Battery Management Systems for Large Lithium-Ion Battery Packs

Battery Management Systems for Large Lithium-Ion Battery Packs

Davide Andrea

ARTECH
HOUSE

BOSTON | LONDON
artechhouse.com

Library of Congress Cataloging-in-Publication Data
A catalog record for this book is available from the U.S. Library of Congress.

British Library Cataloguing in Publication Data
A catalogue record for this book is available from the British Library.

ISBN-13 978-1-60807-104-3

Cover design by Vicki Kane

685 Canton Street
Norwood, MA 02062

10 9 8 7 6 5 4 3 2 1

To Ann

Contents

Preface

As of this writing, lithium-ion (Li-Ion) cells have been the workhorse of small batteries for consumer products (such as cell phones and laptop computers) and are now starting to supplant lead-acid batteries and NiMH cells in large packs for applications such as vehicle traction packs and land-based distributed energy storage.

Compared to other chemistries, Li-Ion cells perform wonderfully, but only if treated well; hence, they require an effective battery management system (BMS).

This book is intended as an aid to the engineer or manager tasked with selecting, specifying, designing, deploying, or using a battery management system (BMS) for a large Li-Ion battery pack.

Over the past 6 years, while developing a few BMSs for large battery packs using Li-Ion cells, I have accumulated some understanding about their requirements, challenges, and solutions. I have shared some of that knowledge by giving talks, publishing white papers, and answering inquiries. With this book I hope to be able to share that knowledge in an organized and comprehensive way. I am fully aware of the fact that my understanding on this subject, while extensive, is by no means complete, making me dread the thought of committing to paper certain incorrect statements. For that I apologize, and ask you to please contact me at http://book.LiIonBMS.com, to correct me, so I may include clarifications and corrections in an online errata sheet, published on that same site.

This book is about control systems and electronics, and not about chemistry: cells are seen as black boxes and only in terms of their equivalent electrical circuit. In general, it is written for the reader with a basic understanding of physics and technology; Chapter 5 is intended for a reader with a good understanding of electronics and software algorithms.

This book is divided into six chapters, starting from general concepts, and then progressively getting into deeper, more practical details.

- Chapter 1 introduces Li-Ion cells, BMS concepts, and the need for a Li-Ion BMS.
- Chapter 2 discusses BMS options: functionality, technology, and topology.
- Chapter 3 explains the functions that may be found in a BMS.
- Chapter 4 explores commercially available BMS solutions.
- Chapters 5 delves deep into BMS electronics and algorithms (should you decide to design your own BMS).
- Chapter 6 guides you during the deployment of a Li-Ion BMS.

CHAPTER 1

Introduction

1.1 Naming Conventions

1.1.1 Cells, Batteries, and Packs

There is some confusion in the terms used to describe the various components of a battery pack, probably due to the fact that we say "batteries" when referring to alkaline cells, and that we tend to forget that a car starter battery is really made up of six cells.

In this book we use the following terms:

- *Cell*: the most basic element of a battery [providing 3V to 4V in the case of lithium-ion (Li-Ion)];
- *Block*: a collection of cells wired directly in parallel, also providing 3V to 4V;
- *Battery*: a collection of cells (or blocks) wired in series, and constituting a single physical module, providing a higher voltage (for example, a battery module that uses four cells in series to provide 12V nominally);
- *Pack* (or battery pack): a collection of batteries, arranged in any series and/or parallel combination.

1.1.2 Resistance

When cell manufacturers list *resistance* in their specs, they usually are talking about AC impedance (see Section 1.2.7). What the user needs to know is the DC series resistance, not AC impedance, because DC is what flows through the cells. Therefore, throughout this book, the term resistance refers specifically to the internal DC series resistance of a cell or battery.

1.2 Li-Ion Cells

Li-Ion rechargeable cells have the highest energy density, and among the highest power densities, of any cell commercially available today. They are capable of amazing performance (Figure 1.1), and are the standard choice for many consumer

Figure 1.1 The KillaCycle, the world's quickest electric motorcycle, burning rubber. (Courtesy of Bill Dubé.)

electronic products, such as laptop computers and cell phones. They are also fast becoming the choice for traction packs in vehicles.

1.2.1 Formats

Li-Ion cells are available in four basic formats (Figure 1.2 and Table 1.1): cylindrical (small and large), prismatic, and pouch.

Some of these formats are far easier to use than others, making them more appropriate for small projects. Cylindrical cells inherently retain their shape against expansion due to chemical processes when fully charged, while, with the other formats, you must provide an overall battery enclosure to retain their expansion.

Additionally, K2 Energy assembles a set of small cylindrical cells in a prismatic case, forming a battery that has the mechanical and thermal advantages of small cylindrical cells and the convenience of prismatic cells.

Figure 1.2 Li-Ion cell formats: small and large cylindrical, pouch, and prismatic.

Table 1.1 Comparison of Cell Formats

	Small Cylindrical	*Large Cylindrical*	*Prismatic*	*Pouch*
Shape	Encased in a metal cylinder, usually 65-mm long	Encased in a metal or hard plastic cylinder	Encased in semihard plastic case	Contained in a soft bag
Connections	Welded nickel or copper strips or plates	Threaded stud for nut or threaded hole for bolt	Threaded hole for bolt	Tabs that are clamped, welded, or soldered
Retention against expansion when fully charged	Inherent from cylindrical shape	Inherent from cylindrical shape	Requires retaining plates at ends of battery	Requires retaining plates at ends of battery
Appropriateness for mall projects	Poor: high design effort, requires welding, labor intensive	Good: some design effort	Excellent: little design effort	Very poor: design effort too high
Appropriateness for production runs	Good: welded connections are reliable	Good	Excellent	Good: high performance
Field replacement	Not possible	Possible but not easy	Easy	In general not possible
Notes	Best for retrofits, as small shape can be fit in all available space	Not widely available	Best availability, very little design effort required	High energy/power density (by themselves); significant design effort required: only appropriate for large production runs

1.2.2 Chemistry

Lithium-Ion is the name given to a class of rechargeable cells that use lithium intercalation reactions in both electrodes; lithium ions travel between the two electrodes in a so-called rocking chair framework.

Among Li-Ion cells, lithium polymer (LiPo) cells use a polymer or gel electrolyte, while all other cells use a nonaqueous, liquid electrolyte.

Many Li-Ion chemistries are available. They are usually named according to the composition of the cathode. They include:

- $LiCoO_2$: Standard lithium-cobalt-oxide;
- LiMnNiCo: Lithium-manganese-nickel-cobalt;
- $LiFePO_4$ and Li_2FePO_4F: Nano-phosphate/lithium-iron-phosphate/lithium-ferro-phosphate;
- $LiMnO_2$: Lithium-manganese-oxide;
- $Li_4Ti_5O_{12}$: Lithium-titanate;
- $LiMn_2O_4$: Lithium-manganese-oxide;
- $LiNiO_2$: Lithium-nickel-oxide.

The nominal voltage, energy, and power density of these cells varies with their chemistry. Some are considered safer and are more appropriate for large traction packs (especially $LiFePO_4$ and lithium-titanate) compared to standard ($LiCoO_2$) Li-Ion cells.

1.2.3 Safety

Li-Ion cells perform magnificently, but are rather unforgiving if operated outside a rather tight safe operating area (SOA), with consequences ranging from the annoying to the dangerous.

In most cases the only effect is simply that the life of the cell is reduced, or that the cells are damaged, with no safety issues. However, abusing a Li-Ion cell in particular ways can be extremely dangerous and can result in physical damage (piercing, crushing) and/or overheating (from over-voltage, over-current, or external heat).

I have witnessed in horror the result of making a short circuit across a "safe" $LiFePO_4$ battery: the cells expelled their contents very violently, in succession, like firecrackers (Figures 1.3 and 1.4). Thanks to safety procedures in place at the time, the battery was on a cart and next to an exit, which allowed the technicians to quickly wheel it outside.

I also have performed a forensic analysis on a Prius converted to PHEV (Figure 1.5), which caught on fire due to arcing inside it Li-Ion traction pack (see Section 6.1.1.5).

No one got hurt in either event. In both cases, human error was at fault: poor mechanical design and bad manufacturing processes (no quality control and distractions present in the manufacturing area). In neither case was a BMS to blame.

Those are extreme examples. However unlikely such events may be, let them serve as a warning of how dangerous large Li-Ion batteries can be. Be safe when working on such a battery.

Figure 1.3 Putting out a Li-Ion battery on fire after a direct short circuit across it.

Figure 1.4 The same battery from Figure 1.3, after the fire.

Figure 1.5 Remains of a Prius PHEV after its Li-Ion traction pack caught fire.

- Think clearly, and categorically tell people nearby to be quiet—chitchat and Li-Ion do not mix.
- Wear proper safety equipment: goggles and insulated gloves for working with batteries above 40V.
- Do not place metal objects on a plane that is higher than unprotected cells, as gravity will invariably cause those objects to fall on the cells. This means no screwdrivers, no meter probes, no socket wrenches, no paint cans, no calipers. If it is not firmly in your hand, place it beneath the battery.
- Design batteries properly, and build them according to the design, using tight quality control.
- Prepare a procedure to follow in case a battery should ignite (cable cutters, quick ejection out of the building, fire suppression); before you work on a bat-

tery, know that procedure by heart, and be prepared to complete it within 10 seconds of the starting event.

1.2.4 Safe Operating Area

The SOA of Li-Ion cells is bound by current, temperature, and voltage.

- Li-Ion cells will be quickly damaged and may burst into flames if overcharged above a certain voltage.
- Most Li-Ion cells will be damaged if allowed to be discharged below a certain voltage.
- The lifetime of Li-Ion cells will be drastically reduced if discharged outside a certain temperature range, or charged outside an even tighter temperature range.
- Li-Ion cells can experience a thermal runaway and ignite if allowed to exceed a safe temperature; even cells that are not prone to thermal runaway may contain an organic electrolyte, which will fuel flames.
- Li-Ion cells' lifetime will be reduced if discharged at too high a current, or charged too fast.
- Li-Ion cells may be damaged if operated at high pulse currents for more than a few seconds.

These limits vary considerably with the chemistry of the cell. For example, standard Li-Ion cells (LiC_2) without any additional protection mechanisms will go into thermal runaway at a relatively low temperature, while $LiFePO_4$ cells are intrinsically immune to thermal runaway. These limits vary to a certain extend with the manufacturer. For example, A123 and K2 manufacture very similar small cylindrical cells, yet the A123 cells can be discharged down to 0V with no ill effects, while the K2 cells may not be discharged below 1.8V (Figure 1.6).

1.2.5 Efficiency

A significant advantage of Li-Ion cells compared to other chemistries is that they are particularly efficient in energy and charge.

1.2.5.1 Energy

The resistance of a Li-Ion cells is quite low (especially in a so-called power cell), which means that little I^2R power is wasted in heat inside of them. For example, the M1 26,650 cells from A123 (used in power tools and PHEV conversion), have a typical resistance of 10 mΩ; when loaded at 1C (2.3A), they waste $P = 2.3^2 * 10$ mΩ = 53 mW, while delivering $P = 2.3\text{A} * 3.2\text{V} = 7.6\text{W}$, which is an efficiency of 99.3% (98.6% round trip when considering both charging and discharging).

At higher currents the energy efficiency of a cell goes down. More energy is wasted in heat inside the cell (across its resistance) leaving less energy available to do work outside the cell.

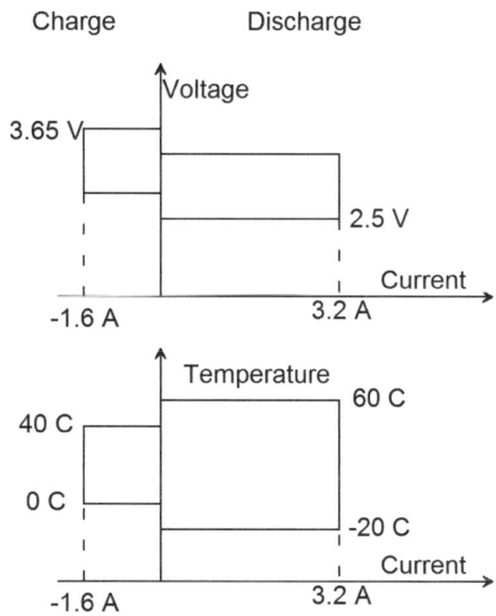

Figure 1.6 Safe operating area of a $LiFePO_4$ 26,650 energy cell.

Maximum power can be extracted from a cell when the load's resistance is equal to the cell's resistance. Half the power will be wasted in heat inside the cell, and half of the power will do work outside the cell.

An M1 26,650 cell from A123 can generate 150A and 500W this way, 250W of which is in heat inside the cells. This can only be sustained for a very short time (less than 10 seconds), as the heat in the cell will quickly degrade it and increase its temperature to dangerous levels. Still, that is something to consider for a race vehicle, for which breaking records is more important than saving the life of the cells, and for which the occasional fire is acceptable.

1.2.5.2 Charge

From a charge standpoint, Li-Ion cells are practically 100% efficient (as long as a charge and discharge cycle is completed within a time that is short enough that self leakage is not an issue). That is, essentially every electron that goes into a cell while charging from completely empty to completely full is available to come back out while discharging back down to completely empty, regardless of the rate of charge or discharge. Note that I didn't say that all the *energy* can be recovered; I said that all the *charge* can be recovered. The cell voltage during discharge is lower than during charge, so the energy discharged is lower than the energy during charging, even though the charge in and out is the same.

If you take exception to this statement, because a specification sheet shows a reduced charge out of a cell at higher current, I would like to point out that I specified complete charge and discharge, while the curves in specification sheet are at constant current, and stop when the cell voltage under load drops down to a certain

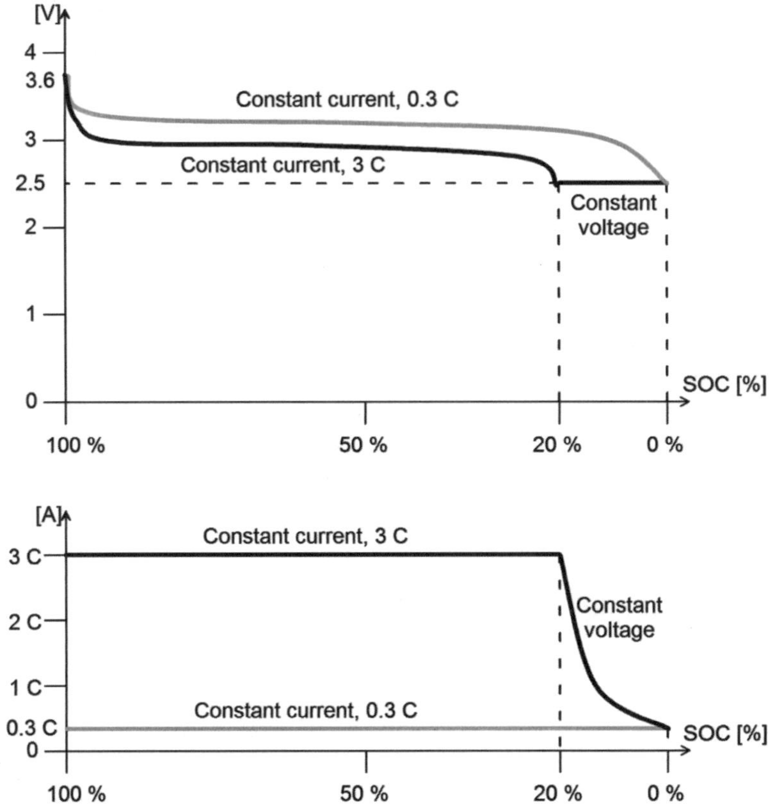

Figure 1.7 The charge that Li-Ion cells can deliver does not depend on rate. After a high current discharge, the remaining charge can be still recovered at a lower current.

level.[1] At that point the cell is not yet fully discharged. You can still extract the remaining charge by discharging the cell at a lower current, for example by discharging it at a constant voltage equal to the cutoff voltage (Figure 1.7). By the time the current drops to zero, the total charge from the cell adds up to essentially the same value, regardless of the rate of discharge (whether the discharge was at a low current or at a high current finished off at a low current).

Of course, in many applications (such as backup power), the load must operate at a high current, and will not be able to extract the last bit of charge out of the cell. For these applications, the fact that all the charge could be extracted at lower current is purely academic.

Some applications do have some flexibility. For example, an EV can go to a "valet" mode at reduced torque, to let the driver "limp home." Such applications are able to use the entire charge from a battery.

1. To those of you familiar with lead acid batteries and the Peukert constant concept, it may be interesting to know that Li-Ion cells have a Peukert constant of about 1.05 (compared to lead acid batteries, which have a Peukert constant between 1.1 and 1.3).

1.2.6 Aging

Li-Ion cells have a longer life than other chemistries, but still have a limited cycle life and may have a limited calendar[2] life.

1.2.6.1 Calendar Life

Standard Li-Ion cells have a relatively short calendar life (Figure 1.8). Whether or not they are cycled, they lose capacity (as users of cell phones and laptops can attest). This is due to a chemical process that occurs when the cells sit fully charged at a voltage above 4.0V. Other Li-Ion chemistries (notably $LiFePO_4$ cells) operate at a lower voltage so that a chemical effect does not take place, and therefore they do not appear to have any calendar life limitations.

1.2.6.2 Cycle Life

You have certainly seen plots of Li-Ion cell capacity versus number of cycles; the capacity decreases linearly with the number of charge and discharge cycles, at a rate that depends on the discharge current used. You may have not seen plots of resistance increase versus cycles, which drops a little after a few cycles, then increases until a few hundreds of cycles, and then continues increasing, though more rapidly.

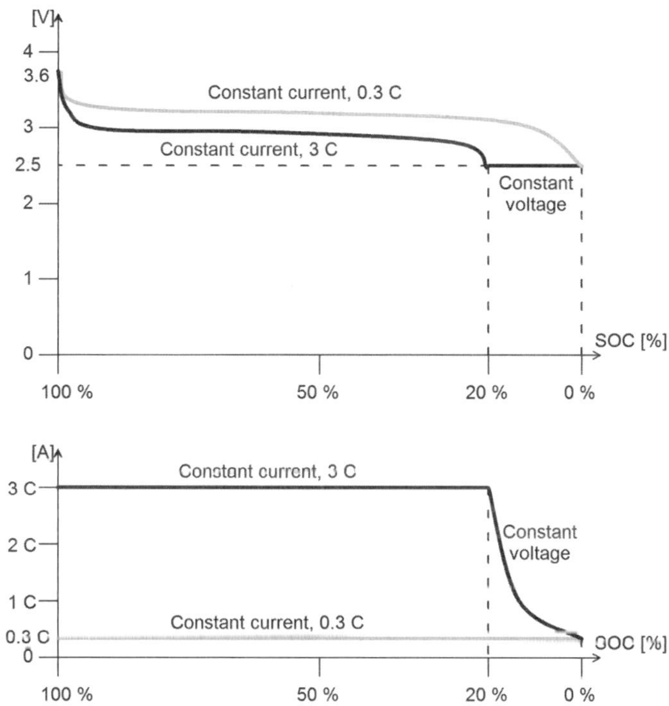

Figure 1.8 Calendar life of Li-Ion cells.

2. Note the spelling of calendar, which is often misspelled "calender." A calendar is a device used to mark the passage of time, very different from a calender, which is a device used in paper making.

What only a few people realize [1] is that these two plots are related. While a portion of the capacity loss is indeed due to a loss of active material inside the cell, the rest is not really lost, but is just simply unused; the cell is undercharged and underdischarged due to the increase of resistance and the fixed cutoff voltages specified by the cell manufacturers. By using fixed cutoff voltages, the cell manufacturer's test equipment charges and discharges cells less and less each time as their resistance increases, resulting in an apparent loss of capacity; it's not just that cells lose some capacity as they are used (which they do), it's also that the cell manufacturer's test algorithm is limited in such a way that the cells appear to be losing even more capacity than they really are.

The portion of that capacity that is unavailable due to increased resistance can be recovered by raising the top cutoff voltage during charge and lowering the bottom cutoff voltage during discharge, using IR ($V = I * R$) compensation (Figure 1.9). The very same cell that has "lost 10% of its capacity," in actuality may still have much of its original capacity; to access it, the charge cutoff voltage must be increased and the discharge cutoff must be decreased, to make up for the additional IR voltage drop due to its increased resistance.

A BMS that is able to measure each cell's resistance and compensate its cutoff voltages accordingly can make better use of a battery's capacity.

1.2.7 Modeling

A chemist thinks of a cell in terms of its processes, while an electrical engineer finds it more useful to view a cell in terms of an electrical circuit; this is called the electrical equivalent model. The simplest model of an Li-Ion cell, yet a very effective one when used at a constant current, is as a voltage source and series resistance [Figure 1.10(a)]. This is indeed the resistance that we are talking about when we use the term "resistance" in the remainder of this book.

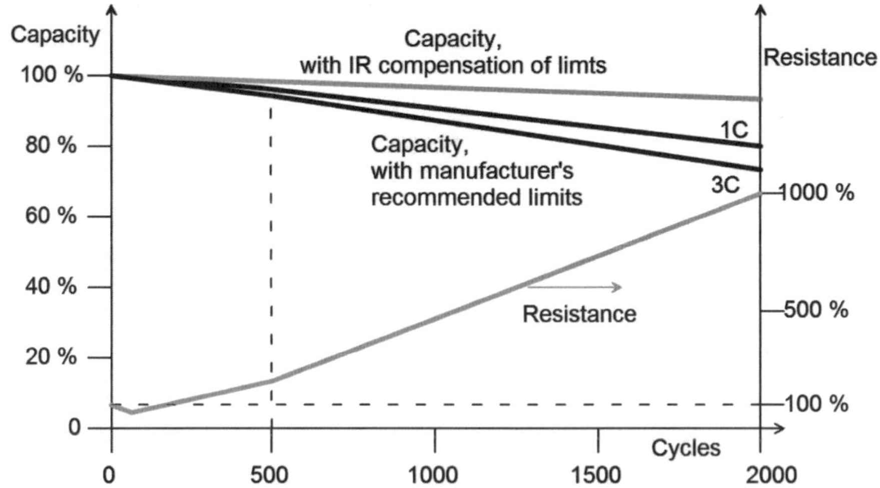

Figure 1.9 Recoverable cell capacity, effective cell capacity, and cell resistance versus cycles.

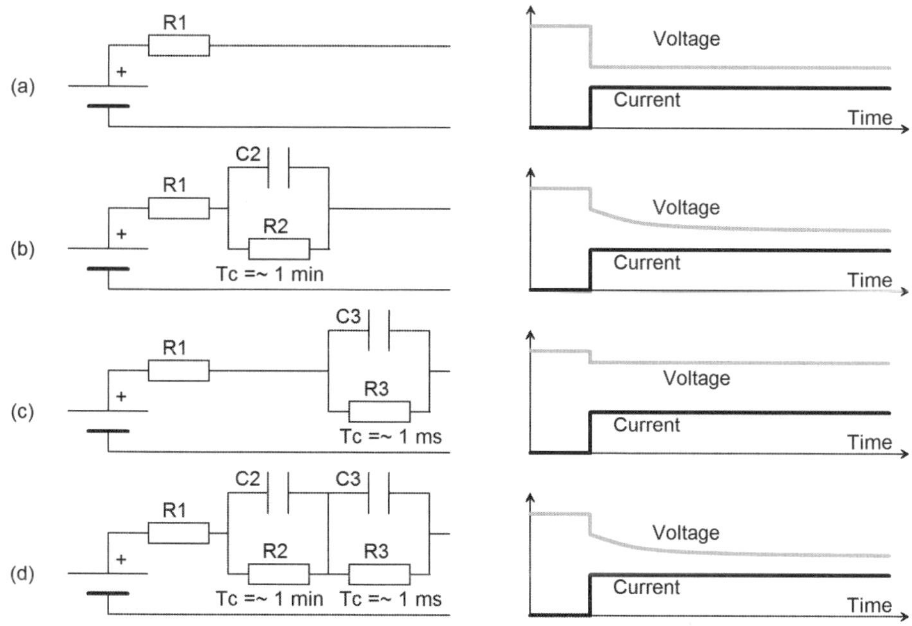

Figure 1.10 Electrical equivalent models of a Li-Ion cell and plots of voltage and current: (a) simple R, (b) with relaxation RC, (c) with AC impedance RC, and (d) with both RC circuits.

In a typical Li-Ion cell, the resistance is on the order of few milliohms (10 to 50 mΩ for a 26,650 LiFePO4 power cell, 0.5 to 5 mΩ for a prismatic cell). That resistance results from the series combination of the effective resistance due to the chemical processes and the bulk metal resistance in the current collectors and terminals. The voltage drop due to this resistance and the cell's current (IR drop) is what chemists call the polarization potential.

Here we are not talking about simple resistance (which is something you can measure with an ohm meter, or calculate as $R = V/I$). We are actually talking about a dynamic resistance, which is different because of the presence of a voltage source in series with that resistance. Dynamic resistance is defined as the ratio of delta voltage over delta current:

$$R = \Delta V / \Delta I$$

Therefore, in order to calculate dynamic resistance, there must be a delta in the current (i.e., the pack current cannot be constant), which results in a delta in the voltage. This resistance varies with (Figure 1.11).

- *SOC:* its value is higher at both low and high SOC levels;
- *Temperature:* its value is higher at colder temperatures;
- *Current:* its value is higher at higher currents and when charging (compared to the same current discharging);
- *Usage:* its value increases over time as the cell is used.

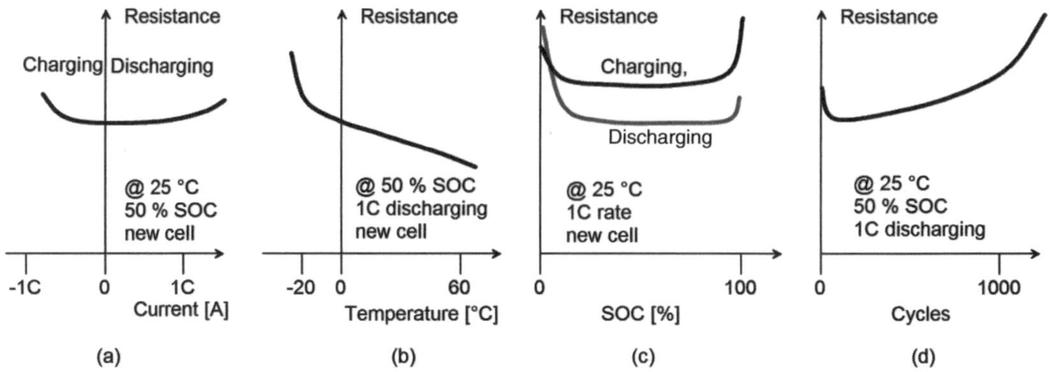

Figure 1.11 Cell resistance variations versus various parameters: (a) current, (b) temperature, (c) SOC, and (d) cycles.

From the point of view of the cell user, the next level of complexity is achieved by splitting the resistance into two resistors and adding a large capacitor in parallel with one of the resistors [Figure 1.10(b)]. This allows the model to correctly emulate the actual behavior of a cell when suddenly loaded. The initial drop in cell voltage is small (just due to $R1$), only to drop further, exponentially, down to the level due to both resistances, with a time constant $T = R2 \times C2$, which is on the order of 1 minute. This effect is what chemists call relaxation.

Unfortunately, cell manufacturers, use a different model [Figure 1.10(c)], which at first may look the same, but instead uses an RC circuit ($R3$ and $C3$) whose time constant is on the order of 1 ms. This allows the model to correctly emulate the AC impedance (at frequencies of the order of 1 kHz) seen by the test equipment available to the cell manufacturers. Cell manufacturers measure this parameter at no load, on a brand new cell, and at 1 kHz, which is a set of conditions that is very different from how the cells are used in the real world. Manufacturers use that measurement method because test equipment to measure 1-kHz impedance is readily available, impedance at 1 kHz remains pretty constant over the life of the cell, and, frankly, their chemists may not be very clear on the concept of DC resistance under load. However, this model is useless to the cell user. If this model were accurate, the voltage drop when a cell was initially loaded would be practically instantaneous and would be much more shallow. Cell manufacturers report the real part of the AC impedance at 1 kHz as the cell's resistance. This is entirely misleading to the user who mistakes that value as the value of the cell's DC resistance. Combining the two models results in a more accurate model with two RC circuits [Figure 1.10(d)], which serves both the cell manufacturer and the user [2].

1.2.8 Unequal Voltages in Series Strings

In a small battery with just a few cells in series, the charger voltage is divided nearly equally among the cells. For example (Figure 1.12), when charging a standard lead-acid starter battery for a car, a constant voltage of 13.5V is applied to it, and each of the six cells within it sees about 2.25V. If any cell is charged more, its voltage

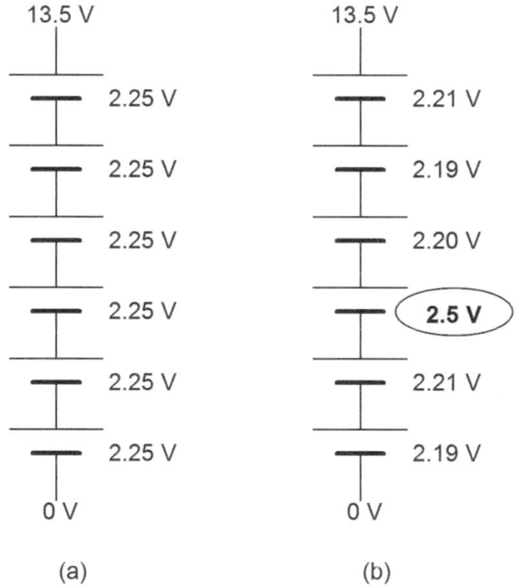

Figure 1.12 A lead acid starter battery: (a) balanced and (b) unbalanced.

will be a bit higher, taking away some voltage from the other cells. For example, if one cell is at 2.5V, the other cells will be, on the average, at 2.20V. That delta voltage among cells is perfectly acceptable; lead acid cells are much more tolerant to variances in their voltage.

For another example (Figure 1.13), a small LiPo battery for a consumer product may have two cells in series. When charging with 8.4V, if the cells are balanced, each cell sees 4.2V. If the cells are out of balance, in the worst case the most discharged cell will be at 3.3V, leaving 4.9V on the most charged one. 4.9V is above the maximum rating for a LiPo cell (4.2V), but it is still low enough that it is not going to go in the thermal runaway and catch fire.

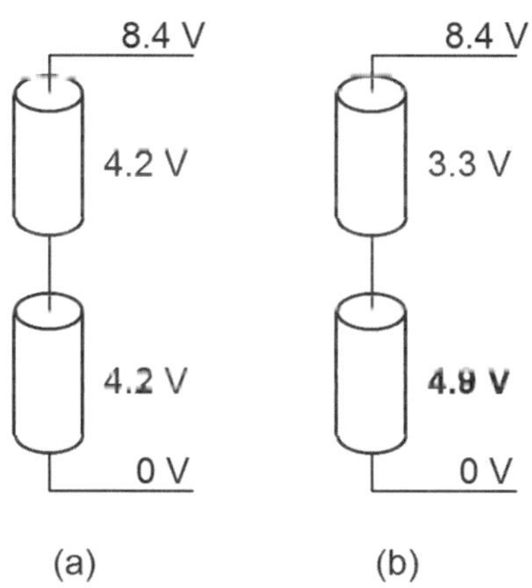

Figure 1.13 Cells in series in a two-cell LiPo battery: (a) balanced and (b) unbalanced.

In a high-voltage battery with many cells in series, though, there is a much greater chance that the overall pack voltage is not evenly divided among its cells. (This is true for any chemistry.)

Consider a four-cell LiPo battery, charged up to 16.8V. If the cells are perfectly balanced, the total voltage will be equally divided into 4.2V per cell [Figure 1.14(a)]. In practice, the cells will be unbalanced, and one will be the first to be fully charged and then be overcharged. Li-Ion cells do not deal well with overcharging. Once charged, they cannot take more current as the other cells in series get their needed charge. Instead, their voltage rises rapidly, possibly to dangerous levels. In this example [Figure 1.14(b)], the second cell is overcharged to 6.3V, while the other ones are around 3.5V. Despite the fact that the total voltage is 16.8V, three of the cells in this battery are not fully charged, and one of its cell is in danger of thermal runaway. Therefore, a system that relies on the total battery voltage to determine when to stop charging the battery (such as a CCCV charger) gives the user a false sense of security; that system *will* overcharge some cells, and *will* create a safety issue as some cells with be overcharged to dangerous levels. It is therefore essential that a BMS monitor such a battery, first and foremost to prevent any cell from being overcharged, and optionally to balance the battery to maximize its performance.

Now consider the same battery, after it has been discharged down to 12.0V. If the cells are perfectly balanced, the total voltage will be equally divided into 3.0V per cell [Figure 1.15(a)]. In practice, the cells will be unbalanced, and one will be the first to be fully discharged, and then be overdischarged. To varying degrees, Li-Ion cells do not deal well with overdischarging. If their voltage is allowed to drop below a certain threshold, irreversible damage may occur. In this example [Figure

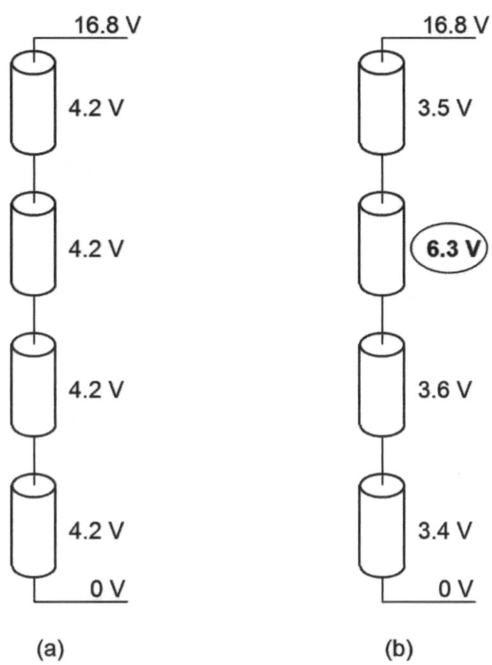

Figure 1.14 Four-cell battery, charged: (a) balanced and (b) unbalanced. One cell is in danger of thermal runaway.

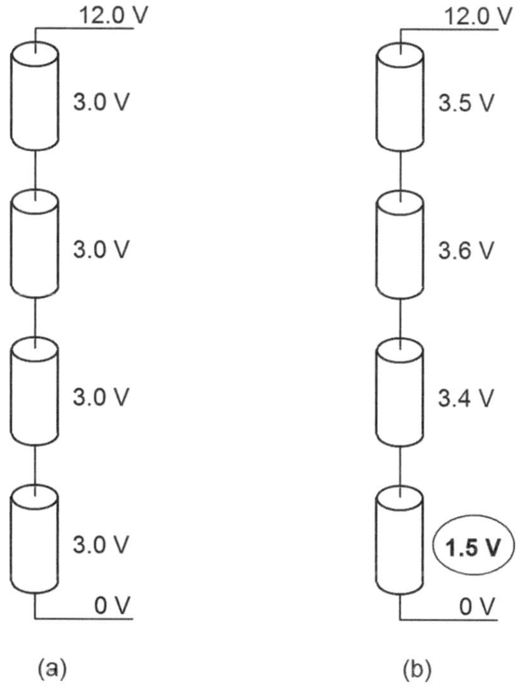

Figure 1.15 Four-cell battery, discharged: (a) balanced and (b) unbalanced.

1.15(b)], one cell is overdischarged down to 1.5V, while the other ones are around 3.5V. Despite the fact that the total voltage is 12V, three of the cells in this battery are not fully discharged, and one of its cells is being damaged. Therefore, a system that relies on the total battery voltage to determine when to stop discharging the battery (such as a motor controller with a low voltage cutoff) gives the user a false sense of security; that system *will* overdischarge some cells, damaging them. It is therefore essential that a BMS monitor such a battery to prevent any cell from being overdischarged and damaged.

1.3 Li-Ion BMSs

In the previous sections we saw how abusing Li-Ion cells may reduce their life, result in damage, and can even be a safety issue. Having analyzed the problems with Li-Ion cells, let us look at Li-Ion BMSs for solutions. It is the job of a BMS is to ensure that the cells in a battery are operated within their SOA. This is particularly important for large Li-Ion battery packs because:

- Li-Ion cells are so much more unforgiving of abuse than other chemistries.
- Large battery packs, with many cells in series, are more prone to be charged and discharged unevenly due to unbalance among cells. Li-Ion cells must not be overcharged or overdischarged.

1.3.1 BMS Definition

There is no unique definition of what a BMS is and does, and sometimes other terms [such as voltage management system (VMS) or protection circuit module (PCM)] are used for what is in effect a BMS. Here I take the wide view that a BMS is any product or technology used with the intent of taking care of a battery in one way or another. This may include any of the following functions:

- To monitor the battery;
- To protect the battery;
- To estimate the battery's state;
- To maximize the battery's performance;
- To report to users and/or external devices.

1.3.2 Li-Ion BMS Functions

For the sake of safety, and for the sake of the cells, a Li-Ion BMS must, at the very least (in order of importance), do the following:

- Prevent the voltage of any cell from exceeding a limit, by stopping the charging current, or requesting that it be stopped. This is a safety issue for all Li-Ion cells.
- Prevent the temperature of any cell from exceeding a limit by stopping the battery current directly, requesting that it be stopped, or requesting cooling. This is a safety issue for Li-Ion cells that are prone to thermal runaway.
- Prevent the voltage of any cell from dropping below a limit by stopping the charging current or requesting that it be stopped.
- Prevent the charging current from exceeding a limit (which varies with cell voltage, cell temperature, and previous level of current) by requesting that the current be reduced or stopped, or by stopping the current directly.
- Prevent the discharging current from exceeding a limit, as described in the previous point.

A BMS is essential when charging a Li-Ion battery. As soon as any cell reaches its maximum charged voltage, it must turn off the charger (Figure 1.16). A BMS may also balance the battery to maximize its capacity. It may do so by removing charge from the most charged cell until its voltage is low enough that the charger may come back on and give the other cells a change to be charged. After many cycles of this process, all the cells will be at the same voltage, fully charged, meaning that the pack will be balanced. A BMS is also essential when discharging a Li-Ion battery. As soon as any cell reaches a low cutoff voltage, it turns off the load (Figure 1.17).

1.3.3 Custom Versus Off-the-Shelf

A major goal in this book is to help you decide whether you should use an off-the-shelf or custom BMS. There are a few specific needs that can only be met by a custom BMS: the need for intellectual property (IP) or tight specifications. Other-

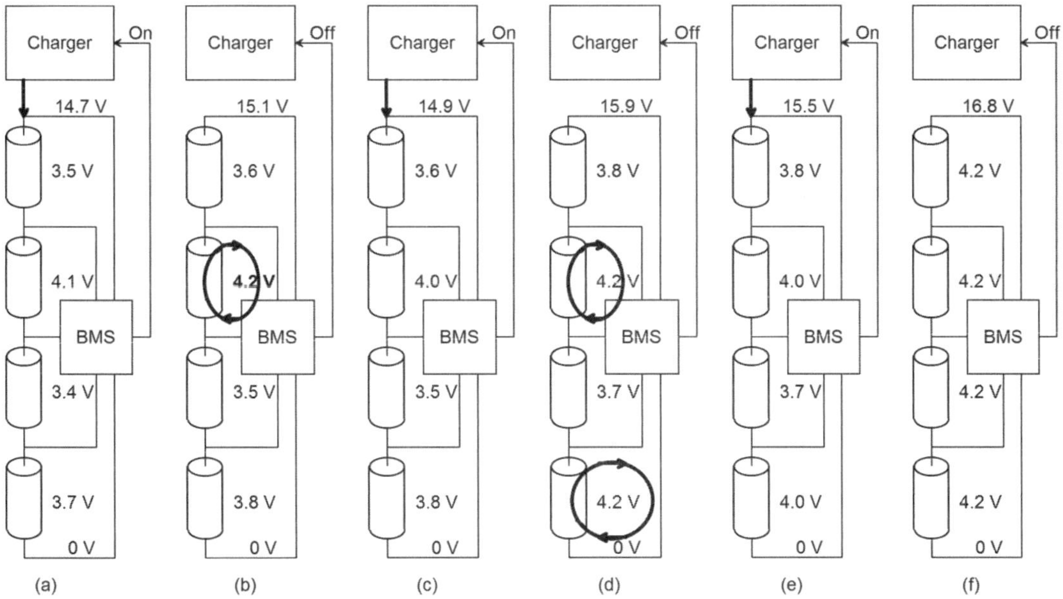

Figure 1.16 Charging with a balancing BMS controlling the charger: (a) charging; (b) charging stops when any one cell reaches the top cutoff voltage; (c) charging restarts after that cell's voltage is slightly reduced by balancing; and (d) the process repeats until (e, f) the pack is balanced.

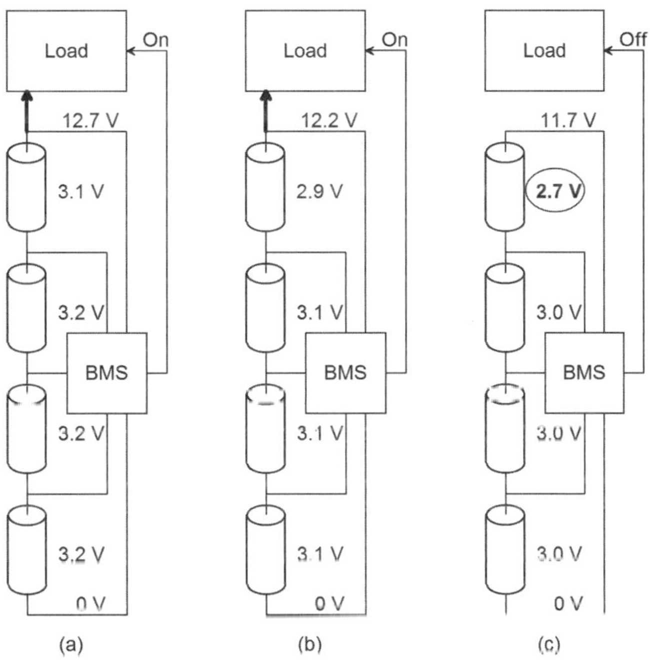

Figure 1.17 Discharging with a BMS controlling the load: (a, b) discharging; and (c) discharging stops when any one cell drops to the bottom cutoff voltage.

wise, a commercially available BMS will get you there much faster for much less money, with fewer resources and a higher likelihood of success. Simply put:

- *Custom:* you own, you control;
- *Off-the-shelf:* fast, easy, cheap.

Entities that will probably benefit from a custom design include:

- Big auto manufacturers who need full control over their products;
- Large electronics corporations who want to enhance their product line;
- Companies poising themselves for acquisition, and wishing to increase their perceived value;
- Hobbyists who want the learning experience.

Conversely, these entities will probably benefit from using an off-the-shelf product:

- Companies designing EV, PHEV, and HEV passenger cars;
- Small to medium manufacturers of specialized vehicles (utility, heavy duty, public transportation);
- Design services companies and engineering consulting firms;
- Vehicle integrators;
- Companies developing land-based back-up systems, utilities;
- Cell manufacturers and battery assembly houses;
- EV converters;
- Efficient companies that are results driven.

Certain cell manufacturers sell a BMS together with their cells (see Section 4.1.3). Their BMS is not available by itself, and, in some cases, the cells are not available by themselves. If using such cells, the cell manufacturer is in effect making the custom versus off-the-shelf choice for you.

Please consider carefully the pros and cons of designing your BMS (Figure 1.18) before you embark on such project, lest you should end up saying: "I wish I had known how complex this was going to be before I started!"

1.4 Li-Ion Batteries

1.4.1 SOC, DOD, and Capacity

The state of charge (SOC) of a cell or a battery at a given time is the proportion of the charge available at that point, compared to the total charge available when it is fully charged. It is expressed in percent, from 100% when full to 0% when empty. The SOC evaluation function is also known as the *fuel gauge*, especially in EVs, because of its analogy to a gas car's fuel gauge. It is essential to understand that each cell in a battery has its own SOC, and that the battery itself has its own, separate SOC.

The depth of discharge (DOD) of a cell or battery is a measure of the charge removed from it. It is expressed in amp-hour (Ah). DOD can be expressed in a percentage as well, and is commonly done so in Lead Acid batteries. It is really more

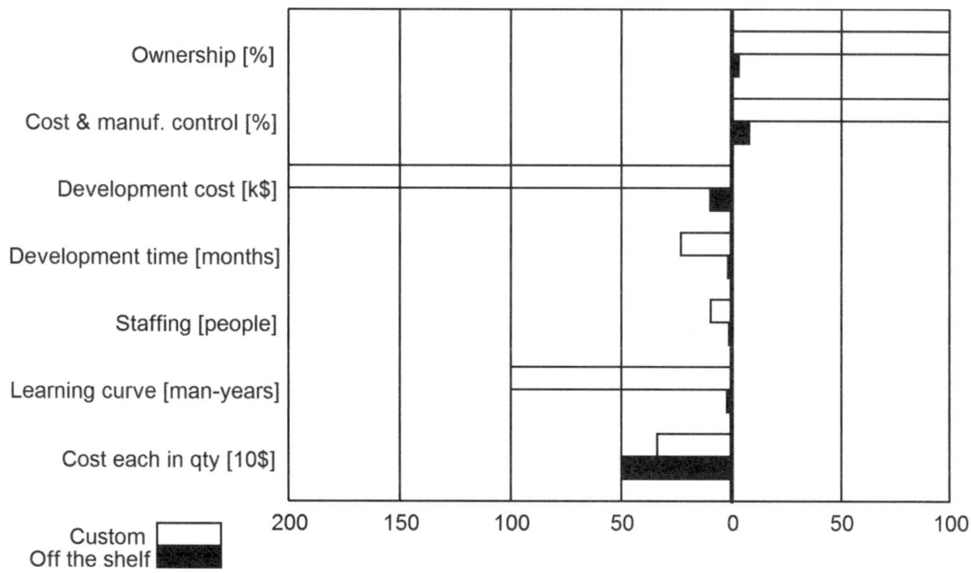

Figure 1.18 Pros and cons of off-the-shelf versus custom BMSs.

useful to express DOD in Ah, so that the combination of SOC (in percent) and the DOD (in Ah) conveys more information than would be the case if both were expressed in percentages. This becomes apparent with a cell that has more capacity than nominal (say, a 100-Ah cell that is actually 105 Ah). When a charge equal to the nominal capacity has been taken from the cell (100 Ah), its SOC is 0%, and the DOD could be either expressed as 100% or as 100 Ah. If you were to extract all of the charge from the cell, the SOC would remain stuck at 0% (because it is not allowed to go negative) and the DOD expressed in percent would remain stuck at 100% (because it is not allowed to go higher than that). However, the DOD expressed in Ah would go up to 105 Ah, which would be correct. Knowing that a cell has a DOD of 105 Ah is far more useful than knowing that it had reached 100% DOD some time ago and is still at 100% DOD even though we extracted more charge from it since. One reason that expressing the DOD in Ah works well is that the DOD does not depend on the discharge rate.

While on a first look one may think that SOC is just the inverse of DOD (one goes up while the other one goes down), that is not really true (see Table 1.2). Not only should their units be different, but also what happens when the battery is empty. Note that if a 100-Ah battery loses half its capacity, the SOC still goes from 100% down to 0%, but the DOD (which used to go from 0 Ah up to 100 Ah) now goes from 0 Ah up to 50 Ah (Figure 1.19).

The actual capacity of a cell or battery (expressed in Ah) is equal to its DOD (also expressed in Ah) when it is completely discharged. The nominal capacity of a cell is specified by the cell manufacturer.[3]

A significant limitation of the effective cell capacity is the point at which charging and discharging is stopped. Cell manufacturers instruct users to stop discharg-

3. As the saying goes: "There are three kinds of liars: liars, damns liars, and battery manufacturers."

Table 1.2 Comparison of SOC and DOD

	SOC	*DOD*
Units	%	Ah
Reference	Two points: empty and full	Only one point: full
Full Reference	100%	0 Ah
Empty Reference	0%	Not applicable
Rate of Change	Not applicable	Proportional to battery current
Past Empty	Won't go below 0%	Continues increasing

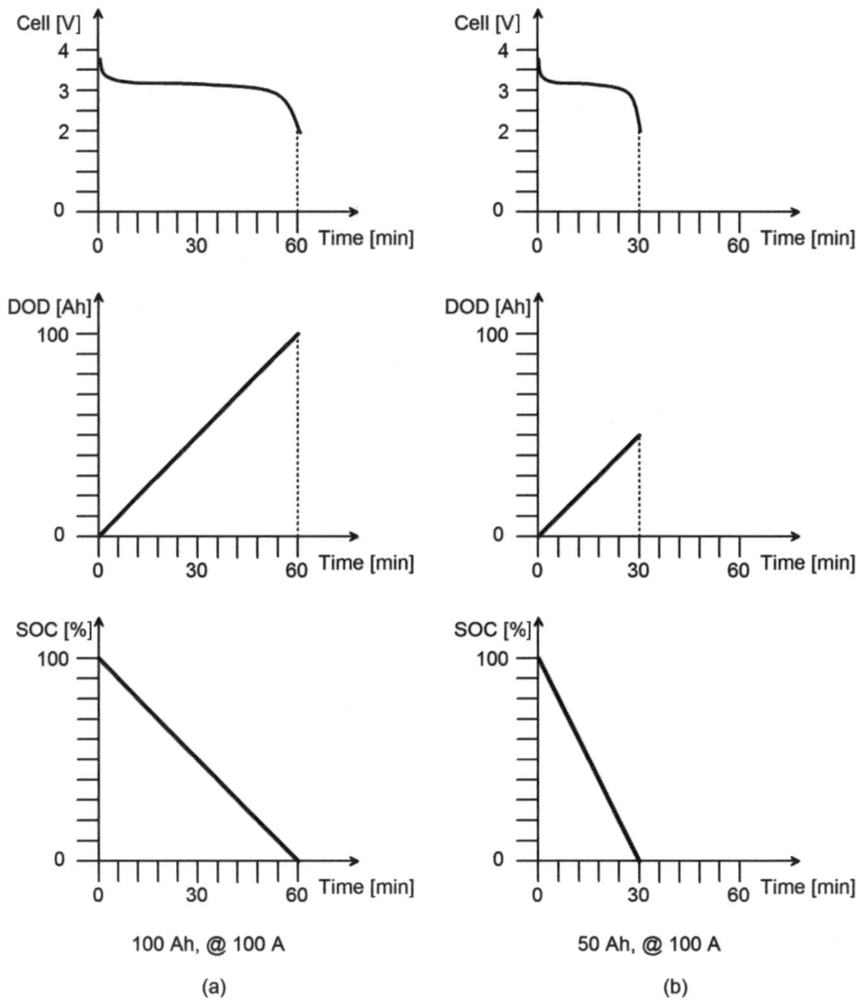

Figure 1.19 SOC and DOD plots: (a) at nominal capacity and (b) after the battery loses 1/2 of its capacity.

ing at a given terminal voltage. That results in the apparent useful cell capacity to vary with the charge and discharge current. At a minimal discharge current, the cell's terminal voltage is essentially the same as the cell's internal or open circuit voltage (OCV), which is the best external indicator of SOC. When the cell reaches the low cutoff voltage, the cell is truly fully discharged. At a high discharge current, the cell's terminal voltage is significantly lower, due to the IR drop across the cell resistance. The cell terminal voltage will reach the low cutoff point earlier, even though the cell is not yet fully discharged. From that, the casual observer may conclude that the actual cell capacity was reduced by the high discharge rate.[4] However, this is untrue. The actual capacity has not changed, only the useful capacity was affected by the high discharge current. Should you then continue to discharge the cell at a lower rate, the cell will be able to generate the entire charge.

An application that is able to measure a cell's resistance may make more use of the charge in that cell by doing IR compensation of its terminal voltage to estimate the OCV and stop the discharge later, when the estimated OCV reaches the low cutoff voltage, instead of when the terminal voltage reaches it (Figure 1.20). This only works up to a point; if the cutoff is lower than 1/2 of the OCV cutoff (i.e., about 1.25V), you get past the maximum power point, and into an area of diminishing returns (see Section 1.2.5.1). Worse yet, if the cell voltage is allowed to go to 0 and reverse, the cell will be permanently damaged.

At the top end, cell manufacturers specify that the cell be charged at constant current until a top cutoff voltage, and then continue to be charged at that constant voltage until the current drops below a certain level. This is a good algorithm.

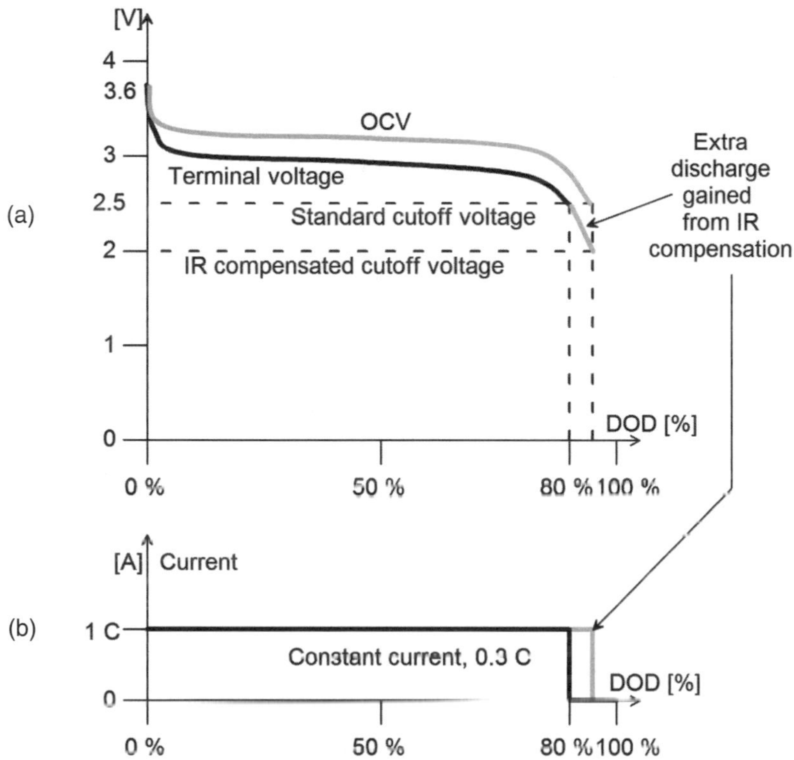

Figure 1.20 Discharging algorithms: (a) as defined by cell manufacturer and (b) IR compensated.

4. "The problem with expressing DOD in Ahr is that it depends on how fast (what rate) you will be using it. There is much less Ahr available at high rates than low rates" [personal communication with Bill Cantor, 2010].

The IR drop limitation described above will occur during the constant current stage: the cell terminal voltage will reach the top cutoff voltage sooner than its internal voltage will, so the constant current stage will be ended prematurely. However, during the constant voltage stage, as the charging current approaches 0, the error due to the IR drop will also approach 0. By the time the low current threshold is reached and the charger is turned off, the cell's OVC will indeed be at the top cutoff voltage, indicating that the cell is fully charged.

If it is possible to measure the cell's resistance as it is being charged, an improvement to that charging algorithm, which will result in faster charging, is to do IR compensation of the cell terminal voltage. This will result in the constant current stage ending later, when the OCV (not the terminal voltage) reaches the cutoff voltage (Figure 1.21).

1.4.2 Balance and Balancing

According to Tom Wicker, "Why do you not include the effect of differences in charging/discharging due to differences in cell internal resistance? ... It seems this is a

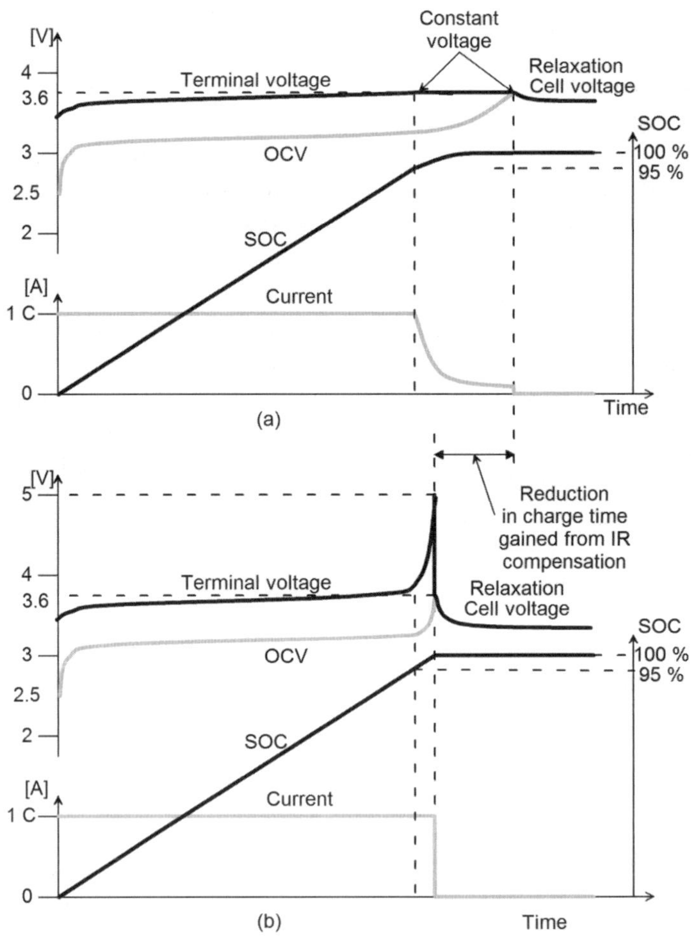

Figure 1.21 Charging algorithms: (a) as defined by cell manufacturer and (b) IR compensated.

potentially larger source of imbalance, requiring larger balancing currents" [Tom Wicker, personal communication, July 11, 2009]. Because of unit-to-unit variations and different charging history, the various cells in a battery may be unbalanced in four ways:

- SOC;
- Leakage (self-discharge current);
- Resistance;
- Capacity.

In general, the term "balance" could refer to the degree by which any of the parameters listed above match each other among the cells in a battery. Here, the term refers to just one of those parameters—SOC.

Balancing is the term used for the process of bringing the SOC levels of cells in a battery closer to each other, in order to maximize the battery's capacity. The balancing process only addresses the first item—it equalizes the cells' SOC. In doing so, it also compensates for the second item—cell-to-cell variations in leakage. Its job may be somewhat hindered by the third item—variations in cell resistance. What balancing does not do is take care of the fourth item—capacity. That's something that can only be done by a different technique—redistribution (see Section 3.2.4).

There is a myth is that balancing is required due to variations in cell resistance.[5] That is not true—balancing does not compensate for such variations. Consider the example of a battery that is composed of ideal, fully charged cells, except that they have differing internal resistance. When sitting on a shelf, the cells will remain at 100% SOC forever, and, as no current is going through the internal resistance, the internal resistance has no effect on balance.

Now consider the same battery, except that the cells have identical leakage. All the cells will be slowly discharged in equal amounts, and their SOC levels will remain in lockstep. The leakage current flows through a portion of the cell's internal resistance, but this resistance is in series with the far larger resistance of the separator, so variations in the cell resistance are insignificant.

Finally, consider the same battery being discharged. During the discharge, the terminal voltage of each cell will differ, due to the varied IR drops across each cell resistance. Once the current is stopped, the open circuit voltage of the cells will be perfectly matched again, and a charge cycle will bring each cell back to 100% SOC in lockstep, with no need for balancing.

An all too common misunderstanding is that balancing results in the entire charge in the battery being made available.[6] That is not true. In a balanced battery, all cells (other than the limiting cell) will still have excess charge available when the battery is fully discharged, and/or the ability to accept additional charge when the battery is fully charged. A balanced battery is one in which, at some point in its cycle, all the cells are exactly at the same SOC.

5. Is a battery "unbalanced" or "imbalanced"? For consistency's sake, I chose unbalanced because it has a corresponding verb and noun (unbalance), but imbalanced only has a corresponding noun (imbalance) and not a verb.
6. "With balancing, the available capacity is increased because the pack will accept more charge and will fully utilize the capacity of every cell" [personal communication with Bill Cantor, 2010].

1.4.2.1 Unbalanced over Time

A battery becomes more and more unbalanced over time due to variations in the internal leakage of its cells. This process is not dependent on cell resistance, nor on cell capacity, nor on absolute leakage; it is only dependent on the *delta* in the leakage. Some examples may help understand this.

Visualizing the cell unbalance process is not easy; you can find a helpful interactive tool online (http://book.LiIonBMS.com/). However, an interactive tool is not possible in a book. The best thing that a book can do is to show 3-D graphs. The following graphs describe a battery with four cells. The *X* axis is time, the *Y* axis (vertical) is the SOC of individual cells, and the *Z* axis (away from you and to the right) shows the four cells.

First consider an ideal, fully charged, balanced battery, on a shelf. All of the cells start at 100% SOC and remain at 100% forever (Figure 1.22). Now consider the same battery, but made of real cells that do have a self discharge (internal leakage), perfectly identical (same capacity and same leakage). All the cells would lose charge at the same rate, and their SOC would decrease in lockstep (Figure 1.23), so the battery would remain balanced (though more and more discharged).

Then consider the same battery, but with nonidentical cells (different capacity and different leakage). The leakage will discharge some cells more rapidly than others, and the battery will quickly become unbalanced; the cells' SOC levels diverge (Figure 1.24).

Given a battery that started balanced and that was unbalanced due to leakage, bringing such a battery back into balance requires maintenance balancing (see Chapter 3). Small amounts of charge must be moved in and/or out of individual cells to make up for the *difference* between the leakage currents of the cells. For example, if the leakage in one cell results in a loss of 2 Ah, and in another cell in a loss of 3 Ah,

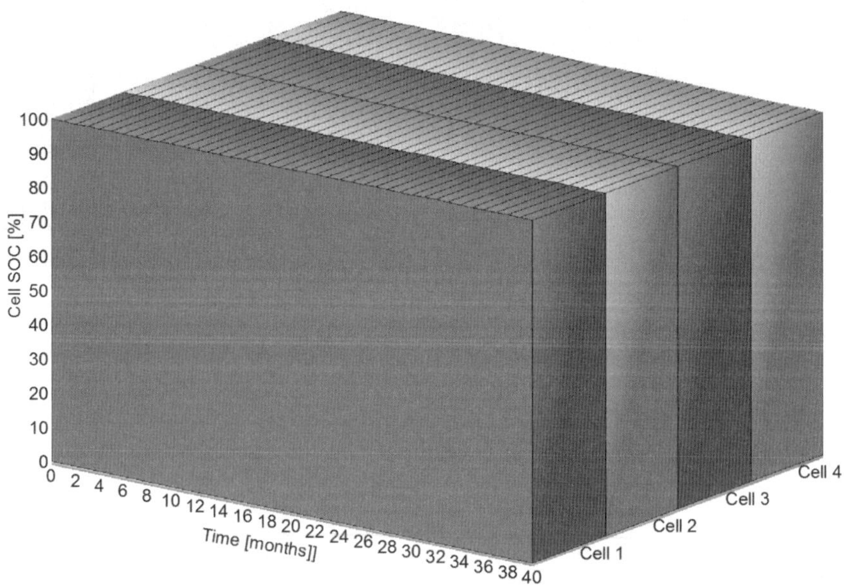

Figure 1.22 An ideal battery on the shelf. It maintains its SOC indefinitely.

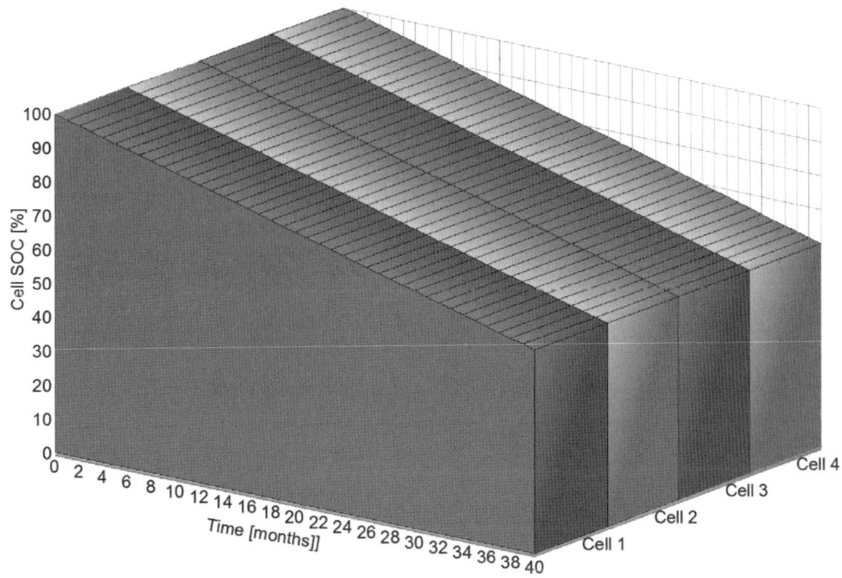

Figure 1.23 A battery on the shelf made up of identical cells. It remains balanced over time, although its SOC decreases due to leakage.

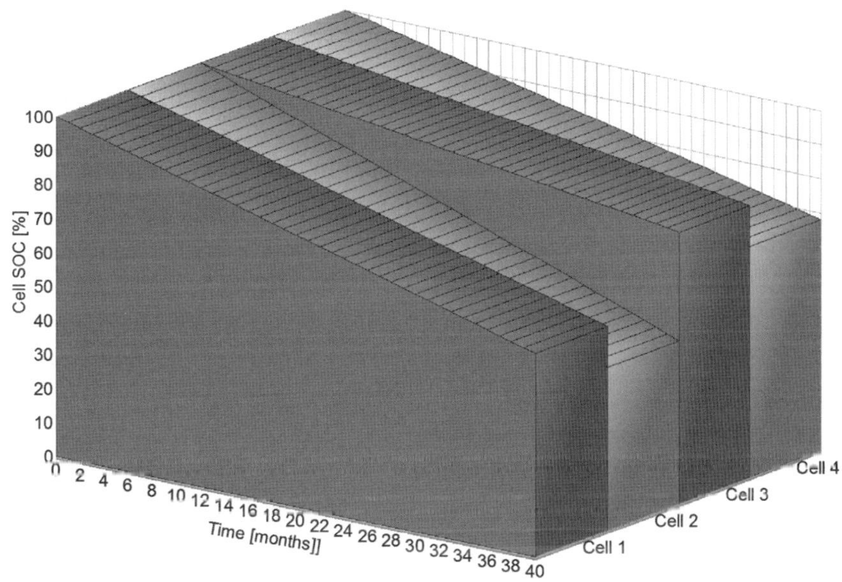

Figure 1.24 Real-world battery: the leakage of each cell is different, so it becomes unbalanced over time.

then all balancing needs to do is to remove 1 Ah from the first cell, so that it too is down by 3 Ah. In a battery that started balanced, maintenance balancing does not compensate for leakage—just the *difference* in the leakage.

1.4.2.2 Starting Unbalanced

In the previous examples we assumed that all the cells started fully charged and balanced (all the cells are at 100% SOC). In reality, when a battery is manufactured, that is rarely the case (Figure 1.25). Not only do batteries start unbalanced, but also differences in leakage make them more unbalanced over time. (I suppose that it is possible that difference in leakage happens to be just right to bring an unbalanced battery closer to balance, but that is statistically quite unlikely.)

To bring a battery that started out as unbalanced back into balance requires gross balancing (see Chapter 3). Large charges must be moved in and/or out of individual cells to make up for that initial unbalance.

1.4.2.3 Battery SOC Versus Cell SOC

While charging, the ability of a battery to accept charge is limited by the cell that is the first one to become fully charged. While discharging, the ability of that battery to release charge is limited by the cell that is the first one to become fully discharged. These two points determine the battery capacity.

Before a battery is balanced, the cell that limits charging and the cell that limits discharging are different cells. When fully charged, the first cell cannot accept any more charge, but the second one can; when fully discharged, the first cell can still be discharged further, but the second one cannot (Figure 1.26).

Balancing increases the battery capacity by expanding the separation between the point in which a cell limits charging, and the point in which a cell limits discharging. After a battery is balanced, the capacity of a battery is limited at both ends by the same cell. This is normally the cell with the lowest capacity (though, in extreme cases, it is the cell with the highest resistance). Again, a balanced battery is one in

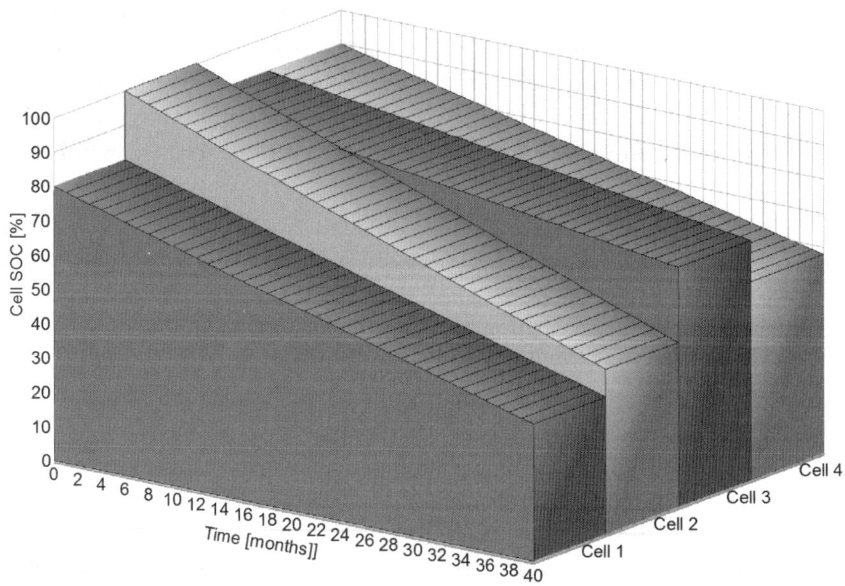

Figure 1.25 Real-world, unbalanced battery: it may become even more unbalanced over time.

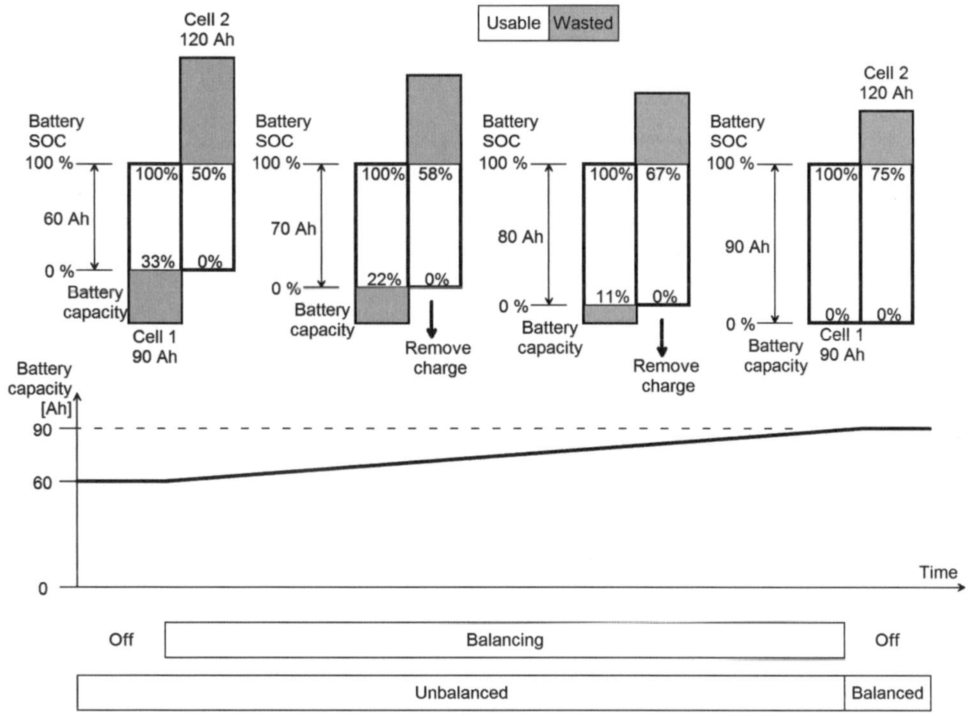

Figure 1.26 The balancing process maximizes battery capacity.

which, at some point, all the cells are exactly at the same SOC. That point can be any SOC level; however, for reasons explained later in Chapter 3, balancing is usually done at 100% SOC.

The previous 3-D graphics were plots of SOC versus time. The following ones graph cell SOC versus battery SOC. Time is not shown, so please visualize it in your mind by considering a plane perpendicular to the *X* axis that moves to the left when charging and to the right when discharging. The instantaneous SOC levels are marked by the intersection of that plane with the curves shown in the graphs.

Let us start from an ideal, balanced battery (Figure 1.27). Note that the SOC levels of the cells are identical to each other and to the battery SOC. When the battery is full (on the left), all the cells are at 100% SOC, as is the battery. When the battery is empty (on the right), all the cells are at 0% SOC, as is the battery.

Now let us look at the same battery, but unbalanced (Figure 1.28). Note that the range of the battery SOC is reduced; it occupies a smaller portion of the *X* axis.

When cell 4, the most charged cell, reaches 100% SOC, no more charging is possible; that point is the fullest the battery can be (by definition, 100%). At that point, the other three cells are less than full (for example cell 0 is at 62%), but they cannot take any more current because that same current would have to flow through cell 3 and would overcharge it.

Conversely, when cell 1, the least charged cell, reaches 0% SOC, no more discharging is possible; that point is the emptiest the battery can be (by definition, 0%). At that point, the other three cells still have some charge left in them (for example,

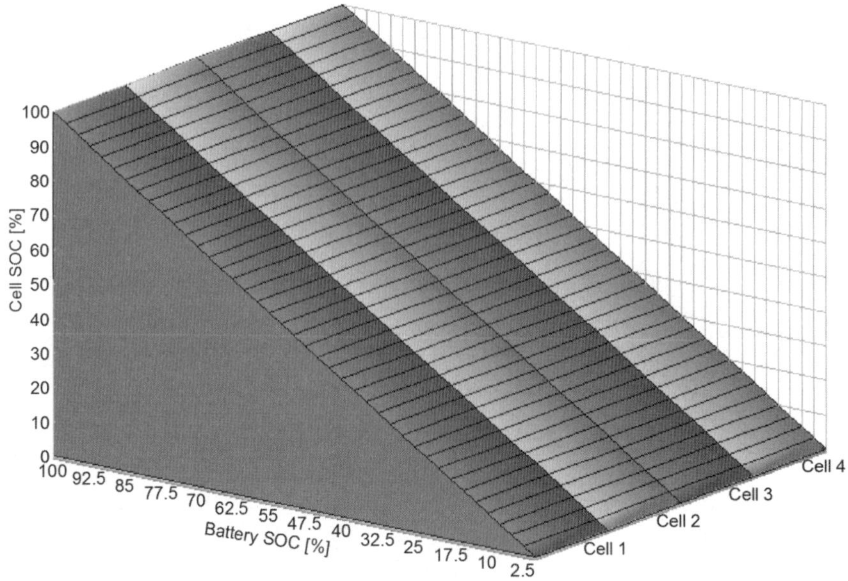

Figure 1.27 SOC levels in an ideal, balanced battery.

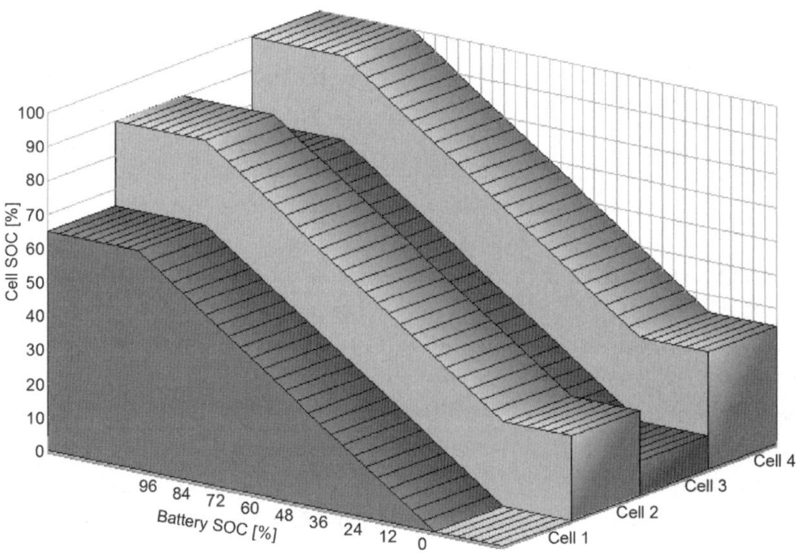

Figure 1.28 SOC levels in an ideal battery, though unbalanced.

cell 4 is at 45%), but they cannot give any more current because that same current would have to flow through cell 0 and would overdischarge it.

You can see how unbalance in the cells' SOC levels results in a reduced battery capacity. If all the cells had exactly the same capacity, once the battery was balanced it would remain balanced at any SOC (Figure 1.27). In a real-world battery, the capacity of each cell is different, which means that a battery that was balanced at a given SOC will be unbalanced as soon it is charged or discharged (its SOC changes)

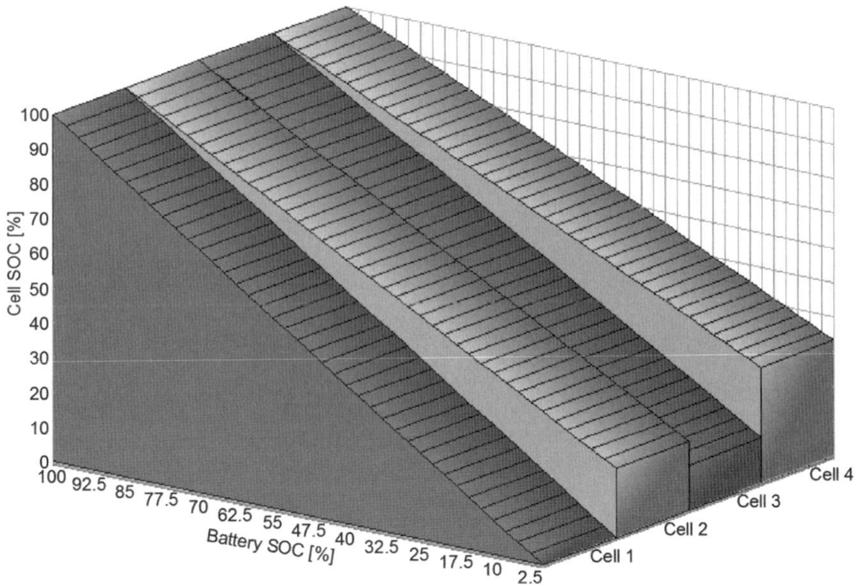

Figure 1.29 SOC levels in a real-world battery balanced at 100% SOC.

(Figure 1.29). Let us look at such a battery, whose cells have different capacities yet are all balanced at 100% SOC. Note that the slope of each cell's curve is different. Cell 4, with the largest capacity, has the flattest slope (because a given current in and out of it has the least effect on its SOC), while cell 1, with the smallest capacity, has the steepest slope (the same current in and out of it has the most effect on its SOC). The cells are balanced at the top (on the left). At that point, each cell is at 100% SOC, as is the battery. However, as the battery is discharged, the SOC of cell 1, the cell with the lowest capacity, drops the fastest, until it reaches 0% SOC. At that point the battery is empty (because cell 1 prevents further discharge), even though the other cells still have charge left in them. Therefore, that sets the battery capacity, which cannot be any higher no matter how the starting SOC of the individual cells is manipulated through balancing.

In the previous example the battery was balanced at 100% SOC. Let us now look at the same battery, balanced at 50% (Figure 1.30) and 0% SOC (Figure 1.31). Again, cell 1, the cell with the least capacity, is the limiting factor. Note that, regardless of where the cells are balanced (100%, 50%, or 0%) the battery capacity did not change. All that changed is the state of the other cells (2, 3, and 4) at the two ends of the charge/discharge cycle. So, on a first order, there is no advantage in balancing a battery at full or empty, or any other point. Later, in Chapter 3, we'll see that top balancing (at 100%) is best.

Finally, let us look at the same battery, but unbalanced (Figure 1.32). Note that the slope of each curve is unchanged (because it's a function of each cell's capacity), but the battery capacity has been reduced. Once cell 3 becomes full, no more charge is possible, and that sets the 100% SOC point for the battery. Similarly, once cell 1 becomes empty, no more discharge is possible, and that sets the 0% SOC point for the battery. The result is that the battery capacity is reduced to about half of the capacity of the same battery when balanced.

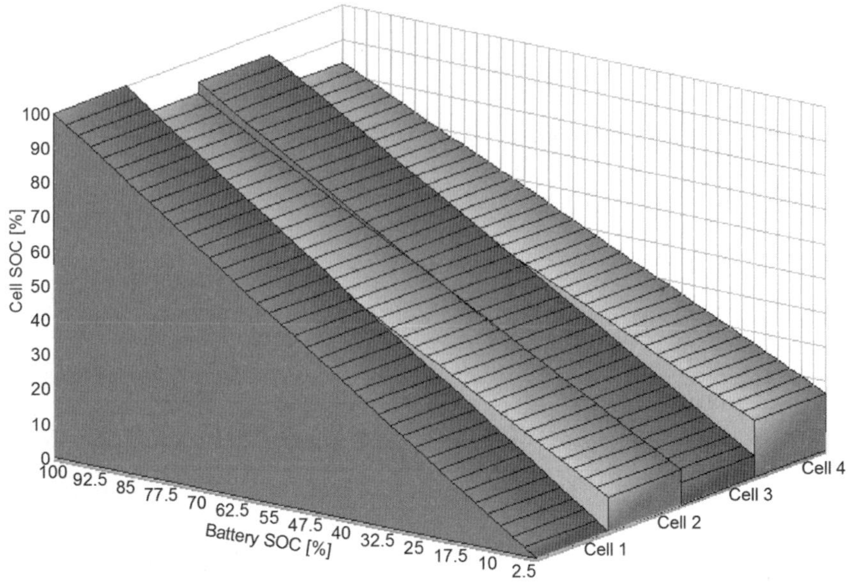

Figure 1.30 SOC levels in a real-world battery balanced at 50% SOC.

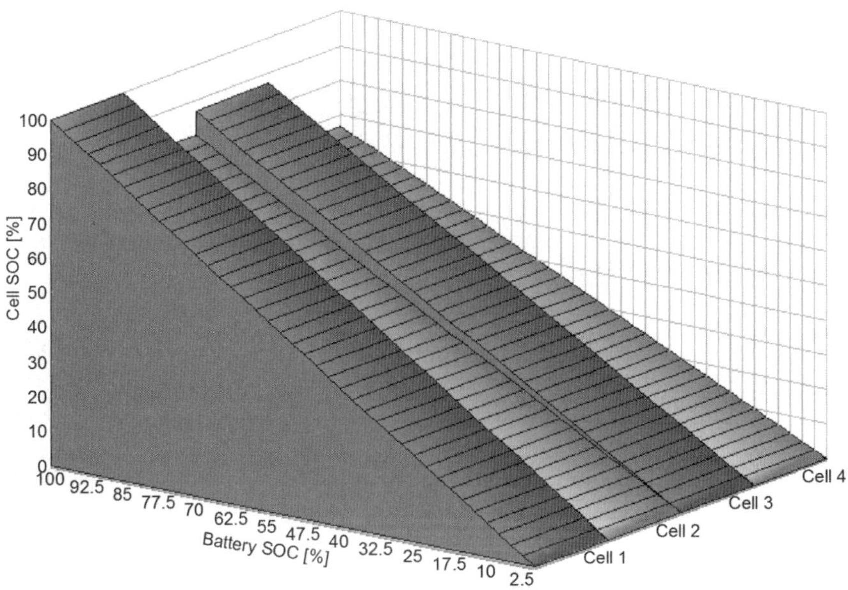

Figure 1.31 SOC levels in a real-world battery balanced at 0% SOC.

In extreme cases, the battery could be so severely unbalanced that it becomes nearly useless (Figure 1.33). Charging just a little bit will fill cell 3 all the way, so charging must stop immediately; discharging just a bit will empty cell 0 all the way, so discharging must also stop immediately. In other words, the battery has effectively 0 capacity.

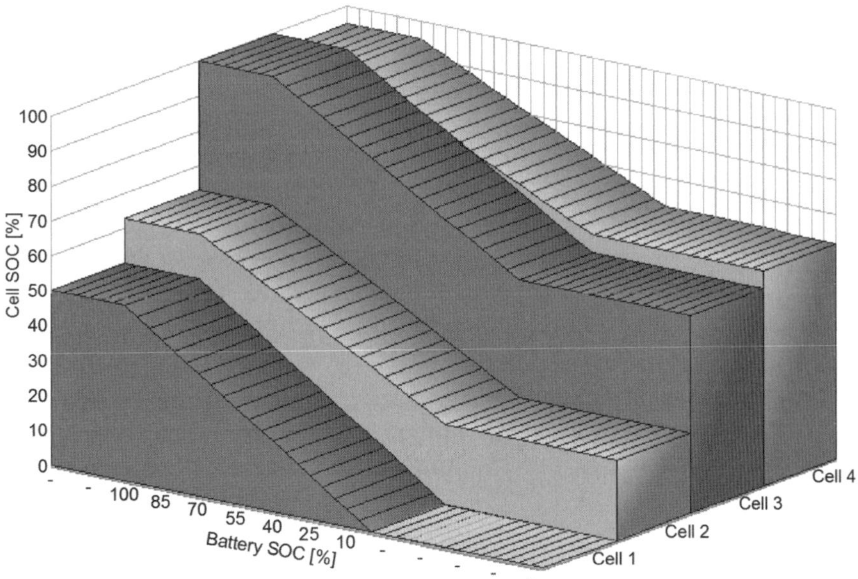

Figure 1.32 SOC levels in a real-world battery, unbalanced.

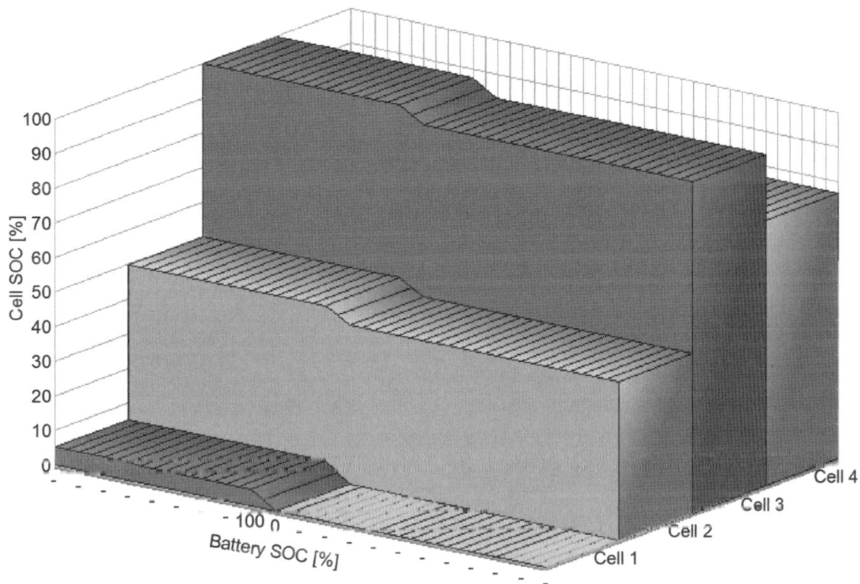

Figure 1.33 SOC levels in a real-world battery, severely unbalanced.

1.4.3 SOH

State of health (SOH) is an arbitrary figure of merit of the actual condition of a battery pack, compared to its nominal condition, expressed in percent points. An SOH of 100% means that the battery's conditions match its specifications. As the battery's conditions worsen, the SOH value is reduced. Below 100%, the meaning of

SOH is arbitrary, and is defined differently by various manufacturers. Additionally, the SOH level below which a battery is considered no longer suitable for a given application is also arbitrary, and is defined differently by various users.

The SOH has no function within the battery and its BMS, but may be useful to a user and to the external system, which can compare the reported SOH to a threshold specific to the application, so that:

- It may determine whether the battery in its present conditions can be used in the application.
- It may estimate the battery's useful lifetime in the application.

SOH evaluation is arbitrary because it does not measure a specific physical quality. A BMS may use a combination of one or more of the following parameters, with arbitrary weight factors, to evaluate the SOH: increase in cell resistance, decrease in actual capacity, number of charge/discharge cycles, self-discharge rate, and passage of time. Sometimes "ability to accept charge" is listed as one of the parameters, but that can be translated to cell resistance, so it is not a separate parameter. The actual formula used by a given BMS to evaluate SOH is often a trade secret.

To overcome the lack of an accepted definition of SOH in the industry, please allow me to offer a definition: Given a set of cells or batteries of the same nominal specification, the performance of a number of such batteries in parallel that do not meet the specification, whose SOH values add up to 100%, is equal to the performance of a single such battery that does meet the specifications.

So, for a simple example, given a set of cells with a nominal capacity of 100 Ah, a BMS would evaluate a 100-Ah cell at 100%, a 70-Ah cell at 70%, and a 30-Ah cell at 30%. Placing the 30-Ah and 70-Ah cells in parallel will result in a 100-Ah battery, which the BMS would evaluate at 100% SOH, which is indeed the sum of the SOH values of the individual cells [Figure 1.34(a)], or, given a set of cells with a nominal resistance of 1 mΩ, a BMS would evaluate a 1-mΩ cell at 100% and a 2-mΩ cell at 50%. Placing two 2-mΩ cells in parallel will result in a 1-mΩ battery, which the BMS would evaluate at 100% SOH, and it is the sum of the SOH values of the individual cells [Figure 1.34(b)].

When calculating SOH, BMS designers must define the interaction of the various parameters, to handle the case in which more than one parameter is below nominal.

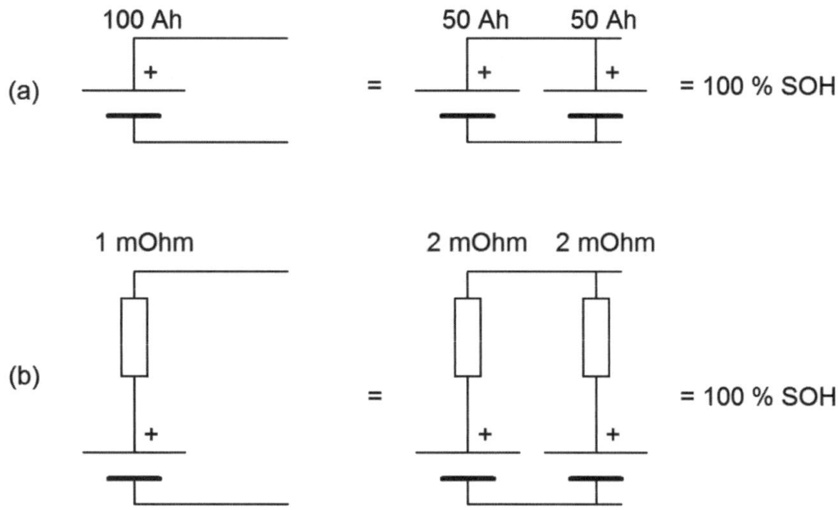

Figure 1.34 SOH of two cells in parallel: (a) two cells of lower capacity than nominal, and (b) two cells of higher resistance than nominal.

References

[1] Dubarry, M., et al., "Capacity and Power Fading Mechanism Identification from a Commercial Cell Evaluation," *Journal of Power Sources*, Vol. 165, 2007, pp. 566–572.

[2] Dubarry, M., and B. Y. Liaw, "Development of a Universal Modeling Tool for Rechargeable Lithium Batteries," *Journal of Power Sources*, Vol. 174, 2007, pp. 856–860.

CHAPTER 2

BMS Options

The wide range of BMS technologies can be broken down by categorizing them based on functionality (what it does), technology (how it does it), topology (how it is physically arranged), and balancing (whether and how it balances a battery).

2.1 Functionality

You can find BMSs with a wide range of functionality, from products that do little or nothing to manage cells (which is actually counterproductive as it gives the user a false sense of security) to complex systems that monitor and protect batteries in every conceivable way.

In this young industry, there often isn't an established terminology for various types of BMSs based on their functionality. Some of the terms I use in this section are industry standard (CCCV, regulator, protector), while others I have chosen to be as descriptive as possible (meter, monitor, balancer). I categorize BMSs by functionality, in increasing order of complexity:

1. Contant current/constant voltage (CCCV) chargers;
2. Regulators;
3. Meters;
4. Monitors;
5. Balancers;
6. Protectors.

2.1.1 CCCV Chargers

Some users of Li-Ion packs rely just on a CCCV charger as their BMS. As we saw in Section 1.2.8, that is an inappropriate choice.

CCCV chargers are standard, regulated power supplies, which are used to charge batteries. They are limited in two ways (Figure 2.1), corresponding to two charging stages:

- CC: When first charging a pack, they produce a fixed constant current and allow the pack voltage to rise as it is charged.

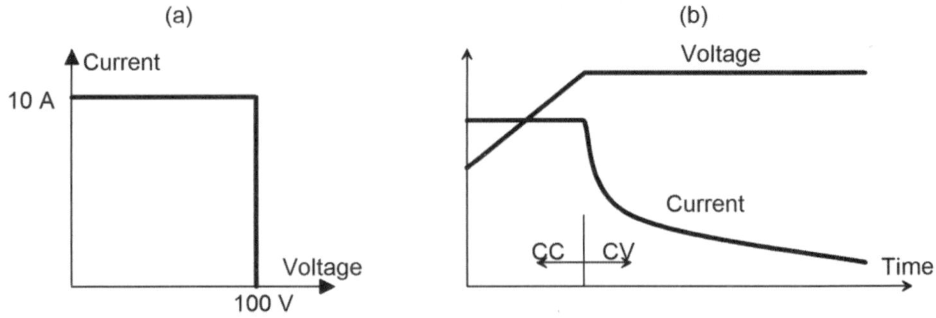

Figure 2.1 (a) CCCV charger characteristics and (b) plot of charge voltage and current.

- CV: When the pack is nearly full, and its voltage reaches the constant voltage, the charger maintains that voltage and the pack current decays exponentially as the pack gets a finishing charge until the pack is full.

CCCV chargers have been used in lead-acid applications, as some feel that they provide sufficient protection. That may be so, but, unlike Li-Ion cells, lead acid batteries are not ruined by a complete discharge, and can handle a certain amount of overcharge.

Some CCCV chargers for large packs are advertised as having a BMS function, because they use a charging profile designed not to overcharge a Li-Ion pack. Expecting a CCCV charger (by itself and with no knowledge of individual cell voltages) to provide protection of a large Li-Ion pack is imprudent. We saw why that simply doesn't work (Section 1.2.8), and why it is imperative to monitor and manage the voltage of each cell in a Li-Ion pack, or else cell voltages can reach dangerous levels. Not only does a CCCV charger provide no protection, but also the false sense of security it provides to the unsophisticated user actually increases the chances of pack damage.

While it would be possible to integrate some BMS functions in a charger (it would need to include many tap wires to sense the voltage of individual Li-Ion cells), doing so just to prevent overcharge would seem a bit ineffective: one might as well use a charger without a BMS function, plus a real BMS, to offer full protection, not just overcharge protection.

A CCCV charger can be useful as part of a system that includes a BMS (Figure 2.2). The BMS turns off the charger when the most charged cell is full, regardless of pack voltage. If the BMS includes balancing, charging can restart when the most charged cell has been discharged a little, so the rest of the cells may receive more charge. Once the pack is balanced, all the cells will reach their maximum voltage simultaneously, with a total voltage exactly equal to the CV of the charger. Finally, the charger is able to do its intended job: complete the charge at a constant voltage and exponentially decaying current, until all cells are equally and completely charged.

Interestingly, allowing a BMS to control a charger has the effect of overriding the charger's built-in intelligence, which means that using a charger with a specialized profile is not required, and may actually be harmful, as the BMS and the smart

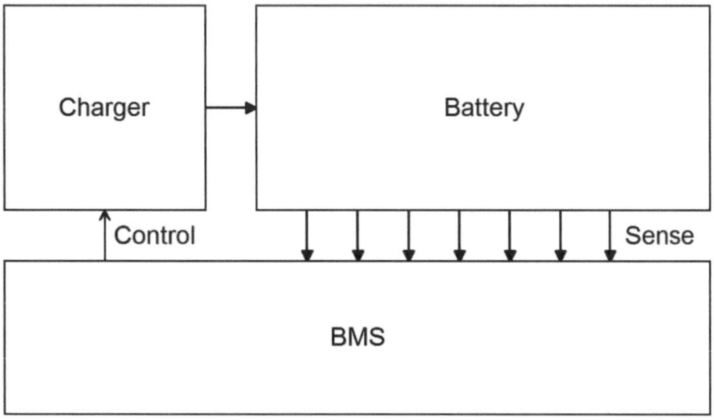

Figure 2.2 Charger controlled by BMS.

charger battle for control. As a matter of fact, when used in conjunction with a BMS, pretty much any charger will do, even a brute force (nonregulated) charger, as long as the BMS can turn it on and off.

In summary, a CCCV charger is not sufficient as a BMS because:

- By itself, it does not prevent overcharging of individual cells.
- It does not prevent overdischarging of cells.
- It does not balance the pack.

2.1.2 Regulators

Some users rely on regulators to balance a Li-Ion pack. While that is a useful function, by itself it provides no protection. Regulators are inexpensive, easy to understand, and therefore are commonly chosen by hobbyists. A regulator is a shunt placed across a cell to bypass some or all of the charging current when the cell is fully charged. In its simplest form, a regulator is a voltage clamp (Figure 2.3), that conducts little or no current up to the full charged voltage of the cell (such as 3.6V for $LiFEPO_4$ or 4.2V for LiPo and standard Li-Ion), at which point it turns on, drawing current trying to maintain that voltage.

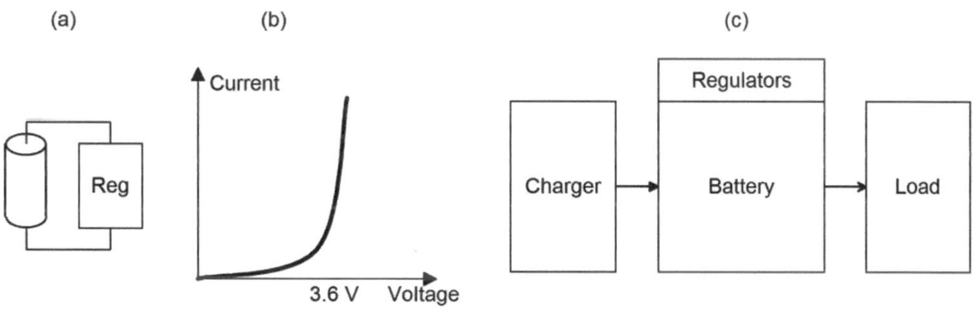

Figure 2.3 (a) A regulator across a cell. (b) Regulator characteristics. (c) A system block diagram using a regulator.

Once all the regulators in a pack are on, all the cells are close to the same voltage and, to a first degree, are all balanced. To a second degree, variations in individual cell resistance mean that the cells will not be exactly balanced, an issue that is discussed later (see Chapter 3).

A problem with regulators is that they can handle only so much current. If the charger can source more current than the regulator is designed for, two things can happen:

- If the regulator is current limited, the excess current goes into the cell, overcharging it.
- If the regulator is not current limited, the excess current goes though it, damaging it.

The solution is to match the charger and the regulators, in one of three ways:

1. Increase the regulators' current capacity to the charger's maximum current (larger, more expensive regulators).
2. Decrease the charger's maximum current to the capacity of the regulators (longer charging time).
3. Program the charger so that, by the time any regulator may be on [based on the pack voltage, but that cannot be relied on (see Section 1.2.8)], its output current is reduced.

By using a matched charger, regulators not only balance the pack, but also prevent overcharging, not by turning off the charger, but by bypassing charging current once all cells are fully charged.

In summary, all that regulators can do is balance a pack. Use of regulators alone is not sufficient as a BMS because:

1. It does not prevent overcharging of individual cells.
2. It does not prevent overdischarging of cells.

Even when used in combination with a matched charger, use of regulators is not sufficient as a BMS because it does not prevent overdischarging of cells.

2.1.3 Meters

Meters simply monitor parameters but do not actively control charging or discharging. They may meet the needs of a hobbyist or a researcher, who may be satisfied with simply knowing the individual cells voltages and doing something manually if anything is amiss. Such a device:

- Measures each cell voltage;
- May also measure other parameters such as pack current and temperature;
- Compiles this data;
- May also compute or estimate pack status, such as SOC;
- Reports them in a display;

- May also include a way to warn the user of abnormal conditions (with a light or a buzzer).

For lack of a better term, we will call these devices "meters" (Figure 2.4).

Users of such meters may think that they have a true BMS, but what they may not realize that they themselves are an integral part of the system in the BMS. Without the user, the system is broken. If the user leaves and lets the pack keep on charging, the control loop in the system is broken, and the pack is likely to be damaged from overcharging (a CCCV charger is of no help, as explained in Section 2.1.1).

In summary, a meter is not sufficient as a BMS because:

1. It does not prevent overcharging of individual cells.
2. It does not prevent overdischarging of cells.
3. It does not balance the pack.

Even if constantly attended by a human, a meter is not sufficient as a BMS because it does not balance the pack.

2.1.4 Monitors

Monitors (Figure 2.5) are like meters in the sense that they measure each cell voltage, but they do close the control loop. Should anything be amiss, they don't rely on a user being nearby and taking corrective action, but take action themselves, by controlling indirectly the charger and the load. A monitor may not optimize a pack's performance (it cannot balance it), but will protect it from operating outside its safe area, and do so autonomously. Monitors are often chosen by researchers to test their Li-Ion packs.

A monitor does not include internally the means of interrupting the pack current. All it can do is to send a request to other devices (charger, load) to reduce or interrupt the pack current. If those other devices are not designed to receive and abide to such requests, a high-power switch (typically a contactor—a high power

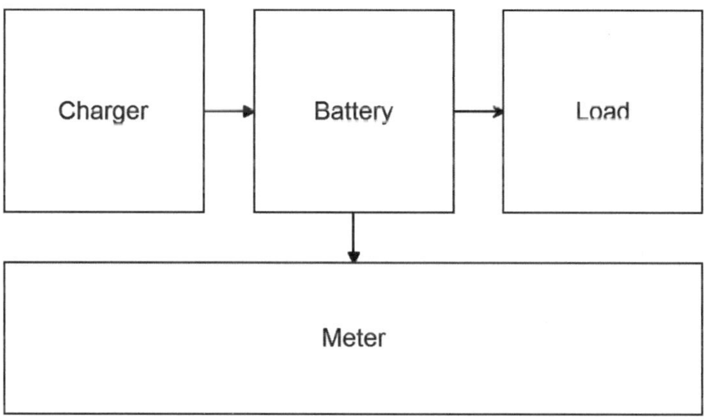

Figure 2.4 System block diagram using a meter.

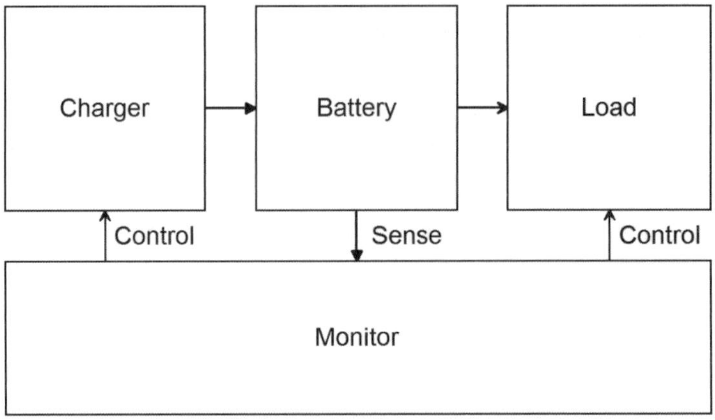

Figure 2.5 System block diagram using a monitor.

relay) must be included in the system, and the balancer must be able to activate it to interrupt the pack current.

A monitor may be standalone (with only a few wires to control the shutting down of the charger and the load), or it may have a display and/or a communications link to report pack data to the rest of the system.

In summary, a monitor provides full protection to the pack, but it does not balance it.

2.1.5 Balancers

A balancer (Figure 2.6) is like a monitor, but it is also able to maximize the pack's performance by balancing the cells. Also, it rarely is a standalone device, as it almost invariably includes a communications link to report pack data to the rest of the system. Balancers are by far the preferred choice by professionals for large Li-Ion packs.

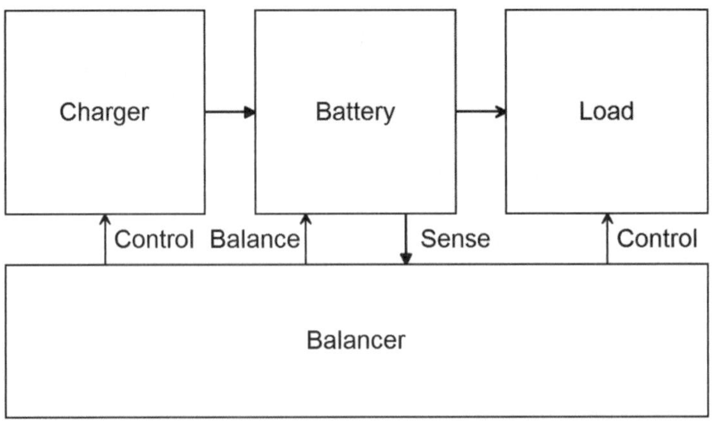

Figure 2.6 System block diagram using a balancer.

A balancer may be physically separated from the batteries, or mounted directly on the cells, or some combination within that range (see Section 2.3), and may use various balancing technologies (see Section 3.2.3), some more effective than others.

In summary, a balancer is sufficient as a BMS as long as it is wired in a way that allows it to control the charging source and the discharging load.

2.1.6 Protectors

A protector is like a balancer, except that it includes a switch to turn off the pack current.

A protector is usually an integral part of a battery, physically located inside the same enclosure, leaving only two power terminals coming out of its enclosure (Figure 2.7). A protector is the industry standard solution for small batteries for consumer products, but it is rarely used in professional, large Li-Ion packs, because its switch is not likely to be able to handle high-power loads. For large packs, a balancer is usually chosen instead (see Section 2.1.5).

The switch in a protector is usually a solid-state type (transistors), able to handle 5A to 50A, and on the order of 50V in both the charging and discharging directions (this requires two sets of transistors in series, one for each direction of the pack current). This power level is usually sufficient just for small batteries.

In summary, a protector is completely sufficient as a BMS for smaller batteries.

2.1.7 Functionality Comparison

As you can see from Table 2.1, full BMS functionality can only be provided by balancers (for large packs) and protectors (for small batteries).

2.2 Technology

There are basically two classes of technologies used to implement a BMS: analog and digital. The distinction between analog and digital is related to how the cell

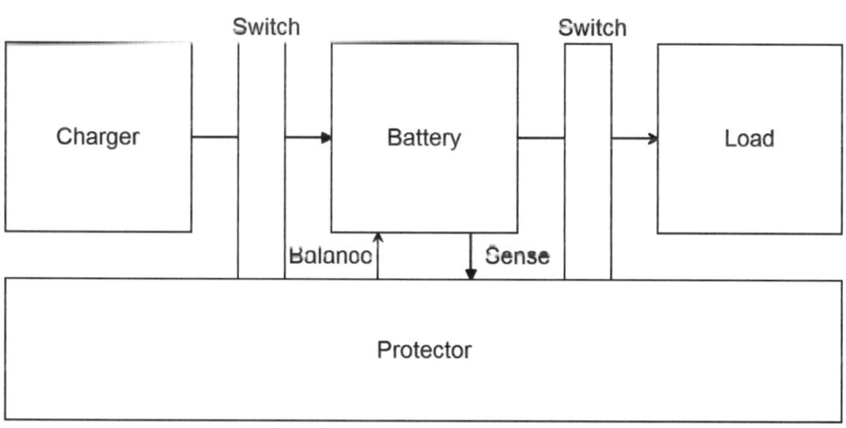

Figure 2.7 System block diagram using a protector.

Table 2.1 Comparison of the Features of the Various BMS Functionalities

					Protect	
	Measure	*Compute*	*Report*	*Balance*	*Charge*	*Discharge*
CCCV Chargers						
Regulators				✔		✔*
Meters	✔	(✔)	✔			
Monitors	✔	(✔)	✔		✔*	✔*
Balancers	✔	(✔)	(✔)	✔	✔*	✔*
Protectors	(✔)	(✔)	(✔)	✔	✔	✔

(✔) = Some may provide this feature
✔* = Requires additional measures.

voltage information is processed. While all systems require some form of analog front end, BMSs that then process the cell voltage with analog circuitry (an analog comparator, op-amp, differential circuit, or the like) is considered to be analog. BMSs that process the cell voltage digitally are considered to be digital.

Alternatively, one could say that there are two classes of complexity in BMS: simple and sophisticated.

Sophisticated BMSs are aware of each individual cell's voltage and can report it, while simple ones do not and cannot. These two sets of BMS classes tend to coincide: simple BMSs are typically analog, and sophisticated BMSs are typically digital. It is not inherent that analog solutions be simple, and digital ones be sophisticated. It is just that designers who are content with simple solutions are more comfortable with analog circuits, and those who have a full grasp of the challenges involved in managing Li-Ion batteries will opt to solve them with digital technology.

2.2.1 Simple (Analog)

The capabilities of analog BMSs are quite limited, and may be barely sufficient to perform the required BMS functions. For one thing, an analog BMS is not aware of individual cell voltages; it may know that one cell is too low, but it will not know which one, or how low. As long as the BMS can shut off the load in case of low cell voltage, not knowing which cell is low, or how low it is, is not a problem. It only becomes a problem if there is a need to analyze a battery without actually opening it up and start poking around with a voltmeter.

The example of an analog regulator may help (Figure 2.8). The regulator may use a power supply supervisor IC, powered by the cell's voltage, and driving a balancing shunt when the cell's voltage exceeds the IC's threshold voltage. Inside the IC there are two components: a voltage reference and an analog comparator, whose output changes state when the cell voltage (divided down) exceeds the reference's voltage. Because of that analog comparator, a regulator is considered an analog device.

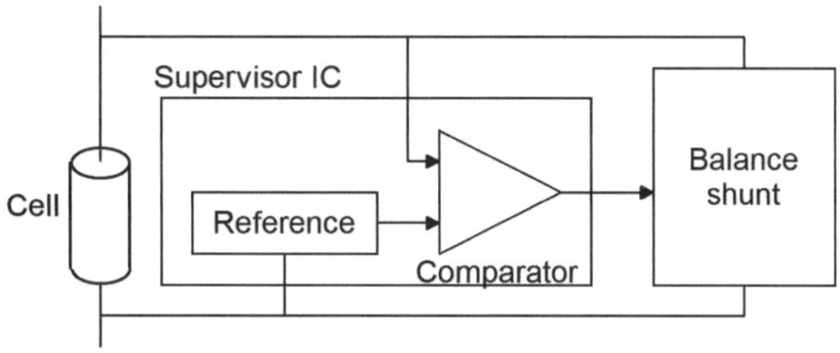

Figure 2.8 Example of an analog BMS: a regulator.

2.2.2 Sophisticated (Digital)

A digital BMS is fully aware of each cell voltage (and possibly more, such as its temperature and its conditions). Therefore, it is able to report this data, which is invaluable in allowing an analysis of the state of the pack. This is normally a requirement for a large, professional Li-Ion pack.

An example of a digital meter may help (Figure 2.9). This device includes an analog multiplexer, able to select and sample the voltage at one of the taps between adjacent cells in series, and send it to an analog-to-digital (A/D) converter. From that point on, the BMS performs all its functions digitally, such as subtracting the reading from two adjacent taps to calculate the voltage of the cell between those two taps.

2.2.3 Technology Comparison

Table 2.2 compares the difference between analog and digital BMS technology.

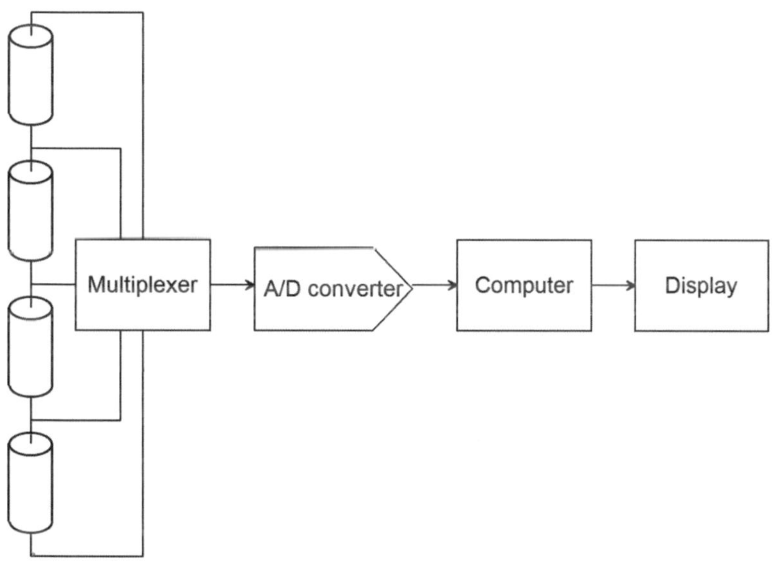

Figure 2.9 Example of a digital BMS: a meter.

Table 2.2 Comparison of the Difference Between Analog and Digital BMS Technology

	Analog (Simple)	*Digital (Sophisticated)*
Cell Voltages	Knows that some cell is low, but not which one, or how low	Knows each cell's voltage
Threshold Detection	Analog comparator	Digital comparator

2.3 Topology

BMSs can be categorized based on how they are installed: separately and directly on each cell, altogether in a single device, or in some intermediate form. This is a far more important characteristic of a BMS that it would first appear, as it affects cost, reliability, ease of installation and maintenance, and measurement accuracy. Here, BMSs are divided according to functionality as: centralized, master-slave, modular, and distributed.

2.3.1 Centralized

A centralized BMS (Figure 2.10) is entirely located in a single assembly, from which a bundle of wires ($N + 1$ wires for N cells in series) goes to the cells. Using only one assembly has several advantages:

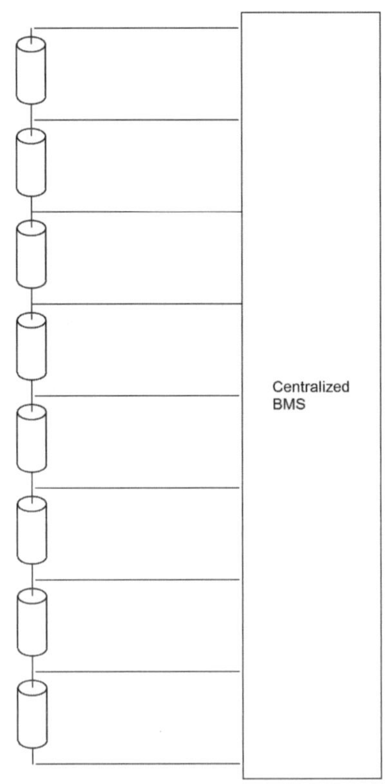

Figure 2.10 Centralized topology block diagram.

- It is compact.
- It is the least expensive approach (it is cheaper to group electronics in a single assembly than splitting it into multiple assemblies).
- If troubleshooting or repair is required, it is easier to replace just a single assembly.

An example of an off-the-shelf, centralized BMS is Convert The Future's Flex BMS48 (see Chapter 4).

2.3.2 Modular

A modular BMS (Figure 2.11) is similar to a centralized one, except that the BMS is divided into multiple, identical modules, each with its bundle of wires going to one of the batteries in the pack. Typically, one of the modules is designated as a master, as it is the one that manages the entire pack and communicates with the rest of the system, while the other modules act as simple remote measuring devices. A communication link transfers the readings from the other modules to the master module.

The modular topology has most of the same advantages as the centralized topology. In addition,

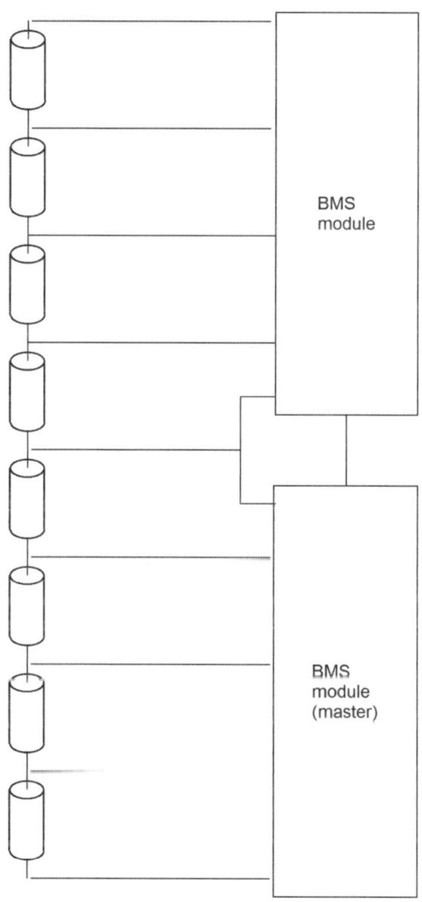

Figure 2.11 Modular topology block diagram.

- The wires to the cells are easier to manage: each module can be placed close to the battery it handles.
- Expansion to larger packs is straightforward: more BMS modules are added.

However,

- The BMS cost is slightly higher than for the centralized topology: slave modules have duplicated, unused functions.
- A few extra tap wires are required, as taps served by two modules need two wires, one to each module.
- As each module can handle a fixed number of cells, it is quite possible that some of its inputs remain unused when it is physically easier to have a few more modules that are better located on the pack, than a lot of long wires that use every available input.

An example of an off-the-shelf, modular BMS is Reap Systems' 14-cell digital BMS (see Chapter 4).

2.3.3 Master-Slave

A master-slave BMS (Figure 2.12) is similar to a modular system, in the sense that it uses multiple identical modules (the slaves), each measuring the voltage of a few

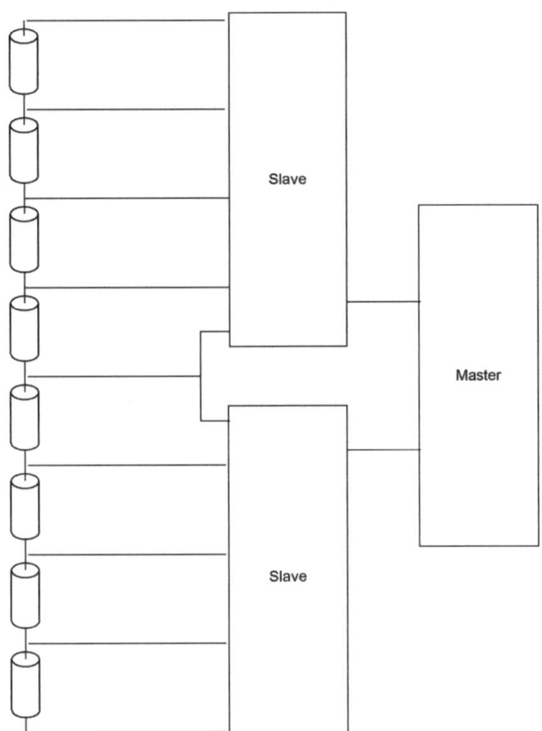

Figure 2.12 Master-slave topology block diagram.

cells. However, the master is different from the modules, and does not measure voltages—it only handles computation and communications.

The master-slave topology has most of the same advantages and disadvantages as the modular topology. In addition, the cost of each slave tends to be less than for the modular topology, as it is optimized just for one job—measuring cell voltages.

An example of an off-the-shelf, master-slave BMS is Black Sheep's BMS_Mini_V3 (see Section 4.2.2.1).

2.3.4 Distributed

A distributed BMS (Figure 2.13) is significantly different from the other topologies (in which the electronics are grouped and housed separately from the cells). In a distributed BMS, the electronics are contained on cell boards that are placed directly on the cells being measured. Instead of many tap wires between cells and electronics, a distributed BMS uses just a few communication wires between the cell boards and a BMS controller. The BMS controller handles computation and communications (a BMS controller is not required in some simpler implementations). An

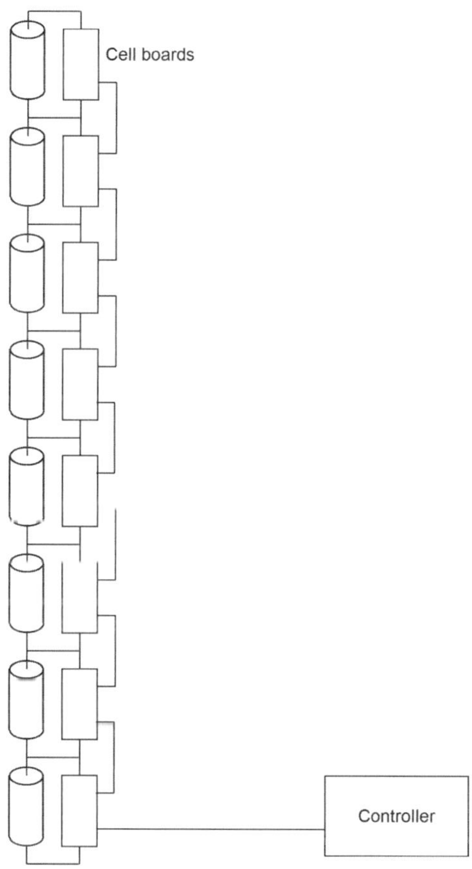

Figure 2.13 Distributed topology block diagram.

example of an off-the-shelf, distributed BMS is EV Power's BMS-CM160-V6 (Fig 4.1.1.2.3).

The distributed topology has significant advantages and disadvantages with respect to other topologies, as discussed in Table 2.3. There is not a clear choice, as it

Table 2.3 Comparison of Distributed and Nondistributed Topologies

	Distributed	*Nondistributed*
Cost	Higher: electronic components are required on each cell board, and many more assemblies are needed.	Lower: fewer assemblies are required, with fewer electronic components.
Connection to Cells	Placing electronics on cells is a novel concept that requires a learning curve on the part of the user.	Using wires between the cells and the measuring electronics is an obvious approach that the user understands easily.
Connection Reliability	Direct connection between cells and measuring electronics requires fewer parts and offers higher reliability. With some cell formats, it is possible to solder or weld the cell boards directly to the cells, increasing reliability further. However, the communication links add some potential failure points.	Tap wires use some five connection points (contacts and crimps), which are all potential failure points. On the other side, there are few or no other connection points for communications.
Installation Ease	More skill required to install cell boards, but no more time. Cell boards that handle multiple cells can be installed very rapidly.	Less skill required to install tap wires, but no less time.
Installation Errors	Less prone to wiring errors, as cell boards can only fit across one cell.	More prone to errors, as tap wires can fit in many spots.
Detailed Troubleshooting	Detailed troubleshooting may be aided by LEDs on the cell boards.	Detailed troubleshooting is not cost effective.
Replacement Assembly Cost	Just the less expensive cell board can be replaced.	Complete, more expensive assemblies must be replaced.
Replacement Labor	High, because cell boards are deep inside the battery.	Low, because the electronic assemblies are not directly inside the battery.
Measurement Precision	Placement of electronics right on the cells allows low-noise, high-resolution voltage readings, unaffected by voltage drops due to relatively high currents during the balancing process.	Measurements cannot be as precise due to errors introduced in the tap wires, especially due to voltage drops during the balancing process.
Temperature Measurement	Placing a sensor on each cell board to measure each cell's temperature is straightforward.	Temperature measurements require additional wires to sensors in the pack (typically just a few sensors: temperature of individual cells is not known).
Battery Electrical Noise Immunity	Cell board electronics are immersed in electrical noise, which may affect their performance. The communication wires are exposed to electrical noise, which may cause errors.	The electronics can be placed away from electrical noise. Any noise picked up by the tap wires can be easily filtered, as they only carry DC.
Expansion Versatility	Adapting a BMS to a given number of cells only involves changing the number of cell boards, and no input remains unused.	Adapting a BMS to a given number of cells, without a redesign to match the number of cells, may leave some inputs unused.

Table 2.3 (continued)

Loss of Isolation	Communication wires are referenced to ground: a short to the high voltage results in a loss of isolation.*	Tap wires are referenced to the high voltage: a short to the chassis results in a loss of isolation.
High Voltage Shorts	Very few wires (the communication wires) run through the high voltage section, reducing the risk of shorts. Each cell voltage is contained in a very small area, minimizing the chance of shorts to other voltages.*	Hundreds of wires (the tap wires) run through the high-voltage section, increasing the risk of shorts considerably.
Plasma Event	In case of a plasma event, the BMS adds a small amount of copper, which can provide paths to the plasma current.	In case of a plasma event, the BMS adds a lot of copper that can provide paths to the plasma current.

*Use of wireless communication links (such as fiber optics) eliminates this risk.

depends a lot on the specific needs of a given application: safety, costs (parts, assembly, and maintenance), and reliability are all affected.

2.3.5 Topology Comparison

Table 2.4 compares the various topologies.

Table 2.4 Comparison of BMS Topologies

	Measurement Quality	*Noise Immunity*	*Versatility*	*Safety*	*Electronics Cost*	*Assembly Cost*	*Maintenance Cost*
Centralized	✔✔	✔✔✔	✔	✔	✔	✔✔	✔
Master-Slave	✔✔	✔✔✔	✔✔	✔	✔✔✔	✔✔	✔
Modular	✔✔	✔✔✔	✔✔	✔	✔✔✔	✔✔	✔
Distributed	✔✔✔	✔✔	✔✔✔	✔✔✔	✔✔✔	✔	✔✔

✔✔✔ = Best; ✔✔ = Better; ✔ = Good

CHAPTER 3

BMS Functions

In this chapter we will explore the various functions of a BMS (Figure 3.1). They may be grouped as:

- Measurement;
- Management;
- Evaluation;
- Communications;
- Logging.

A given BMS may implement any of these functions (Figure 3.1).

3.1 Measurement

The first function of a sophisticated, digital BMS is to gather data about a BMS (a simple, analog BMS does not include this function). These measurements are:

- Cell voltage (and possibly pack voltage);
- Typically, cell temperature, or at least battery temperature;
- Most often, pack current.

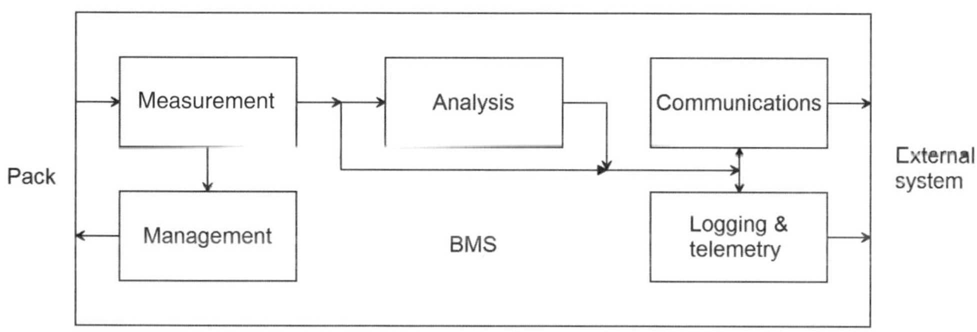

Figure 3.1 BMS functions.

3.1.1 Voltage

A sophisticated, digital BMS measures the voltage of each and every cell in series. It may also measure the total pack voltage, though that is not necessary, as that value can be calculated by adding the individual cell voltages.

3.1.1.1 Methods

A distributed BMS may measure directly the voltage across a cell. (Normally the cell board is powered by the cell itself, as it measures its voltage.) Otherwise, the BMS may measure the voltage of various taps in a battery, and calculates a cell's voltage as the voltage difference between two taps. Or, the BMS may take two measurements simultaneously of the two taps on either side of a cell, and calculate the difference as the cell's voltage. The voltage is sampled by an analog multiplexer, and the reading is taken by an analog to digital (A/D) converter (which may be on the same IC), which then passes the value to a processor (Figure 3.2).

3.1.1.2 Rate

The rate of the readings depends on the application.

- For a backup power application, one reading per minute or per 10 seconds is acceptable.
- For applications where the current varies rapidly (such as vehicles) one reading per second is acceptable.

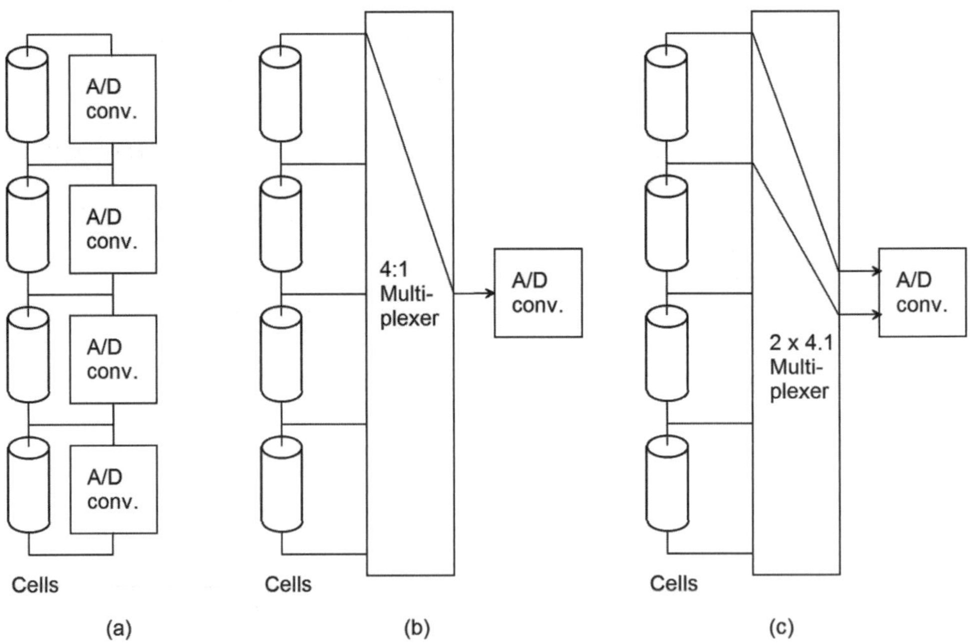

Figure 3.2 Cell voltage measurement methods: (a) discrete, (b) single-end multiplexed, and (c) differential multiplexed.

- For research application, scientists may like to see 10 to 100 readings per second.

If the BMS must calculate the internal resistance of the cells, and the pack current varies rapidly, the voltages of all cells should be measured simultaneously, or almost simultaneously, so they may all be correlated to what the pack current was at the time of measurement. Otherwise, there will be errors in the calculations of resistance for the cells whose voltage was measured at a different time than when the pack current was sampled. As long as the readings were all taken simultaneously, they may be reported sequentially, afterwards.

3.1.1.3 Accuracy

The required accuracy for the reading depends on its use.

- To simply detect when a cell is fully charged or discharged, an accuracy of 100 mV is sufficient, because the OCV versus SOC curve for a Li-Ion cell is very sharp at the top and bottom ends. The voltage rises some 200 mV at the charged end and drops on the order of 500 mV at the discharged end.
- To top balance a battery while charging, an accuracy to 50 mV is sufficient to match the SOC of the cells. Any higher accuracy in the reading may be lost in the errors due to cell resistance.
- To accurately estimate a cell's SOC from its real or estimated open circuit voltage (OCV) at the either end of the OCV versus SOC curve, an accuracy of 10 mV or better is required. Otherwise, the SOC estimate will suffer from errors on the order of 10%.
- To accurately estimate a cell's SOC from its OCV in the central plateau of the OCV versus SOC curve (20% to 80% SOC) an accuracy of 1 mV or better is required. This is a requirement for BMSs that are able to estimate the cell's SOC without knowing its previous history. For example:
  - Products that have a BMS that stays with the product (not with the removable battery);
  - An HEV, whose pack never becomes fully charged or discharged, and therefore is not able to calibrate the SOC at the end of charge or discharge.

Table 3.1 lists the required A/D resolution (based on a 5-V full scale) and total tolerance in the reference and voltage dividers to achieve the accuracies listed previously.

The accuracy of most BMSs is between 10 and 30 mV. Very few BMSs have the high precision electronics that are required to achieve higher accuracy.

3.1.1.4 Isolation

Cell voltages are, of course, at high-voltage potential, which, in a large pack, must be isolated from the low-voltage, ground-referenced signals. Therefore, measuring the cell voltage brings the challenge of doing so without introducing a loss of isolation in the pack. We will explore this issue in more detail when discussing BMS

Table 3.1 Required A/D Resolution and Total Tolerance to Achieve a Given Measurement Accuracy

Accuracy	*Resolution*	*Tolerance*
100 mV	6 bits	1%
30 mV	8 bits	0.25%
10 mV	9 bits	0.1%
1 mV	12 bits	0.01%

designs (Section 5.3.1.1). At this point we just need to emphasize that there needs to be isolation between the pack and the low-voltage signal somewhere in the BMS. A loss of isolation would not necessarily be a problem from the standpoint of the electronics, but would introduce a safety risk, as it would provide a potential current path and a user could provide the rest of that path and receive an electrical shock.

3.1.2 Temperature

Temperature measurement of the pack or, better yet, of individual cells is useful for a few reasons:

- Li-Ion cells must not be discharged if outside a certain temperature range, and not be charged outside an even tighter range, which is a concern in applications that are not temperature controlled, such as mobile applications.
- Should a cell become particularly hot due to internal problems (the cell is bad or is being abused) or external ones (poorly done power connection, localized heat source), it is best to warn the system than to wait for catastrophic failure.
- In a distributed BMS, it is very easy to include a sensor on each cell board, not only sense its cell's temperature, but also to detect whether a balancing load is working.

Digital BMSs may or may not measure temperature, while very few analog BMSs do so. Those that do, do so at the battery level. Distributed BMS may measure each cell's temperature; nondistributed BMS measure the battery or even just the pack temperature (Figure 3.3). If the BMS has just one or a few temperature sensing inputs, the sensors should be placed in strategic places around the battery or pack, such as the spots that are the most likely to be the hottest or the coldest.

3.1.3 Current

Knowledge of the battery current allows a BMS to perform additional functions, which, while not essential, are expected to be offered by a professional product. These are, in order of likelihood that a particular BMS will implement that function:

- Prevent the cells in the battery from being operated outside their SOA in terms of continuous current (analog BMSs that measure battery current usually implement just this one point).

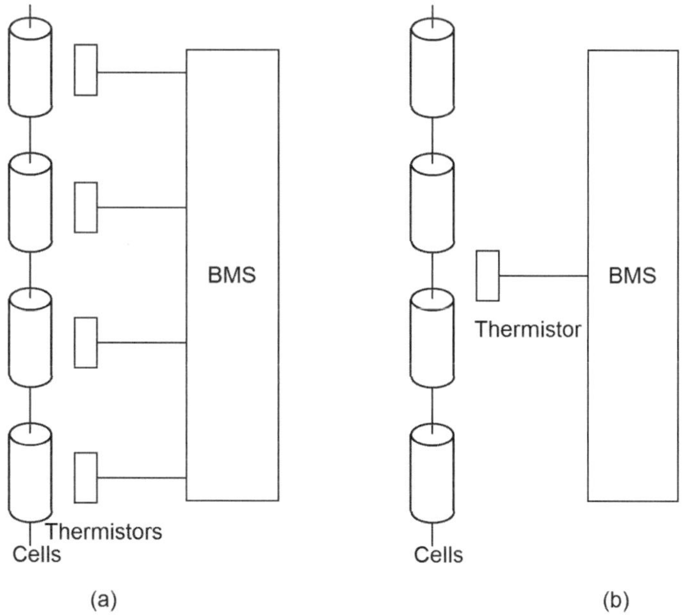

Figure 3.3 Temperature sensing: (a) one per cell and (b) one per battery.

- Use an integral of the current as part of the DOD calculation, to implement a fuel gauge function.
- Simply report the battery current.
- Prevent the cells in the battery from being operated outside their SOA in terms of both peak and continuous current.
- Calculate the cells' internal DC resistance.
- Use the battery current, together with the calculated internal DC resistance, to do IR compensation of the cells' terminal voltage.

There are mainly two ways of measuring high currents:

- Current shunt: a very low-resistance, high-precision resistor;
- Hall effect current sensor.

3.1.3.1 Current Shunt

A current shunt is simply a high precision, low value, high power resistor (Figure 3.4). The pack current is routed through the shunt, which results in a voltage drop across it proportional to the current. Special attention is made to avoid errors due to how the connections are made:

- The end portions of the shunt are particularly massive, compared to the resistive element between them, so that the resistance is barely affected by the way the connection is made;
- The power connections are separate from the sense connections, and further out (called a *Kelvin connection*).

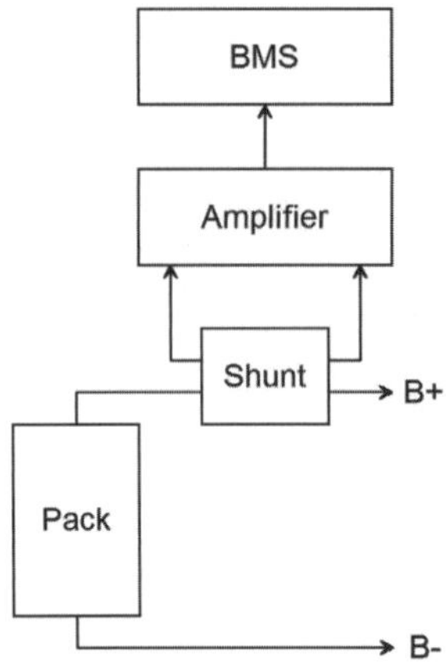

(a)

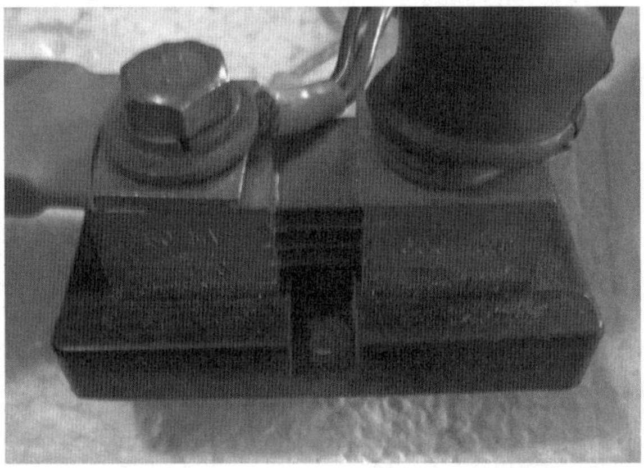

(b)

Figure 3.4 Shunt current sensors: (a) circuit and (b) example.

That voltage across the sense connections can be amplified and measured to derive the pack current. Current shunts are characterized by the following:

- Current shunts have no offset at 0 current, regardless of temperature, so they are good to avoid drift in coulomb counting (though offset is introduced by the accompanying electronics).
- Current shunts are not isolated from the pack. For large battery packs, the BMS must provide some form of isolation.
- The resistance of a current shunt changes with temperature, which introduces errors.
- A shunt sensor introduces some energy losses.
- A shunt current sensor produces a tiny signal (on the order of millivolts full scale). The BMS must provide an amplifier, and any wiring between them must be protected from electrical noise interference, typically by using shielded, twisted pair wiring.

3.1.3.2 Hall Effect Sensor

A Hall effect sensor is placed inside the magnetic field produced by a cable that carries the pack current, and it produces a voltage that is proportional to that current; that voltage can be measured directly (Figure 3.5). High-current Hall effect sensors are modules shaped like a ring, through whose opening a cable carrying the pack current is routed. Low-current Hall effect sensors are ICs with two power terminals, through which the current is routed. Hall effect sensors are characterized by the following:

- The current reported by a Hall effect sensor remains accurate over time and temperature.
- Hall effect sensors are isolated from the pack current and therefore no isolation is needed.
- Hall effect sensors suffer from offset at 0 current, which changes with temperature. So, even if they are zeroed at room temperature, they will report a small current when there isn't one as they get hot or cold. Frequent calibration is possible in applications that have periods of 0 current, such as HEVs.

Hall effect current sensors are modules that includes their own amplifier, so, unlike the signal of current shuts, their output is at a high level. They can be powered by one supply (5V) or two supplies (+/− 12V or +/− 15V), and they can be unidirectional (can only see current in one direction) or bidirectional (can see both charging and discharging current). Based on that, their output can be referenced to ground (0V at 0A), or have an offset (typically 2.5V at 0A). In particular, the output of a two-supply bidirectional sensor is bipolar—it will swing above and below ground.

The analog input on a BMS needs to be compatible with the output voltage of the current sensor: 0- to 5-V output, or −12V to +12V. To make a bipolar current

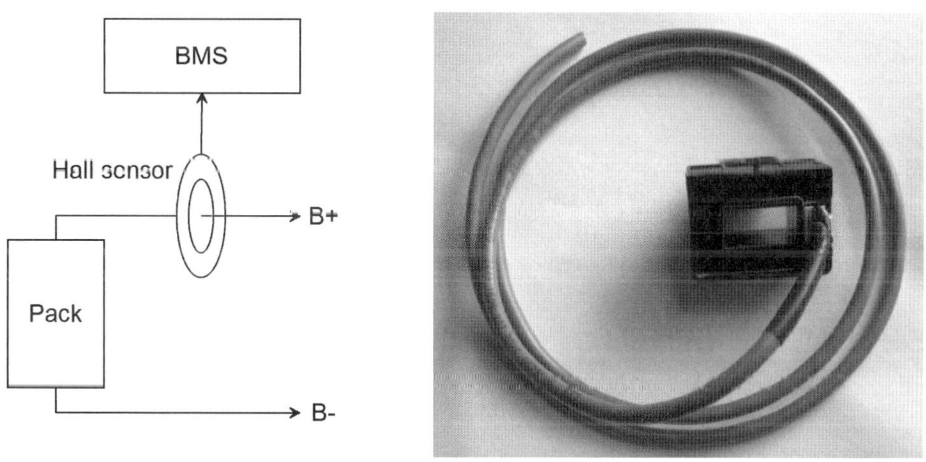

(a) (b)

Figure 3.5 Hall effect current sensors: (a) circuit and (b) cable mount sensor module.

sensor work with a 0- to 5-V input, a 2:1 voltage divider is required, with one resistor in series with the signal and the other resistor between the BMS input and a 5-V supply.

3.2 Management

A BMS may manage a pack in three ways:

- *Protection:* not allow the battery to be damaged by usage outside its SOA;
- *Balancing or redistribution:* maximizing the pack's capacity;
- *Thermal management:* actively attempt to bring the battery into its safe area.

A simple, analog BMS may implement the protection and/or balancing. A digital BMS will implement a fair portion or all of these functions.

3.2.1 Protection

A good BMS will protect its pack by preventing operation outside of the cells' SOA, based on various conditions. Depending on its type, a BMS may do so by interrupting the current or requesting that it be interrupted (on-off control) or reduced (analog current limiting).

3.2.1.1 Monitored Conditions

A good BMS will protect its pack by preventing operation outside of the cells' SOA (Figure 3.6) based on one or more of the following conditions:

- Pack current;
- Cell voltage (plus, possibly, pack voltage);
- Cell (or pack) temperature.

Pack Current

Li-Ion cells have different current limits for charging than they do for discharging, and can handle high peak currents for a short period. Therefore, cell manufacturers specify as many as four parameters:

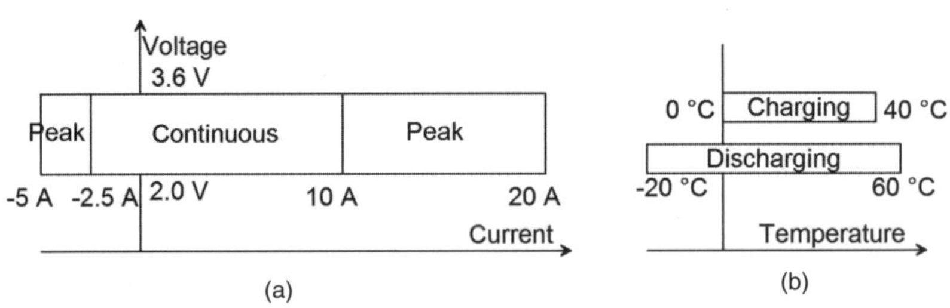

Figure 3.6 Example of a safe operating area for a Li-Ion cell: (a) current and voltage and (b) temperature.

- Continuous charging current;
- Peak charging current;
- Continuous discharging current;
- Peak discharging current.

If a BMS does any protection at all, it is likely to have only a fixed setting for maximum battery current, regardless of direction or duration. Only the more sophisticated BMSs will offer four separately adjustable settings for these four current limits.

A BMS that is able to discriminate between continuous and peak currents will include an algorithm to integrate the excess current during a peak, to determine when it is time to start reducing, or even to interrupt the pack current (Figure 3.7). Such an algorithm will have a nearly instantaneous reaction to extreme current peaks (such as due to a short circuit that has not already blown a fuse) and a more forgiving response to current pulses that are not excessive so they may last for a few minutes (such as from accelerating a vehicle).

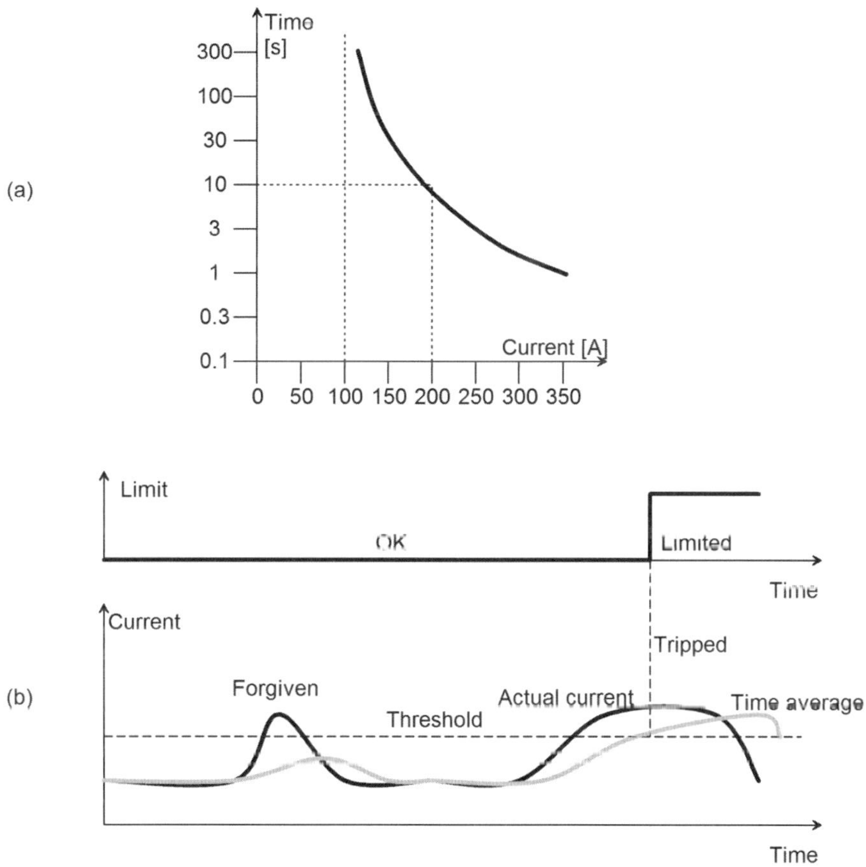

Figure 3.7 Response to continuous and peak current: (a) response time assuming limits of 100A continuous and 200A, 10-second peak; and (b) plot of delayed response to a current peak.

When approaching a limit, the BMS may request that the current be reduced (such as by gradually reducing the torque available from a motor). That creates a feedback loop, where the higher current results in the BMS requesting the current to be reduced, which the system abides to, reducing the current used. The result in some applications is that the current may settle at a compromise value and remain there (depending on the loop gain, it could also result in oscillations).

Cell Voltage

Because each cell's voltage must be held within a certain range, the BMS must have a way of knowing if any of the cells is close to or at one of those voltage limits. Some BMS will simply shut down the battery current if either limit is reached. Others will request a gradual reduction of battery current as the limit is approached.

When approaching the high limit, the BMS may request that the charging current be reduced. Some chargers allow control of the charging current, in which case the cells will be charged at a decreasing current as the pack approaches its fully charged state.

When approaching the low limit, the BMS may request that the discharging current be reduced (such as by gradually reducing the torque available from a motor). In a vehicle, if this were to happen suddenly during a quick acceleration, it would be disconcerting to the driver and even dangerous. Therefore, it is best if the BMS uses an average of the minimum cell voltage to determine the current limit, instead of the instantaneous value. That way, quick accelerations can be forgiven, but long lasting climbs up a mountain result in a request that the current be reduced (Figure 3.8).

Pack Temperature

The temperature of a Li-Ion pack must be within a range for discharge, and an even smaller range for charge. Some BMS will simply shut down the battery current if the pack is too cold or too hot; others will request a gradual reduction of battery current as the limit is approached; and some will try to bring the temperature back into range, through thermal management (see Section 3.2.2).

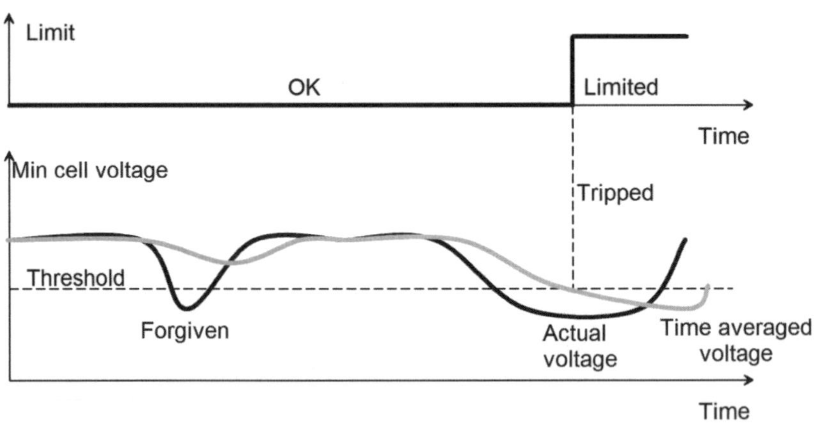

Figure 3.8 Delayed reaction to low cell voltage: a short pulse doesn't result in a fault, a long pulse does.

Pack Voltage

A BMS does not need to monitor the pack voltage for the sake of the pack itself. However, the pack voltage may matter to the rest of the system. For example, a fully balanced pack with 100 LiPo cells in series may reach 420V, yet a motor controller connected to the pack may be rated for up to 400V. While the BMS will prevent any individual cell from exceeding 4.2V, it may also be able to prevent the pack voltage from exceeding 400V. If the pack is not fully balanced, when the BMS stops the charger at a pack voltage of 400V, just some cells are up to the full 4.2V, but the average cell voltage is 4.0V. Once the pack is fully balanced, when the BMS stops the charger at a pack voltage of 400V, all cells are at 4.0V.

3.2.1.2 Current Interruption

Depending on its type, the BMS may protect the pack from operating outside these limits in various ways (Figure 3.9):

- *Monitors and balancers:* Sends requests to reduce or stop battery usage;
- *Protectors:* Interrupts the battery current directly;
- *Others:* Do not do anything about it.

Request for Interruption

In case of operation near or at these limits, monitors and balancers protect the pack by requesting the outside system to reduce or stop using the pack. They can do so with dedicated lines and/or with data on a communication link, and with a linearly variable value and/or an on-off signal (Figure 3.10). These signals may be:

- *Discharge current limit (DCL):* The BMS outputs this signal for the purpose of controlling the pack's discharge current. Normally is it at a nominal value. As the BMS sees that the conditions come close to where discharging would be unsafe, it reduces this signal to request that discharge current be reduced, all the way down to 0. This signal can be communicated to the external system in either or both of these ways:

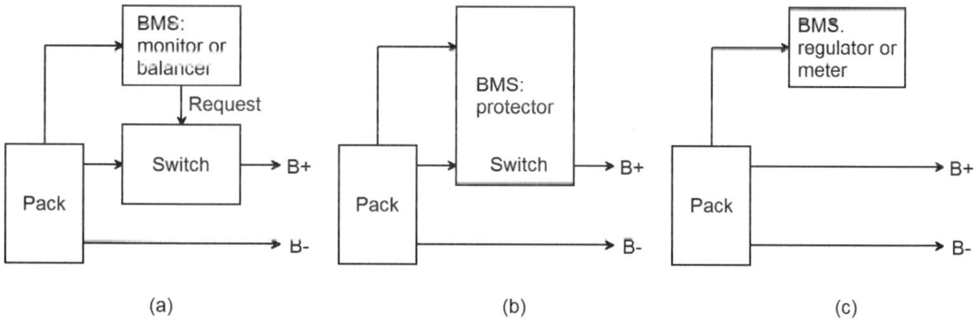

Figure 3.9 Current interruption methods: (a) request for switching off, (b) direct switching off, and (c) neither.

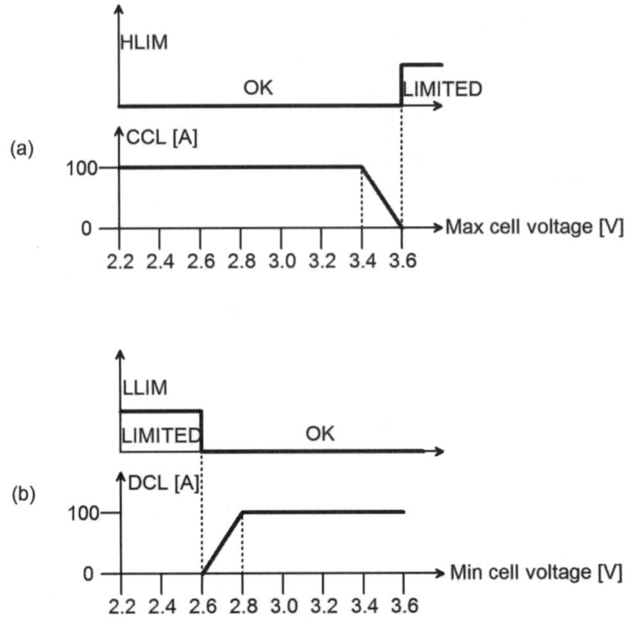

Figure 3.10 Current reduction and interruption requests based on cell voltage: (a) charge limit and (b) discharge limit.

 - *Analog DCL:* A dedicated line with a voltage that is proportional to the DCL (e.g., 5V if no limit, down to 0V if no discharge current is allowed);
 - *Data DCL:* A value on a communication link, either in Amps, or in percentage (e.g., 300A if no limit, down to 0A if no discharge current is allowed).
- *Charge current limit (CCL):* Similarly, the BMS outputs this signal for the purpose of controlling the pack's charge current. Normally is it at a nominal value. As the BMS sees that the conditions come close to where charging would be unsafe, it reduces this signal to request that charge current be reduced, all the way down to 0.
 - *Analog CCL:* A dedicated line with a voltage that is proportional to the CCL;
 - *Data CCL:* A value on a communication link, either in Amps or percentage.
- *Low Limit (LLIM or LVL):* When the BMS sees that the conditions exceed the pack's safe operating area for discharging, it generates this signal to request that that discharge be stopped altogether. That signal can be communicated to the external system in either or both of these ways:
 - *Digital:* a dedicated line, that is open or closed (such as done by a switch), or with a logic level that is high or low;
 - *Data:* A bit on a communication link.
- *High Limit (HLIM or HVL):* Similarly, the BMS generates this signal when it sees that the conditions exceed the pack's safe operating area for charging.
 - *Digital:* A dedicated line with a logic level that is high or low (or that is open or closed);
 - *Data:* A bit on a communication link.

- *Reserve (or Valet Mode, or Limp Home):* When the BMS sees that the conditions approach the pack's safe operating area for discharging, it generates this signal to request that that discharge be limited to a reduced level. Some motor controllers offer a valet input to enable the user to limp home at a reduced power. That signal can be communicated to the external system in either or both of these ways:
  - *Digital:* A dedicated line with a logic level that is high or low (or that is open or closed);
  - *Data:* A bit on a communication link.

Direct Interruption

Protectors, on the other hand, take matters into their own hands by interrupt the battery current themselves in case of operation outside the pack's safe operating area. The battery current is routed through the protector, through a switch that the protector controls. The fact that the protector doesn't need to rely on the rest of the system to shut down the current is advantageous. However, the abrupt shut down can be at best inconvenient and at worst dangerous.

A protector can interrupt the battery current with a solid state switch or a contactor. Regardless of the switching technology, and whether the switch is located in the BMS (in a protector) or outside of it (other types of BMS), it must be able to operate at the maximum pack voltage and current. This switch could be a single point of failure of the pack, though no more or less than any other component in the power circuit.

3.2.2 Thermal Management

The temperature range of Li-Ion cells (e.g., −20 to +60°C) is better than for other chemistries, but is still worse than what is required by many applications (such as the automotive environment: −40 to +85°C). It is completely unacceptable for military applications. Therefore, some applications require thermal management of the pack.

A BMS may control the temperature of its pack through:

- Heating;
- Cooling.

3.2.2.1 Heating

Knowing the pack temperature, the BMS may be able to control a heater to keep the pack above its minimum operating temperature. Usually, this is only done if the pack is able to draw energy from its charging supply (i.e., when a vehicle is plugged into the wall). A trick to heat a pack with a distributed BMS (albeit not as effective as using a heater) is to turn on all the passive balancing loads on the cell boards, which produce some heat (on the order of 1W per cell, which is on the order of 100W for the entire pack).

3.2.2.2 Cooling

Similarly, knowing the pack temperature, a BMS may be able to control a fan or blower to keep the pack below its maximum operating temperature.

Do not think that a fan will be of any help "in a hot day in a parking lot in Arizona." All that a fan can do is to equalize the temperature of the pack to the surrounding ambient temperature. If the ambient is at 60°C, then using a fan may actually *increase* the pack temperature (which may be still cool from the previous night due to the pack's insulation and high thermal mass). A fan will only help if the pack is hotter than the ambient, due either to heat generated by the pack itself during use, or to having previously been in an even hotter environment. The only thing that will help in the Arizona scenario is refrigeration.

The noise from a ventilation system can be a problem, so a BMS may include provisions for variable speed control of a fan, so that it may operate quietly when the pack is not too hot and operate at higher and higher speeds as the pack temperature increases.

3.2.3 Balancing

Balancing leaves room for more charge, without overcharging the most charged cell. Eventually, the balancing process brings all the cells to the same SOC (Figure 3.11). Balancing (see Section 1.4.2) can be performed by the BMS or by a distributed charger (see Section 5.5). If done by the BMS, balancing can be passive (energy is wasted in heat) or active (energy is transferred between cells). Redistribution (see Section 3.2.4) goes a step beyond balancing, and allows the full use of the capacity of each and every cell.

Without balancing, all cells (or blocks of cells in parallel) in series in a pack see exactly the same current, and therefore their DOD (in Ah) changes at exactly the same rate. A BMS balances a battery by making it possible for a particular cell (or block of cells) to see a current that is different from the pack current, in one of the following ways:

- Removing some charge from the most charged cells, leaving room for more charging current and allowing the less charged cells to receive more charge;

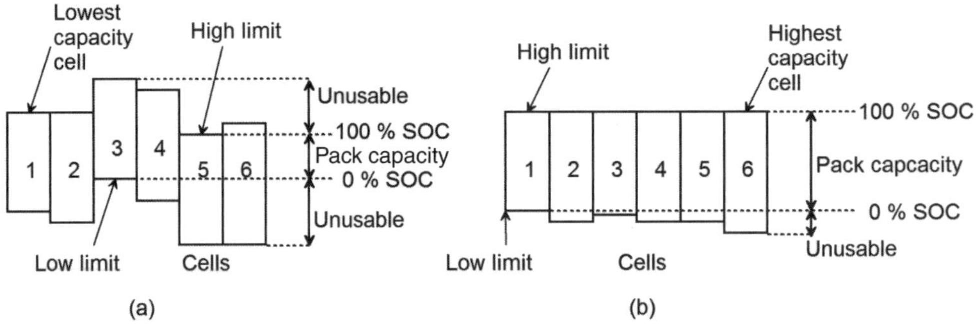

Figure 3.11 The balancing process: (a) unbalanced battery and (b) top balanced battery.

- Bypassing some or all of the charging current around the most charged cells, allowing the charging current to charge the less charged cells further;
- Feeding extra current just to the least charged cells.

Table 3.2 compares the effects of using no balance, with balancing (Section 3.2.3), distributed charging (Section 3.2.5) and redistribution (Section 3.2.4).

The removed charge can either be wasted in heat (passive balance) or transferred (active balance).

As we saw when talking about regulators, bypassing the entire charging current around a cell suffers from certain issues (see Section 2.1.2), and feeding extra current to low cells requires more complex electronics. Therefore, in this section, unless otherwise stated, let's assume that balancing is done by removing some charge from cells and wasting it in heat.

3.2.3.1 Balancing Algorithms

Balancing algorithms can be based on:

- Voltage;
- Final voltage;
- SOC history.

The three balancing algorithms operate on different parts of the OCV versus SOC curve (Figure 3.12).

Table 3.2 Comparison of No Balance, Balancing, Distributed Charging, and Redistribution

	None	*Balancing*	*Distributed Charging*	*Redistribution*	
Method	N/A	Passive	Active	Active	Active
Current Transferred	None	Low: 10 mA to 1A	Medium: 100 mA to 10A	High: 1A to 100A	High: 1A to 100A
Battery Energy Utilization	0~90%	~90%	100%		
Battery Capacity	Reduced over time	Minimum cell capacity	Average cell capacity		
Pack SOC	Unrelated to cell SOC	SOC of cell with least capacity	SOC of all cells		
Cells' SOC	All over the place	At 100% all cells have same SOC	All cells always at same SOC		
Duration	N/A	Top end: at end of charge, SOC history: may be at any time	During charging	Whenever in use	

Top-balancing is assumed.

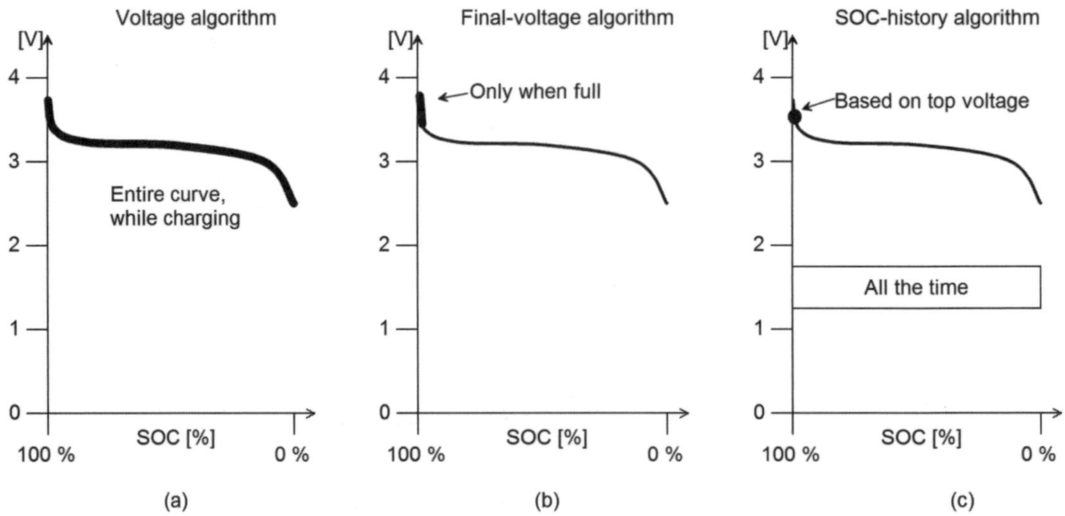

Figure 3.12 Comparison of balancing algorithms: (a) voltage, (b) final voltage, and (c) SOC history.

Voltage Based

Voltage-based balancing is the simplest algorithm, though it is usually counterproductive. It is based on the notion that cells with equal voltage are at the same SOC. That is true, but only if you look at the open circuit voltage (OCV), which is often not the same as looking at the terminal voltage. In any case, the OCV versus SOC curve of Li-Ion cells has a large, flat plateau in the middle, making SOC determination from voltage very difficult in that area.

The algorithm is simple: while charging, charge is removed from the cells with the highest voltages. The problem with this method is that, while charging, a cell's terminal voltage is higher than its internal voltage by the voltage drop across its internal resistance, and that internal resistance changes from cell to cell. If all the cells' terminal voltages were the same during charging, their equivalent OCVs would all be different (due to the difference in their resistances), and therefore so would be their SOC levels.

It is possible to overcome this algorithm's limitation. If the BMS knows the resistance of each cell, it can calculate its OCV by removing the IR drop its terminal voltage. Alternatively, the BMS could periodically stop charging, allow the cell terminal voltages to relax down to the OCV, and measure the OCV directly.

However, another limitation would remain. Except at the very end of the charge, the OCV is a poor indicator of SOC. Other than at high SOC levels or large deltas in SOC, the difference in the OCV of two cells that are at different SOC is minimal and therefore it is nearly useless as a way to determine which cell has more charge (Figure 3.13) [1].

Final Voltage Based

Final voltage–based balancing is the most commonly used algorithm. It works well, but takes time. This method is similar to the previous one, except that, instead of working at all times, it works only at the end of charge (at the top) (Figure 3.14). The

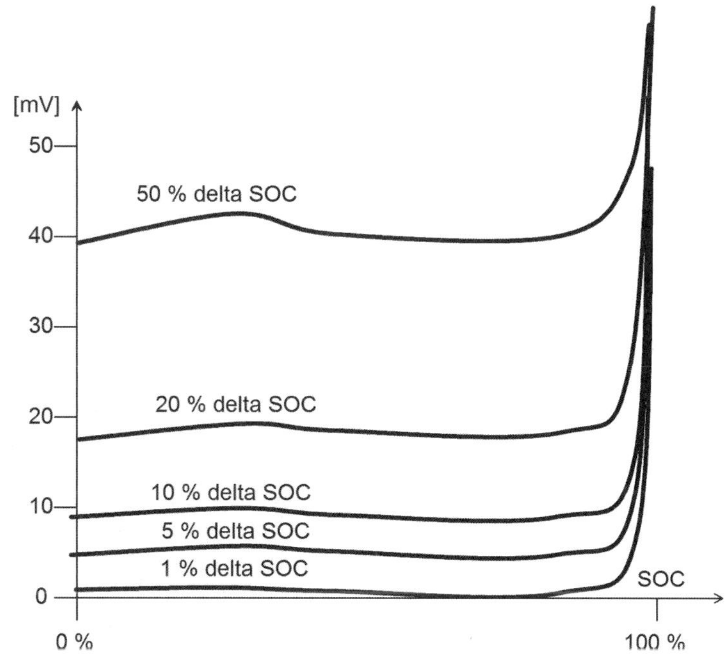

Figure 3.13 Difference in OCVs of two $LiFePO_4$ cells versus SOC, at various delta SOC levels.

algorithm is: energy is removed from the cells whose voltage is above a threshold (such as 3.4V for a $LiFePO_4$ cell).

Later we will analyze balancing in the middle or the bottom instead of at the top; for now let us assume that balancing is always done at the top. (It is not possible to balance both at the top and at the bottom, because real-world cells have different capacities, so they cannot be balanced both at 100% SOC and at 0% SOC.)

The advantage of this method is that it avoids operating in the flat portion of the voltage versus SOC curve, where voltage cannot be relied on as a measure of SOC, operating instead at the ends, where voltage is strongly affected by the SOC. When a Li-Ion cell is close to being fully charged, its voltage increases rapidly as it is charged further. A change of 100 mV indicates a change in SOC of on the order of 1~3%. So, if all the cells are within 100 mV, their SOC are also within 1~3%. That is why this balancing method operates in that area.

One problem with the final voltage algorithm is that usually there is very little time left between the moment the cell voltages are high enough to start using them for balancing, and the moment that the charging source is removed. For example, an EV that is plugged in only part of the day may have 2 hours for charging, leaving only 10 minutes for balancing at the end of charge. A solution to this limitation is to use a much stronger balancing current, to make up for the short time available.

Another problem with the final voltage algorithm is that each cell's terminal voltage during charging is higher than its OCV due to the IR drop across its resistance, and cell-to-cell variations in resistance can easily affect the cell voltage more than the SOC does. Therefore, this algorithm could also be counterproductive, by resulting in the discharging of cells with the highest resistance instead of the ones with the highest SOC, and therefore increasing the unbalance.

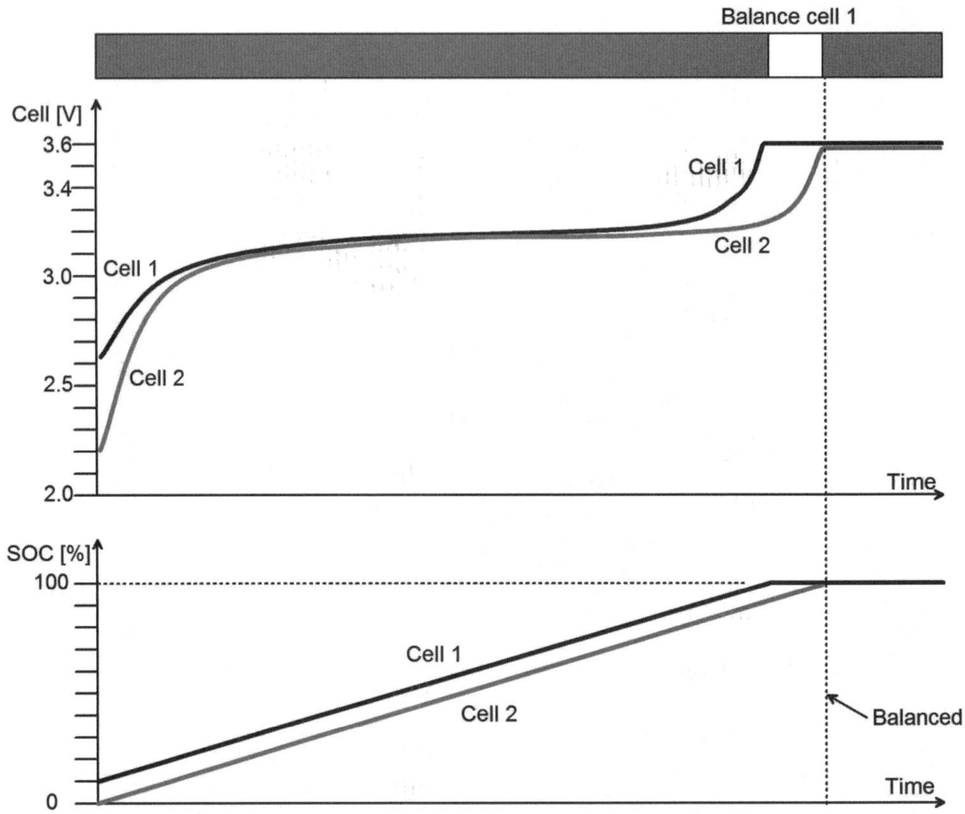

Figure 3.14 Final voltage algorithm used to balance two cells.

There are various solutions to this limitation:

- If the BMS knows the resistance of each cell, it can calculate its OCV (by doing IR drop compensation), from which it can estimate its SOC.
- The BMS can turn off charging during balancing, so that there are no errors due to IR drop across varying cell resistances, and wait for the cell voltage to settle.
- The BMS can tell the charger to reduce the charging current, to reduce these errors.
- The BMS can turn the charger on and off every few minutes, resulting is short bursts of full charging current interspersed by long periods at no current, allowing balancing that is hardly affected by these errors (e.g., a 10-A charging current, with a 1:100 duty cycle—such as 10 seconds on, 1,000 seconds off—will average to 100 mA, leaving 99% of the time for error-free balancing) (Figure 3.15).

SOC History Based

This is the most sophisticated balancing algorithm. It works very well, but it takes significant computing power, as it is based on knowing the past SOC of each individual cell, and calculating how long balancing should last for each cell.

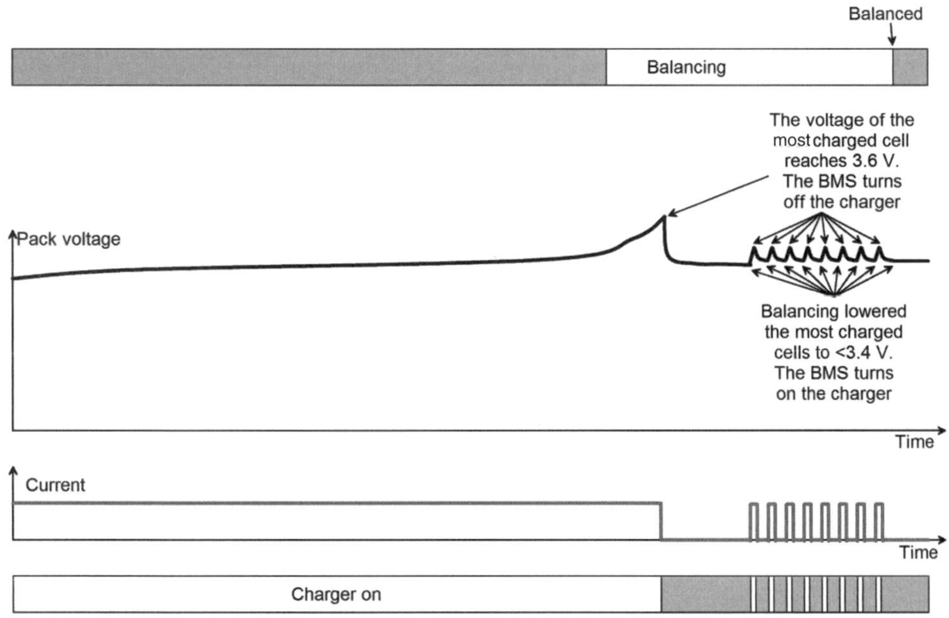

Figure 3.15 Final voltage algorithm, turning the charger off and on.

At the end of a charging cycle, the BMS learns the SOC of each cell (from the OCVs). Knowing the nominal capacity, the BMS converts the SOC of each cell to its DOD.

$$\text{DOD[Ah]} = \text{Capacity [Ah]} * (1 - \text{SOC [\%]}/100\%)$$

For each cell, the BMS calculates the delta in its charge compared to the charge in the least-charged cell. Knowing the magnitude of the balance current, for each cell, the BMS calculates how long that balance current must be applied to that cell to remove its delta charge.

$$\text{Balance Time [h]} = \text{Delta Charge [Ah]}/\text{Balance Current [A]}$$

Then, during the next discharging and charging cycle, the BMS applies the balancing discharge current to each cell (other than the least-charged one), for the exact duration it has calculated for that particular cell. When balancing is completed, the DOD of each cell (in Ah) will be exactly the same. Therefore, once charging is completed, all the cells will be at exactly 100% SOC (Figure 3.16).

This method has all the advantages and few of the limitations of the previous two. By being able to balance all the time, this method is faster than the final-voltage method for a given amount of available balancing current. Conversely, given a maximum balancing time, it can do so with less balancing current. In effect, this method has the effect of increasing the average balancing current with respect to the final-voltage algorithm, for a given BMS balancing current. The rule of thumb is that the increase is by a factor of 1 to 5. Beyond that, the BMS hardware itself has to be changed to handle more balancing current.

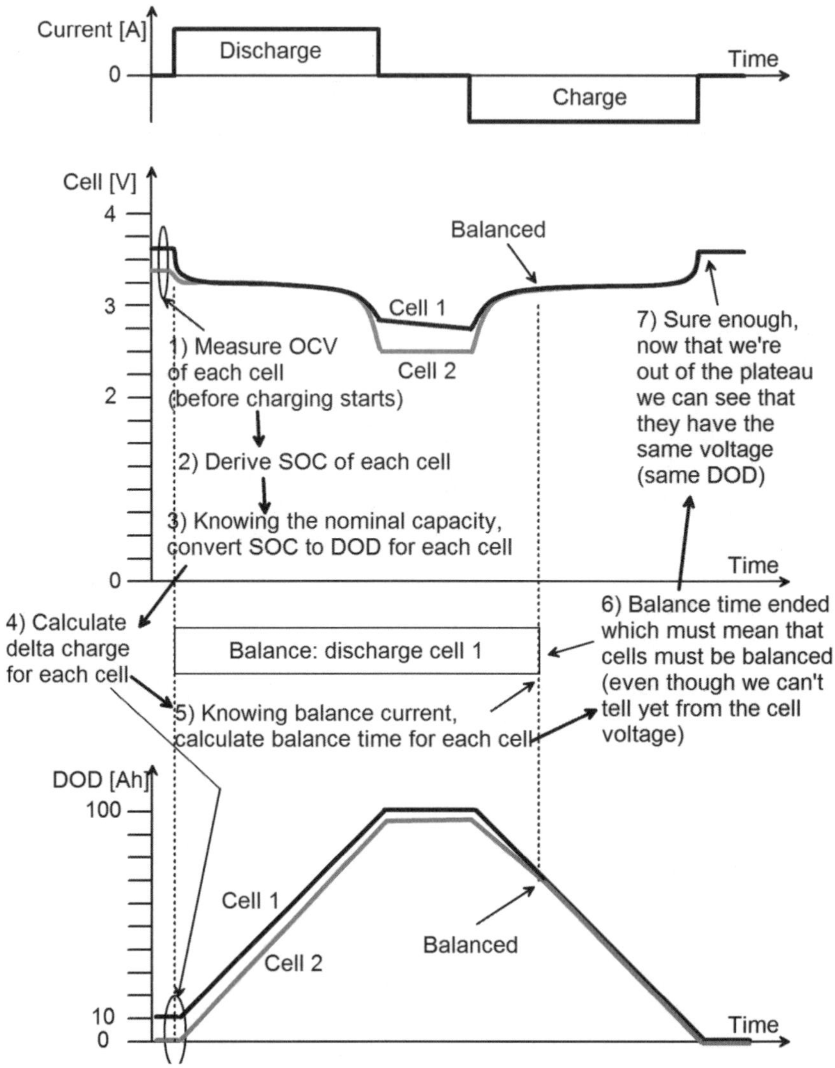

Figure 3.16 SOC history algorithm used to balance two cells.

For example, let us look at an electric vehicle that is driven 4 hours every day, and plugged in for 12 hours every night, during which time charging takes 8 hours, and final voltage balancing takes 4 hours. Let us assume that its BMS can balance at 100 mA. Then, the balance current is 100 mA/(4 hours/24 hours) = 17 mA. If the battery requires a 10-mA average balancing current, the BMS will be able to keep the battery in balance. If, however, the battery requires a 50-mA average balancing current, then the BMS will not be able to do so. One way to increase the balance current is to increase the maximum current that the BMS can handle (say, from 100 mA to 1A). Another way is to increase the time available for balancing. If the BMS were to use the SOC history algorithm to know a priori which cells are likely to need balancing, it can perform balancing whenever it is on (16 hours in this case), for an average of 100 mA/(16 hours/24 hours) = 67 mA, which is sufficient.

Just like the previous balancing algorithms, this one is affected by inaccuracies in the SOC estimation based on cell voltage. It works best if the cell voltage is measured when there is no battery current (to avoid errors due to the cell resistance), after the voltage has settled, and away from the flat plateau at the center of the voltage versus SOC curve.

Algorithm Comparison

Table 3.3 compares the three balancing algorithms.

Top Versus Midbalancing

We saw that batteries can be balanced at any SOC (see Section 1.4.2). So far we assumed that balancing is always done at the top (100% SOC). Top balancing is indeed ideal for packs whose main purpose is to store energy, but not all packs are used to store energy.

Some packs are used as a power source, such as in HEVs, or applications that draw intense power for a very short time (such as a high-power, pulsed laser). Such packs are kept at about 50% SOC, and never reach the top or the bottom. Once balanced at the time of manufacture, such packs may not need any further balancing, as the reduction in pack capacity due to variations in cell leakage may never be noticed for the life of the pack. If power packs are balanced, they either have to be fully charged once in a while to do top balancing, or they may be balanced at 50% SOC.

Table 3.3 Comparison of Balancing Algorithms

	Voltage Based	*Final Voltage Based*	*SOC History Based*
Principle of Operation	Balances whenever charging, regardless of SOC. Strives to match cell voltages.	Balances at high SOC. Strives to match cell voltages.	Balances all the time. Strives to match cell DOD, based on previous history of cells.
Pros	Very simple method.	At high SOC, the cell voltage chances rapidly, so it gives better data on the true SOC. The charging current can be reduced so errors due to the IR drop across the cell's internal resistance are minimized; or the charger is mostly kept off during balancing, so cell resistance is a small factor.	The BMS balancing current can be lower, and balancing can be done in fewer cycles, as balancing can occur all the time. Cell resistance has little effect.
Cons	Using cell voltage as an indication of SOC is not effective because the OCV versus SOC curve is quite flat at mid SOC levels. Strongly affected by cell resistance, because it mostly runs when the terminal voltage is higher than the OCV due to charging current.	Balancing only at the top means there is less time to balance after the battery is charged, until power goes away. Therefore, the BMS has to balance at a higher current level.	Requires more computing power and more memory to store the history of each cell.

Balancing a power pack can be very challenging. Let us look at a Li-Ion traction pack for an HEV. I will present two approaches. The first one uses a standard BMS doing SOC history balance, and the HEV's vehicle control unit (VCU) is in charge of controlling the timing of the process. The second one uses a high-precision BMS, able to measure cell SOC in the middle of the plateau in the OCV versus SOC curve.

The first approach is to balance at the top, because we have a BMS that knows how to do that. Now we need a good strategy to bring the HEV traction pack to 100% SOC in a way that remains transparent to the driver. When the traction pack is fully charged, regenerative braking is not possible, so the handling feels different. Balancing at the top has an added benefit, as the HEV will be able to calibrate the DOD at the same time.

A good strategy is to balance the pack during a long trip on a long stretch of freeway. The HEV's VCU may notice that the HEV has been traveling at more than 50 mph for a long time, so it assumes the HEV is on a freeway. It then starts a balancing process (Figure 3.17), composed of the following steps:

1. Use more energy from the engine to charge the traction pack (the mpg will be reduced) until a cell is fully charged.
2. Clear the pack DOD to calibrate it.

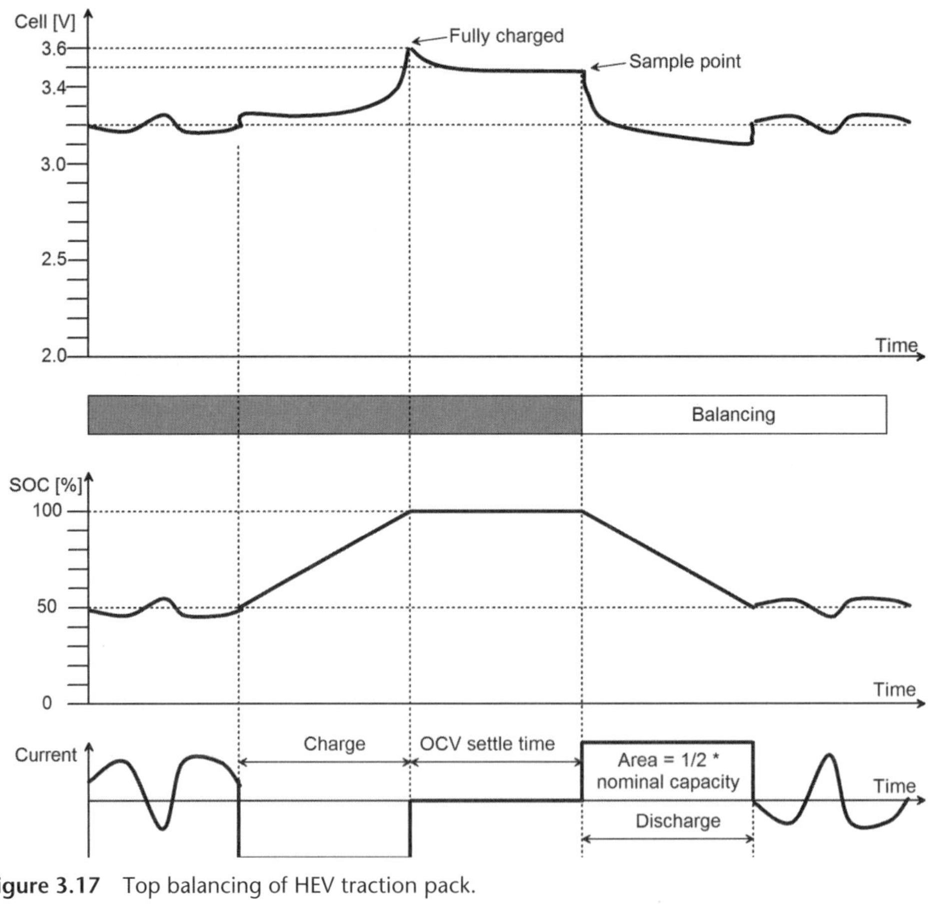

Figure 3.17 Top balancing of HEV traction pack.

3. Stop the battery current for about 10 to 30 minutes (to allow the cell voltages to settle).
4. Measure each cell's voltage (the OCV), convert to SOC, then to DOD, then to delta charge.
5. Use less energy from the engine and start using energy from the traction pack until 50% SOC (the mpg will be increased).
6. Restore normal HEV operation.
7. Balance the pack at 50% SOC using the SOC history algorithm.

Note that, in theory, the cells would have to be ranged from fully charged to fully discharged to allow the BMS to measure their capacity, and therefore to convert DOD to 50% SOC. In reality, the variation in cell capacity is not enough to justify that extra effort.

If the BMS is not able to do SOC history balancing, then balancing must occur while the battery is at the top (see item 4 in the previous list). Yet, the HEV wants to minimize the time when the battery is not at 50% to reduce the chance that the driver is affected by the process. That means that the BMS needs to balance in a hurry, using a high level of balancing current.

The second approach is to use a BMS whose accuracy is sufficient to estimate the cell's SOC in the middle of the plateau in the OCV versus SOC curve on the order of 1 mV. Then, the BMS may wait for a long period without any significant battery current (for example, stopped, waiting for a train to go by), take high accuracy cell measurements, convert them to SOC, and then use the SOC history algorithm to balance the cells at 50% (Figure 3.18).

Top Versus Bottom Balancing

Bottom balancing a pack may be used in applications that do not have a BMS to protect individual cells from undercharging (as long as individual cells are protected

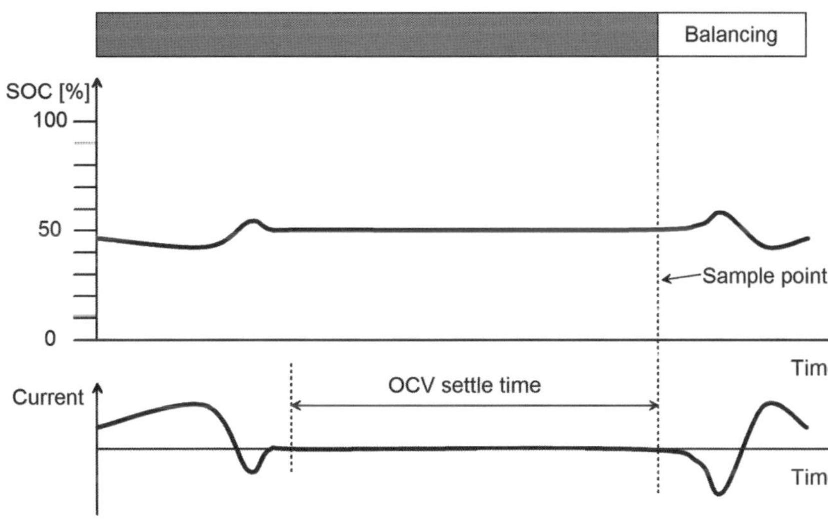

Figure 3.18 Midbalancing of HEV traction pack with a high-precision BMS.

from overcharging). If all the cells are perfectly balanced when fully discharged, they will all be at the same low voltage at that point. Then, monitoring the pack voltage will be sufficient to detect that the pack is fully discharged, at which point the load is turned off before any cell is overdischarged. A BMS is still required to monitor each cell's voltage to detect when a cell is fully charged, as bottom balancing makes cell more likely to be overcharged. Overcharging is more of a concern than overdischarging; an overdischarged cell simply dies—an overcharged cell may catch on fire!

The consensus among battery experts is that bottom balancing makes no sense. To address the dissent from a very small but vociferous minority who insist that top balancing is bad[1] [2], here is a list of why top balancing is far superior to bottom balancing.

1. *Top balancing allows batteries to store more energy.* Because the cell voltage decreases with SOC (granted, not as much in Li-Ion than other chemistries), a battery stores more stored energy when top balanced than when bottom balanced. Another way of saying that is that the energy density of cells decreases with SOC, so it is better to start with every cell fully charged (where their energy density is higher) that to end with all cells fully discharged (where their energy density is lower).

   Let's look at a simple example of a battery formed with two standard Li-Ion cells is series, one with a capacity of 10 Ah and one with a capacity of 7 Ah.

   When top balanced, one cell stores 4.2V * 10 Ah = 42 Wh of energy, and the other one stores 4.2V * 7 Ah = 29.4 Wh of energy, for a total of 42 Wh + 29.4 Wh = 71.4 Wh. After taking 7 Ah from the battery, the first cell still stores 3.7V * (10 – 7) Ah = 11.1 Wh (which is not available for use) and the second cell is totally discharged (0 Wh). So, the useful energy storage of the battery is 71.4 Wh – 11.1 Wh = 60.3 Wh.

   Now let's look at bottom balancing. When bottom balanced, both cells are empty, for a total energy of 0 Wh. After adding 7 Ah to the battery, the first cell stores 3.7V * 7 Ah = 25.9 Wh (the remaining 3-Ah capacity is not available) and the second cell is totally charged, storing 4.2V * 7 Ah = 29.4 Ah. So, the useful energy storage of the battery is 25.9 + 29.4 = 54.8 Wh.

1. Jack Rickard actively campaigns against balancing circuits by calling them a dangerous fire hazard that add needless expense and do the opposite of what they are meant to do. He bottom balances his pack manually once in a while. He argues (correctly), that balancing the cells at the top unbalances them at the bottom, as a consequence of their differing capacity. When his battery pack is empty, the lowest capacity cells drop out first, and are damaged by too low a voltage or even a voltage reversal. He therefore concludes (incorrectly) that balancing is the reason his cells are killed. The real reason is that he is not using a BMS to detect when a cell is low, relying instead on monitoring the entire pack voltage to know when to stop driving. While bottom balancing does allow him to prevent cell overdischarge without a BMS, unfortunately he exposes his pack's cells to overcharging, which is a far more dangerous prospect. As we saw (in Section 1.2.5), monitoring the total voltage of a long series of cells offers no indication of the state of each individual cell. If he performed bottom balancing instead of top balancing, as he proposes, he may be able to prevent damaging his cells when driving (by stopping when the pack voltage is low). However, cell damage would occur during charging with a CCCV charger (see Section 1.2.5), which would be worse: an overdischarged cell will just die, but an overcharged cell can be a fire hazard.

As you see, the battery stores 10% less energy when bottom balanced than when top balanced.

Therefore, top balancing allows batteries to store more energy.

2. *Top balancing allows load to be run longer.* In applications that have a high-power load (which is generally the case), it is essential that the battery resistance be as low as possible and as long as possible. During discharge, the cell resistance is low from 100% SOC until it starts getting close to empty.

   In a bottom-balanced battery, near the end of discharge all the cell resistances increase simultaneously while all the cell voltages drop simultaneously; that results in a simultaneous, large increase in the pack resistance and large drop in pack voltage. Either one of these effects by itself would prevent the battery from powering the load at full power. Both of them simultaneously do so to an even greater degree.

   In a top-balanced battery nearing the end of charge, only a few cells will have reached a high resistance and low voltage. The total battery resistance and total battery voltage will be hardly affected by those few cells, allowing the load to draw full power until the very end of charge. Therefore, top balancing allows a high-power load to be powered longer.
3. *Only top balancing is acceptable for Thundersky cells.* Certain cell chemistries (e.g., Thundersky cells) require that each cell be brought to the top voltage regularly to maintain an internal chemical balance (not to be confused with battery balance) and maximize the life of the cell. Therefore, with such cells, top balancing is the only acceptable choice.
4. *Top balancing is more accurate.* Assuming that the discharge current is higher than the charge current (which is generally the case), the errors in the cell terminal voltages due to IR drops are lower when charging.
5. *Top charging makes the most use of the charge in a pack.* A CCCV charger is able to completely charge all the cells in a top balanced pack, in three stages:

- Constant current until one cell is full;
- On and off as all the cells are top balanced;
- Constant voltage as the current is gradually reduced, until all the cells are equally and completely charged (see Section 6.2.3).

   I don't believe that any device exists that is able to do the analogous function at the bottom end. If it existed, it would be some sort of constant current/constant voltage load, able to discharge the pack at constant current, until the voltage drops down to a certain level, at which point it reduces its current draw to maintain that voltage until the current gets to a low threshold. Not many loads could do any useful work under such conditions, so, only top charging can make full use of the pack charge.
6. *Top charging is compatible with unscheduled pack usage.* The previous two points assumed that bottom balancing would occur at the end of discharging. That is not the only option: bottom balancing can also be done at the beginning of charging. When charging, starting from a discharged pack, a BMS would turn on balancing loads across the cells that still have some charge in them, to discharge them fully. At that point, the pack would be bottom balanced, and the BMS would be able to turn on the charger.

Bottom balancing is likely to take hours, starting from the moment the system is first allowed to start charging (say, when an EV is plugged into the wall outlet). Some hours later, when the pack may be needed again, it may be still balancing, meaning that it will be actually *more* empty than when charging started. Therefore, bottom balancing before charging is not appropriate when the pack could be needed at a moment's notice.

7. *Top charging is more forgiving if the BMS is overridden "to get home."* A BMS should stop the discharge when any cell is low. But, at that point in some applications the BMS may be overridden (so that an EV can get home, even at the risking of killing some cells). In a top-balanced pack, only a few cells are at risk, but in a bottom-balanced pack every cell is at risk. Therefore, top balancing is better in applications where the user is able to override the BMS in an emergency.

3.2.3.2 Balance Current Requirements

Is a 100-mA current sufficient to balance a Li-Ion pack? What about 1A? 10A? The answer depends mostly on what the job is:

- Gross balancing, to bring into balance a pack that was built or repaired using cell with a mismatched SOC;
- Maintenance balancing, to continue keeping a balanced pack in balance.

A pack should be built to be balanced right from the start, so that the BMS is not required to provide an initial, gross balancing. This can be done in a couple of ways:

- Balance before build: start with cells that are all fully charged;
- Balance after build: while the pack is still open and there is access to the cells, use a power supply to charge each cell completely.

That way, the pack starts balanced and all the BMS has to do is keep it in balance (maintenance balancing). If the pack is built or repaired with no consideration for the SOC of each individual cell, the BMS will have to do gross balancing.

Gross Balance

If a BMS is expected to do gross balancing of a large pack in a reasonable amount of time, it will have to use a relatively high balancing current (Figure 3.19). The maximum length of time required to do gross balancing on a pack that is completely out of balance will depend on its capacity, and on how much balancing current the BMS can provide:

Gross Balancing Time [hours] = Pack Capacity [Ah]/Balancing Current [A]

Figure 3.19 shows that:

- A BMS with a 1-A balancing current will take almost 1 week to balance a 100-Ah pack that is completely out of balance.

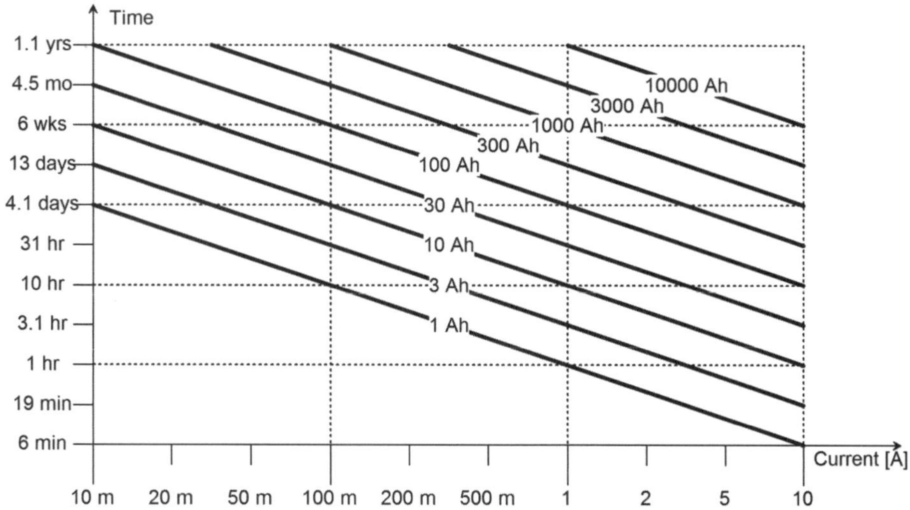

Figure 3.19 Time required to balance a grossly unbalanced pack versus balance current for various pack sizes.

- A balance current of 10 mA will not be able to balance a 1,000-Ah pack within the lifetime of its owner.
- Conversely, a balance current of 10A would be an overkill for a 1-Ah battery, as it will balance it in less than 6 minutes.

Maintenance Balance

If a battery starts balanced, keeping it in balance is a far easier job than gross balancing. All that is required is to compensate for the variation in leakage in the cells (self discharge). For example, if all the cells have exactly the same leakage, then no balancing is required. The SOC of all the cells will slowly drop exactly by the same amount, so the battery will remain in balance

If all the cells have exactly the same leakage, except for one cell whose leakage is 1 mA less, then the BMS has to discharge that one cell at an average current of 1 mA to make up for the difference in leakage (either that, or add 1 mA to all the other cells).

In the example above, the average balancing current was 1 mA. If the BMS can do balancing nonstop, then 1 mA is the actual balancing current. However, in many applications, the BMS is turned on only part of the time (yet, leakage discharges cells all the time). In such applications, the balance current has to be higher, in inverse proportion to how much time is available for balancing. For example, if the BMS is only able to balance 1 hour every day, the balance current must be 24 mA to achieve an average of 1 mA.

Of course, it is OK if the BMS is capable of a higher balancing current. If so, the BMS can achieve the required balance current by one of two methods:

- Reduce the balancing current down to the required level;
- Turn the balancing current on and off with a duty cycle such that, on the average, the balancing current is down to the required level.

Therefore, the current required for maintenance balancing is proportional to the difference in the leakage currents and to the portion of time available for balancing:

Balance Current [A] = (Max Leakage [A] − Min Leakage [A])/
Balance Time Proportion [−]

Typically, Li-Ion cell manufacturers do not specify leakage current. A few do specify how long cells can be kept in storage at room temperature. That value is of the order of years. From that value, one can estimate the leakage current.

Leakage Current [A] = Capacity [Ah]/Storage Time [hours]

Assuming a self-discharge time of 4 years, the leakage works out to:

Leakage Current [mA] = 0.03 * Capacity [Ah]

From that, we can plot the required balancing current for various size batteries (Figure 3.20).

Figure 3.20 shows that a 100-mA balancing current should be enough for:

- A 3,000-Ah battery, if the BMS is able to balance 24/7 (such as for backup power);

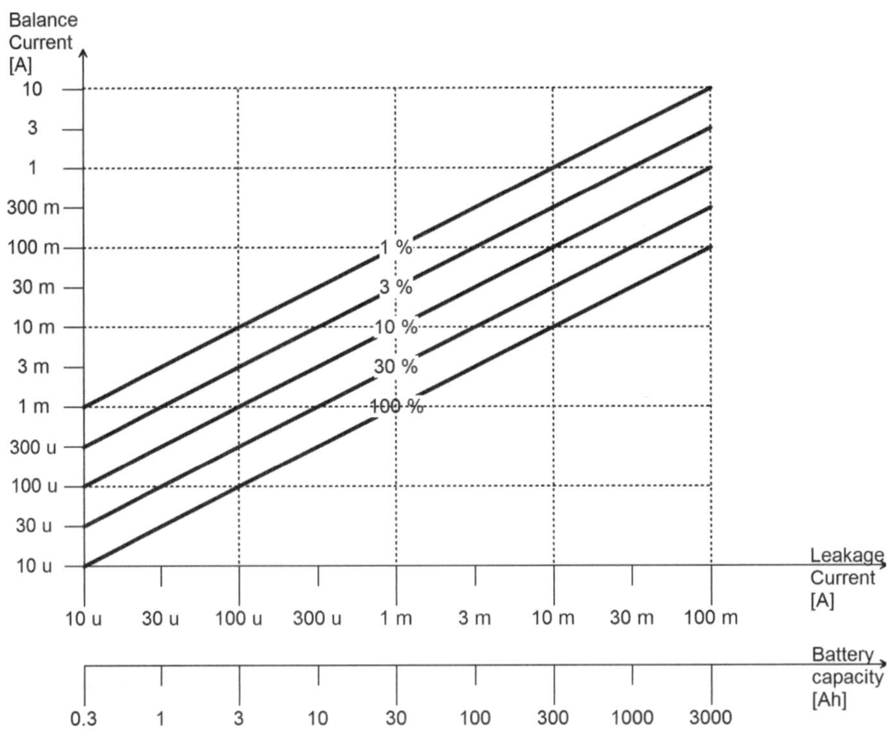

Figure 3.20 Required balancing current versus delta leakage current for various time availability for balancing.

- A 300-Ah battery, if the BMS is able to balance 10% of the time (such as an EV);
- A 30-Ah battery, if the BMS is able to balance only 1% of the time.

How much balance current is required for a Li-Ion battery, during normal operation? My findings have shown:

- The current level actually required tends to be less than one may have guessed;
- 10 mA is sufficient for small backup supply applications (10 kWh), 100 mA for large applications (100 kWh);
- 100 mA is sufficient to handle any automotive application (10 kWh, plugged in nightly);
- 1A is sufficient for large battery applications, other than backup (> 100 kWh, cycled daily).

Conclusions

Here is a summary of the points discussed in this section on balancing.

- Balancing compensates for the SOC of individual cells. It does not compensate for capacity unbalance (which is what redistribution does).
- A balanced battery is able to supply the maximum charge, limited only by the cell with the least capacity (or highest internal resistance, in some extreme cases).
- A battery should be balanced when it is built, so that the BMS is not required to do gross balancing.
- If the battery is balanced at the factory, the BMS only needs to be able to handle a balancing current sufficient to compensate for cell-to-cell variations in self-discharge current, during normal operation.
- There is no reason to specify a BMS that can handle more balancing current that is required by the battery in the worst case.
- It makes little economic and engineering sense to specify a BMS that can balance a grossly unbalanced battery just once in its lifetime. That BMS will be more costly and bulkier and produce far more heat than a BMS that is designed just for the job that it must do 99% of the time—keep a previously balanced battery in balance.
- A BMS that provides 100 mA of balancing current is sufficient for most Li-Ion applications.
- Using a balancing algorithm based on SOC history can increase a given BMS balancing capability by a factor of 2–5 times.

3.2.3.3 Active Versus Passive Balance

Balancing can be:

- Passive: energy is removed from the most charged cell and is wasted in heat; or
- Active: energy is transferred between cells and therefore it is not wasted.

The disadvantages of passive balancing are obvious:

- Wasting energy is costly.
- At high-balancing currents, the wasted energy is converted to heat, which can affect the operation of the pack.

At a first glance, active balancing is better because it doesn't waste energy. In reality, active balancing does have some disadvantages:

- More components than passive balancing: higher cost, lower reliability, more occupied volume;
- Power wasted in standby may result in greater losses than for the equivalent passive balancing.

It may be useful to compare passive and active balancing in some applications. Table 3.4 lists the assumptions in this analysis, and Table 3.5 compares active and passive balancing in some typical applications using those assumptions.

For these examples, we can compare the waste heat for both methods (Figure 3.21).

Active balancing is recommended in applications in which low current passive balancing would take too long, or in which heat generation due to high current passive balancing could be problematic.

Table 3.6 lists applications for which active balancing may be recommended. Other than these applications, passive balancing is probably preferable.

Table 3.4 Assumptions in Active Versus Passive Balance Comparison

Method	*Assumptions*
Common to Both	Cell voltage when charged: 4V
	12 months self-discharge time for worst cell, 18 months for best cell
	For a 100-Ah cell, that works out to 12 mA for the worst cell, and 8 mA for the best cell, with a delta of 4 mA
	That means that the BMS has to transfer an average of 4 mA/cell
	If able to balance 24/7, that is truly 4 mA; if only 10% of the time is available for balancing, that works out to 40 mA average
	Unbalance: half the cells (meaning that 1/2 the cells have low leakage and require bleeding to match the cell with high leakage)
	Balance maintenance only (no gross balancing)
Passive	$1/cell
	100-mA balancing current (= 0.4W)
	0-mW standby power
	0% efficiency when wasting energy
Active	$10/cell
	3-A balancing current (= 12W)
	50 mW standby power*
	70% efficiency when transferring energy

*Note that the assumption is that the active balancer uses a bit of power whenever turned on, whether or not it is actually transferring energy. This is the case with many but not all active balancers.

Table 3.5 Comparison of Active Versus Passive Balance for Various Applications

Application	*Passive*	*Active*
UPS: 100 Ah (4-mA delta leakage) 15 cells in series Always plugged in the wall	On/off duty cycle during balancing: 4% Power of waste heat: 0.1W Cost: $15	On/off duty cycle during balancing: 0.1% Power of waste heat: 0.8W Cost: $150
Distributed power source: 1,000 Ah (40-mA delta leakage) 300 cells in series Always plugged in the wall	On/off duty cycle during balancing: 40% Power of waste heat: 24W Cost: $300	On/off duty cycle during balancing: 1.3% Power of waste heat: 22W Cost: $3,000
EV, PHEV: 100 Ah (4-mA delta leakage) 100 cells in series Charged daily, plugged in the wall 12 hours a day, 8 hours charge/ 4 hours balancing	On/off duty cycle during balancing: 24% Power of waste heat: 0.8W Cost: $100	On/off duty cycle during balancing: 0.8% Power of waste heat: 2.7W Cost: $1,000
Public transportation EV: 1,000 Ah (40-mA delta leakage) 100 cells in series Charged every 4 hours, plugged in the wall 30 minutes each time	On/off duty cycle during balancing: > 100%* Power of waste heat: 8W Cost: $100	On/off duty cycle during balancing: 100% Power of waste heat: 3W Cost: $1,000
HEV: 10 Ah (0.4-mA delta leakage) 100 cells in series Charged 1/2 the time while driving SOC kept at 50% +/− 20% Balanced once a week for 10 minutes by going to 100% SOC	On/off duty cycle during balancing: > 100 %* Power of waste heat: 0.08W Cost: $100	On/off duty cycle during balancing: 0.8% Power of waste heat: 0.44W Cost: $1,000
E-bike: 10 Ah (0.4-mA delta leakage) 10 cells in series Charged nightly SOC kept at 50% +/− 20% Balanced once a week for 10 minutes by going to 100% SOC	On/off duty cycle during balancing: 2.4 % Power of waste heat: 0.01W Cost: $10	On/off duty cycle during balancing: 0.1% Power of waste heat: 0.25W Cost: $100

*Note that when the balancing, the on/off ratio would have to be more than 100% (which is impossible). The final voltage method will not work, and the SOC history method must be used instead.

Some BMSs use on-chip passive balancing, which is limited to small balancing current (on the order of 20 mA/cell), due to the heat dissipation limitation of the IC. Therefore, chip manufacturers (notably Texas Instruments) are eager to propose active balancing as a way to overcome the limitations of on-chip passive balancing.

Conclusions

- The power handled when maintaining a Li-Ion battery pack in balance is usually minimal—typically on the order of 0.1 to 10W per cell in series. Therefore, in most cases, the question on whether active or passive balancing is better is purely academic.

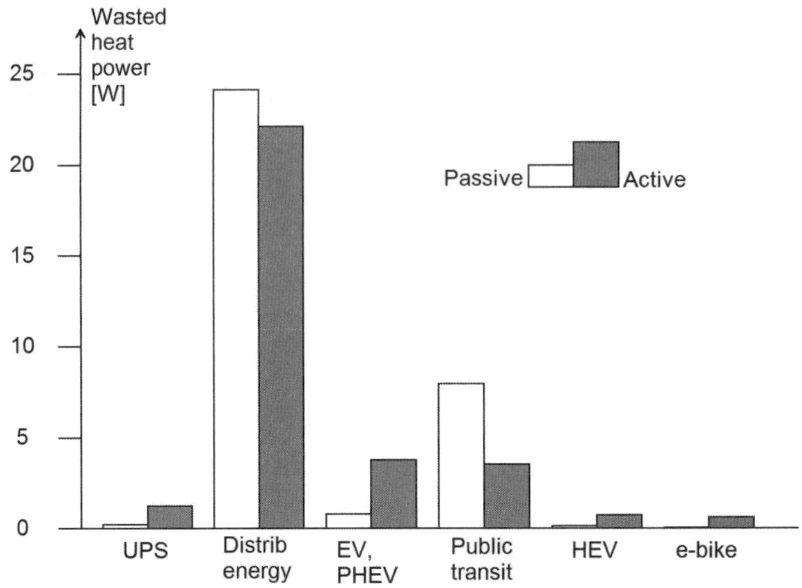

Figure 3.21 Power wasted in heat, passive versus active, for various applications (lower is better).

Table 3.6 Active Balancing Recommended Applications

Application	*Reason*
Large pack built with mismatched SOC cells, or whose cells may be replaced in the field without regard to SOC	Requires gross balancing
Large pack that must be charged in a short time	Little time available for balancing, must be done at high current
Medium to large pack that must be extremely efficient	Must not waste energy as heat
Medium to large pack operating always at high temperatures	Cell leakage (self-discharge current) is very high

- In applications in which the power handled is noticeable (> 10W/cell), it is possible that passive balancing and active balancing generate similar amounts of waste heat (due to the standby power of active balancing circuits).
- In general, the significantly higher cost of active balancing trumps any energy saving arguments, especially when those savings are minimal.
- The argument in favor of active balancing comes down not to its efficiency, but to its ability to transfer a significant amount of charge quickly in applications where only a short time is available for balancing, or frequent gross balancing is required.
- Active balancing is a worthwhile alternative to passive balancing that uses on-chip resistors (which is limited by the IC operating temperature to only tens of milliamperes).

- Algorithms based on SOC history increase the effectiveness of balancing hardware (typically by a factor of 2 to 5), regardless of whether balancing is done actively or passively.

3.2.3.4 Active Balance Techniques

There are four active balancing techniques (Figure 3.22):

- *Cell to cell:* energy is moved between adjacent cells;
- *Cell to battery:* energy is taken from the most charged cells and sent to the entire battery;
- *Battery to cell:* energy is taken from the entire battery and sent to the least-charged cells;
- *Bidirectional:* either cell to battery or battery to cell, depending on the need.

A comparison of these four techniques (Table 3.7) shows that:

- Cell to cell is fine for small batteries.
- Cell to battery is the simplest, and has the highest efficiency.
- Battery to cell is best if using a charger with N outputs for N cells.
- Bidirectional is best for redistributions (see Section 3.2.4).

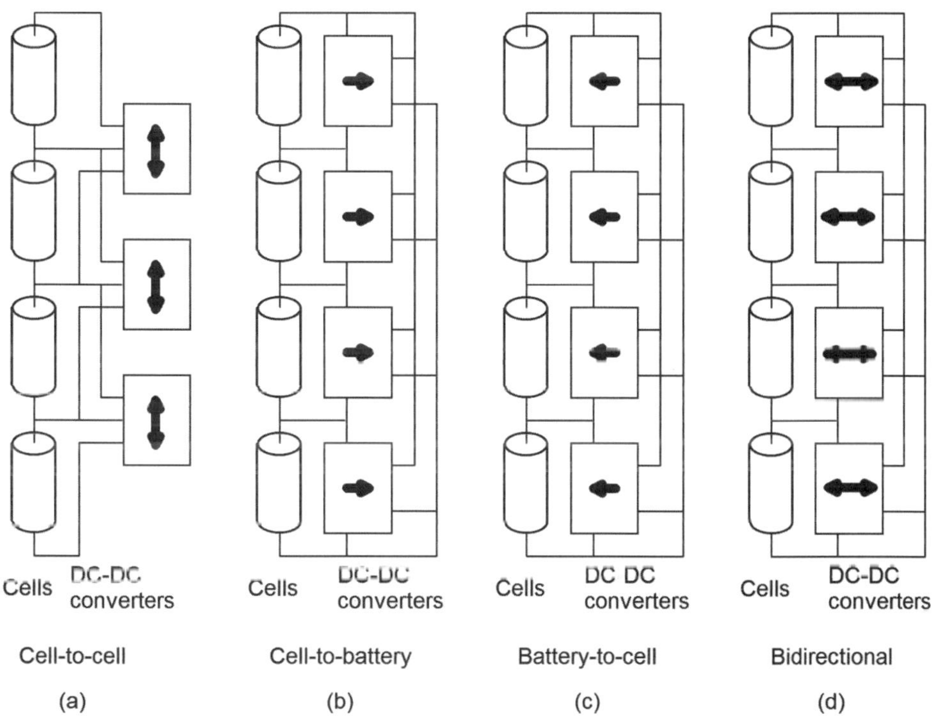

Figure 3.22 Active balancing topologies: (a) cell to cell, (b) cell to battery, (c) battery to cell, and (d) bidirectional.

Table 3.7 Comparison of Active Balancing Algorithms

	Cell to Cell	*Cell to Battery*	*Battery to Cell*	*Bidirectional*
Type of DC-DC Converter	Nonisolated DC-DC converters, low voltage to low voltage	Low voltage to high voltage	High voltage to low voltage (or bulk DC-DC converter with *N* switched outputs)	Bidirectional
Number of Converters, for an *N*-cell battery	$N - 1$	N	N	N
Direction and Operation	Fed by a cell when it has higher voltage than the adjacent cell Feeds the adjacent cell	Fed by a cell when it has excess charge Feeds the battery	Fed by the battery Feeds a cell when it has insufficient charge	Fed by a cell with excess charge, if those are the majority, or fed by battery Feeds the cells with insufficient charge, if those are the majority, or feeds the battery
Pros	Fewer converters Highest efficiency per converter (around 90%) Simpler, less expensive converters All DC-DC converter connections are at low voltage relative to each other	More efficient: high-voltage output rectifiers Simpler: low voltage transistors, controlled from same low voltage side as the cell electronics Best when only a few cells are low capacity—most converters are on	Most effective when most cells are low capacity: the majority of the converters are operating Can be implemented with a single, bulk, high-power DC-DC converter, and many switched outputs to the cells	Effective regardless of whether most cells are low SOC or high SOC. Best for redistribution: energy can go either way
Cons	More wires A midpack opening blows up two converters Takes longer to balance: energy has to go from converter to converter to reach the intended cell Overall efficiency is poor: losses occur at each step, from converter to converter	Not terribly efficient (around 80%)	Requires high-voltage transistors, isolated control from cell side to the drive transistors on high voltage side Inefficient (around 70%)—low-voltage rectifiers (synchronous rectifiers may help, for additional cost and complexity) Parallel charging has difficulties charging those cell with higher resistance	Most complex: switches on both ends Inefficient (around 70%)—low-voltage rectifiers (synchronous rectifiers may help, for additional cost and complexity) Parallel charging has difficulties charging those cells with higher resistance

3.2.4 Redistribution

Redistribution is a technique that shuffles energy in a battery in such way that all of its energy can be used (Figure 3.23). While discharging, additional energy is taken from the cells with the highest capacity, so that the cells with the lowest capacity are no longer the limiting factor in the battery capacity. An effect of redistribution is

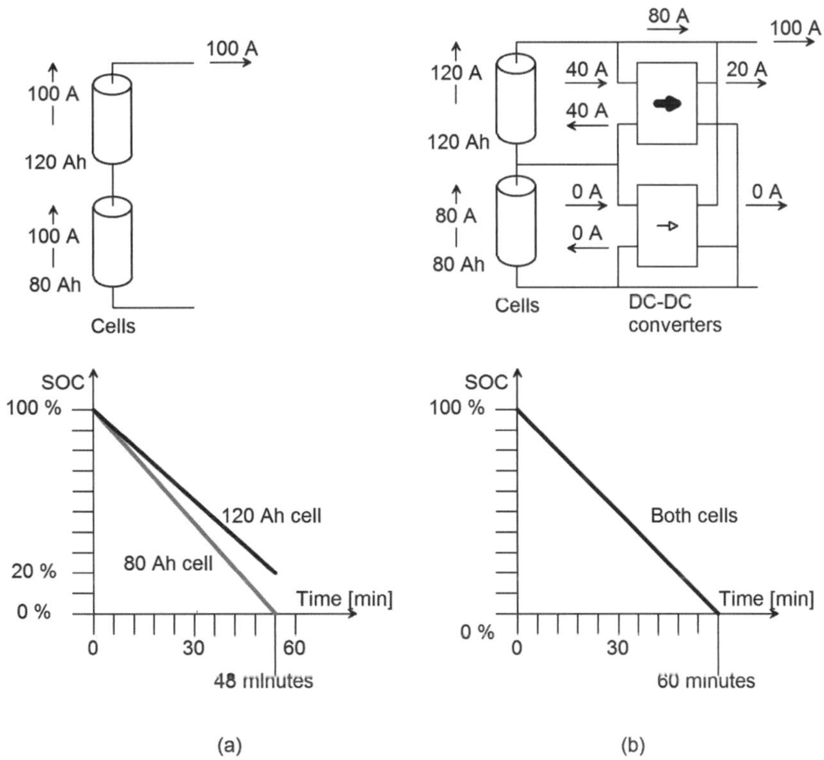

Figure 3.23 Two-cell battery: (a) without redistribution and (b) with redistribution. The charge in both cells is used fully.

that the SOC of the battery and the SOC of each cell are always equal: during discharge, all the cells start at 100% SOC and all end at 0% SOC.

For a simple example, let us look at a two-cell battery. The capacity of one cell is 20% high, and the capacity of the other cell 20% low. Without redistribution, the battery capacity is limited by the second cell, and the battery will be fully discharged in 48 minutes [Figure 3.23(a)]. With redistribution, extra energy is taken from the first cell, to make up for the second cell's low capacity, so that they will both discharge for 60 minutes [Figure 3.23(b)]. The DC-DC converter powered by the larger cell converts its cell voltage to the full battery voltage, while the one across the smaller cell remains off.

3.2.4.1 Comparison Between Balancing and Redistribution

Redistribution is similar to active balancing, except that the DC-DC converters used must handle more power, and the algorithms are a bit more complex. Table 3.8 compares them.

3.2.4.2 Converter Power

High-power DC-DC converters are required for redistribution. Roughly speaking, the power required for of each converter is:

Table 3.8 Comparison Between Balancing and Redistribution

	Balancing	*Redistribution*
Battery Energy Utilization	About 90%	100%
Battery Capacity	Equal to minimum cell capacity	Equal to the average cell capacity
Pack SOC	Equal to SOC of the cell with the least capacity	Equal to the SOC of all the cells
Cell SOC	At 100% SOC, all the cells have the same SOC	All the cells are always at the same SOC
Method	Active or passive Once the battery is balanced, energy is no longer transferred	Active only Dynamically transfers energy during use, during each and every cycle
Current	Low: 10 mA to 1A	High: 10 to 100A

$$P = \text{Average Load Power} * \text{Variation in Cell Capacity}/\text{Number of Cells in Series}$$

By knowing each cell's capacity before hand, each converter is operated as long as necessary (as long as the entire charge or discharge period).

For example, for a 10-kW load, +0/−10% variation in cell capacity, and 100 cells in series:

$$P = 10\text{ kW} * 10\%/100 = 10\text{W}$$

In this example, at first it may be hard to believe that 10-W DC-DC converters can power a 10-kW load, until you note that:

- There are 100 converters, which may transfer up to 990W altogether (at least one converter will remain off).
- Converters can work constantly throughout the discharge period, to transfer only 10% of the charge, giving an additional 10:1 advantage.

Therefore, the total power is $10 \times 990\text{W} = 9.9\text{ kW}$.

3.2.4.3 Redistribution Versus Additional Cells

While the technical advantage of redistribution is obvious, a cost analysis may show that in some cases it is cheaper to add cells to the battery than to do redistribution.

Assuming:

- DC-DC converters cost $1/W.
- Cells cost 0.3/Wh.

Then, if the load discharges the battery in less than 20 minutes, adding cells is cheaper that using redistribution (Figure 3.24).

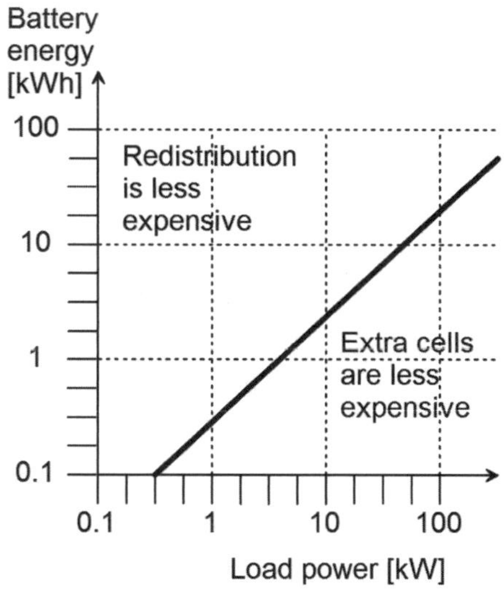

Figure 3.24 Redistribution versus extra cells: for smaller batteries, and at high power, it is cheaper to add cells.

3.2.4.4 Conclusions

In conclusion:

- Redistribution makes all the energy in a battery available for use.
- In a typical application, the required power of the DC-DC converters is 1/1,000 of the load power.
- If the load discharges the battery faster than 20 minutes, it is cheaper to just add cells to the battery.

3.2.5 Distributed Charging

Distributed charging is a feature of chargers, not BMSs, but is discussed in this book because it is an alternative to a BMS's balancing function (see Section 3.2.3). Instead of using a single, large bulk charger [Figure 3.25(a)] and a BMS with balancing, it is possible to use a number of small chargers, one per cell [Figure 3.25(b)]. Each charger will charge its own cell, up to the full voltage, and no further. This approach will inherently result in a pack that is top balanced without any risk of overcharge.

A subtle disadvantage of distributed charging is that it has difficulties charging those cells in the pack that have a higher resistance. This is unlike bulk charging, in which all the cells receive exactly the same current, with parallel charging all the cells receive the same voltage. Cells with high resistance draw less current from that voltage, taking longer to charge.

It is also possible to use a mixed approach. The combination of a large bulk charger plus a charger per cell [Figure 3.25(c)]. The bulk charger will do most of the work, and low-power chargers will each add a little bit of charge to the cells that need it, to complete the charge and balance the battery. The elegance of this approach is offset by the cost and low efficiency of many individual chargers and

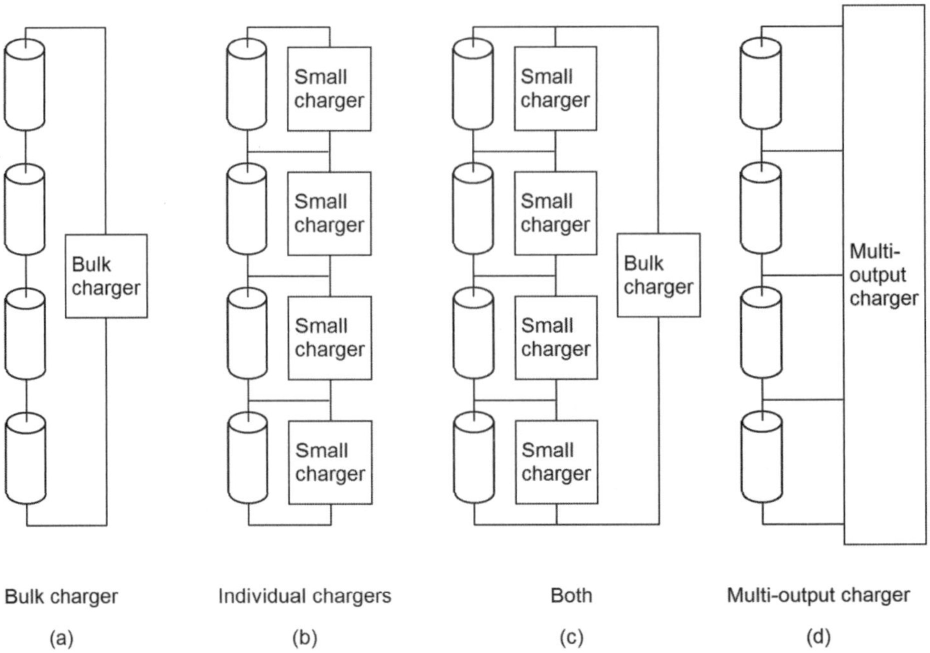

Figure 3.25 (a) Bulk charger, (b) many small chargers, (c) both together, and (d) multioutput charger.

the need for $N+1$ power wires going to N chargers. A less expensive option is to use a single large charger with multiple low-voltage outputs [Figure 3.25(d)].

3.3 Evaluation

From the measured data, the BMS may be able to calculate or estimate certain parameters that relate to the state of the pack. These include:

- State of charge (SOC) and depth of discharge (DOD);
- Resistance;
- Capacity;
- State of health (SOH).

In general, analog BMSs do not implement any of these functions. Most digital BMSs only implement the SOC function. Only the more sophisticated digital BMSs implement all of them.

These evaluation functions are not for the protection of the pack; they are only for the convenience of the user. For example, SOC gives an indication of how much longer the pack can be used, while SOH gives an indication of when it's time to replace the pack, or start using it in a less demanding regime.

Still, a lot rests on these measurements, because of the user's perspective. A BMS that reports a grossly incorrect SOC may be still doing its main function correctly (manage the pack), but will lose the confidence of its user. A BMS that incorrectly reports a low SOH may cost the user dearly when a pack is replaced too soon.

Unfortunately, all of these parameters are indirect estimates, and therefore prone to large discrepancies between the actual, physical reality and the computed estimate. In particular, there is no direct way of measuring SOC of Li-Ion cells (comparable to using a hydrometer to measure the SOC of a vented lead acid cell) and, to date, there is no reliable method to do so indirectly.

3.3.1 State of Charge and Depth of Discharge

The user typically wishes to know the battery's SOC or DOD, for the same reason that a car driver wishes to know how much fuel is left in the car's tank—to estimate how long before it is empty.

For however inaccurate car fuel gauges may be, fluid level measurement is easier to do than SOC estimation. The fuel level in a tank can be measured directly, while a cell's SOC cannot. Estimating SOC and DOD in a Li-Ion pack is, at best, an inexact science, and, at worse, a wild guessing game.

There is no direct way of measuring the SOC of a Li-Ion battery. There are indirect ways of estimating it, but each suffers from limitations. Among them, two commonly used methods are:

- Voltage translation;
- Current integration (coulomb counting).

Both techniques are useful, but each by itself is unable to reliably estimate SOC in a Li-Ion battery. By combining them, a reasonable estimate of SOC is possible.

3.3.1.1 Voltage Translation

With some cell chemistries, the battery voltage decreases more or less linearly as the battery is discharged, so one may consider using a simple voltmeter as an SOC indicator (Figure 3.26). Knowing the relationship of open-circuit battery voltage and SOC (which is the idea behind voltage translation) allows the voltmeter to be calibrated to report an approximate SOC.

A major limitation with this technique is that the battery's terminal voltage is affected by parameters other than SOC. Having prior knowledge of the way those parameters affect the terminal voltage, it may be possible to provide a certain amount of compensation, allowing voltage translation to be a useful way to estimate that battery's SOC.

The usefulness of voltage translation in Li-Ion batteries is limited, because, for most of its SOC range, the voltage of a Li-Ion cell remains rather constant (Figure 3.27). With zero cell current, using a precise voltmeter (with accuracy on the order of 1 mV), and allowing a long time for the cell voltage to settle (the time constant is on the order of tens of minutes), voltage translation is possible in a laboratory (though mostly impractical in a product). Yet, the voltage of a Li-Ion cell does change significantly at both ends of its OCV versus SOC curve. Therefore, voltage

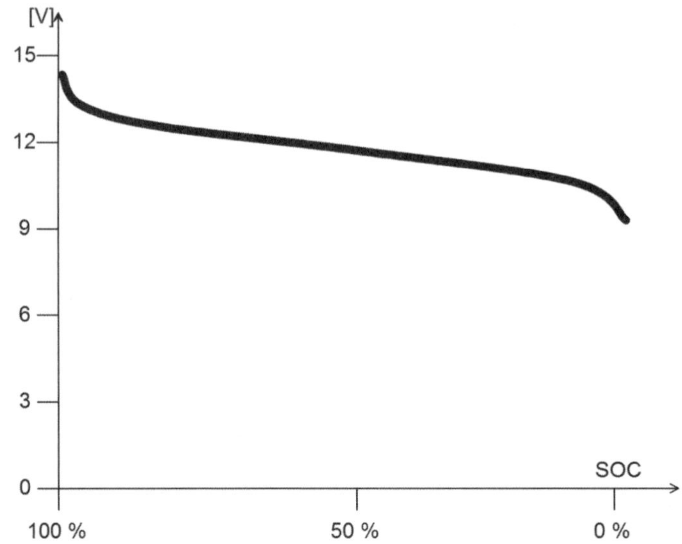

Figure 3.26 OCV versus SOC in a lead-acid battery is somewhat linear.

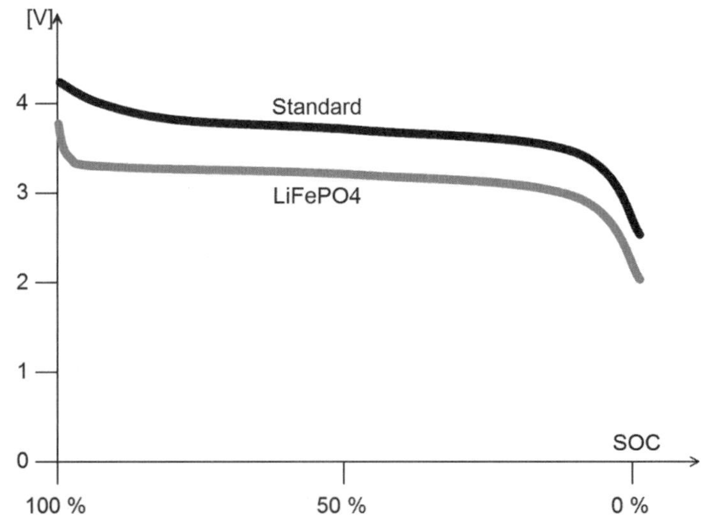

Figure 3.27 The OCV versus SOC curve for Li-Ion has large plateau. Voltage translation is difficult.

translation can be used to estimate the SOC of a Li-Ion cell when it is nearly full or nearly empty.

3.3.1.2 Coulomb Counting

Integrating the current into or out of a battery gives the relative value of its charge, just as counting currency in and out of a bank account gives the relative amount in the account. The operative word here is relative. Like any definite integral, coulomb counting needs a starting point. If the initial charge in the battery is known, from then on coulomb counting can be used to calculate charge. For example, a 2-A current into a battery, for 3 hours, will add 2 * 3 = 6 Ah charge to the battery (Figure 3.28). The battery's DOD will have decreased by 6 Ah. Without knowing the initial DOD, however, we cannot know the final DOD. (The charging process is 100% efficient. See Section 1.2.5.2.)

Coulomb Counting can be a very accurate technique, with two limitations:

- Leakage current within a cell does not go through the current sensor and is therefore not taken into account;
- Offset in the measurement of the battery current will result in the SOC drifting up (or down) over time (in any integration, a nonzero constant in the variable being integrated causes the integral to change over time).

Coulomb counting works well with Li-Ion cells because they have low leakage. Drift remains a major limitation (Figure 3.29) due to the offset in the current sensor, especially Hall effect sensors (see Section 3.1.3.2).

Drift can become significant in applications that, for long periods, use very little battery current or shuttle current back and forth. In particular:

- *Standby batteries:* even if the battery is full, a small offset in the current sensor in the discharging direction will result in the reported SOC drifting all the way to 0% SOC over time.
- *HEV pack:* uses energy from the battery when it needs it, and replenishes it when it can, trying to maintain a 50% SOC. While the reported SOC may very well stay around 50%, over time the actual SOC will drift due to the small offset in the current sensor. Eventually the actual battery charge will approach either the full or the empty state (Figure 3.30).

An HEV can calibrate its current sensor to mostly eliminate SOC drift due to offset in the current sensor. Once in a while, the vehicle control unit (VCU) may

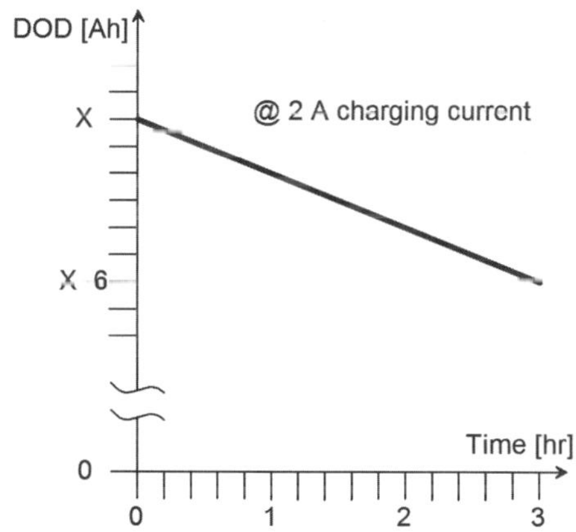

Figure 3.28 Coulomb counting gives only relative DOD.

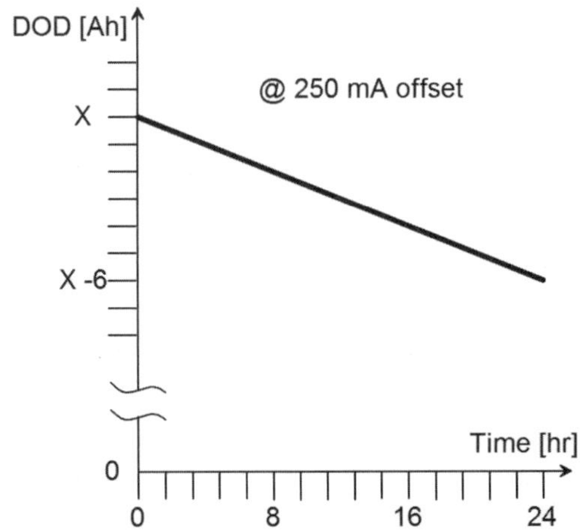

Figure 3.29 Drift in DOD over the course of a day, due to a 250 mA offset in the measured current.

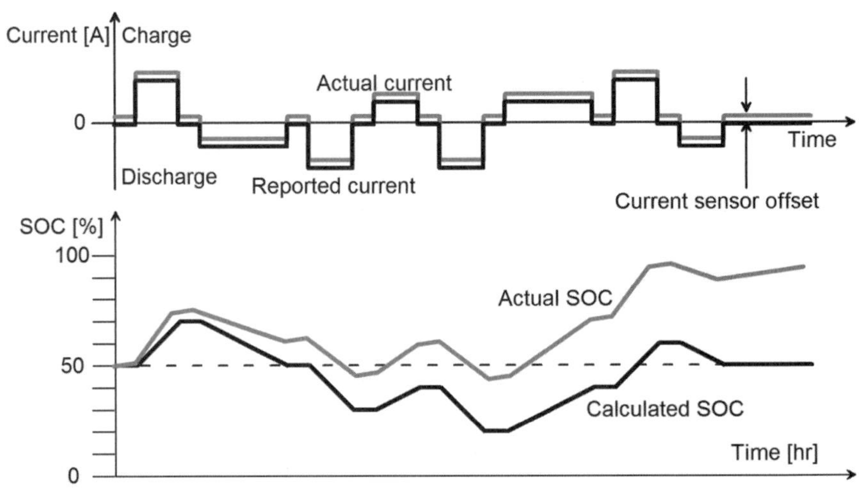

Figure 3.30 Long-term drift in actual SOC with respect to reported SOC in an HEV, due to offset in the current measurement.

turn off the motor AC inverter for a while and notify the BMS that the battery current is 0 so that the BMS may save the current sensor reading as its offset, and later use that offset to correct readings.

3.3.1.3 Combining the Techniques

Coulomb counting can be used to estimate the DOD of a Li-Ion pack as long as there's a way of calibrating it at some point, and often enough to overcome drift. Going back to the bank account analogy: balancing your check book synchronizes the amount you believe is in your account with the amount that your bank says is in that account. Similarly, coulomb counting needs a way to calibrate its result, so that the charge it reports is the actual DOD. Voltage translation provides a way of doing so, just as balancing a checkbook does for a bank account. Combining these two techniques results in a reasonable way of estimating of DOD in a Li-Ion cell (Figure 3.31):

- The battery current is integrated (coulomb counting) to get the relative charge in and out of the battery.
- The battery voltage is monitored, to calibrate the DOD when the actual charge approaches either end.

If the DOD estimated through coulomb counting is uncalibrated (it is not equal to the actual DOD), eventually the battery will be charged or discharged so far that voltage translation can be used to estimate SOC and, knowing the capacity can be translated to DOD, the estimated DOD can be calibrated. For example, if the actual DOD of a 100-Ah Li-Ion cell is 20 Ah but the BMS estimates its DOD to be 50 Ah, the cell may be charged until its voltage reaches a threshold (say, 3.4V), which corresponds to an actual SOC (say, 90%). At that point, the BMS sets the estimated SOC to 90%, calculates the corresponding DOD at 10 Ah, and calibrates the DOD (Figure 3.32).

Going back to the issue of drift, let's see how combining these two techniques affect DOD estimation in the two applications we considered earlier.

- *Standby systems:* the pack is kept full, so voltage translation is used, avoiding the long term drift of coulomb counting.
- *Hybrid car (HEV) traction packs:* when the actual SOC drifts in such way that a cell voltage reaches a threshold at either end, the BMS calibrates the SOC based on that voltage (Figure 3.33). A more sophisticated approach can calibrate the SOC in the process of balancing the traction pack.

For the above method to work, the actual pack capacity must be known so that the conversion between SOC and DOD will be correct. Otherwise, the estimated

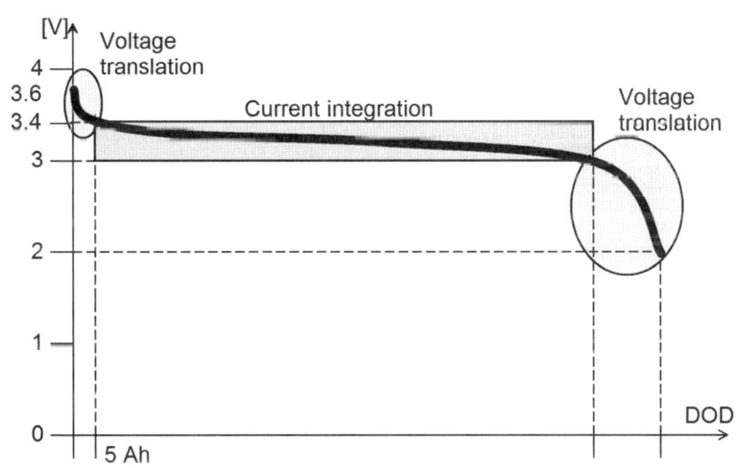

Figure 3.31 Combining coulomb counting and voltage translation to estimate DOD.

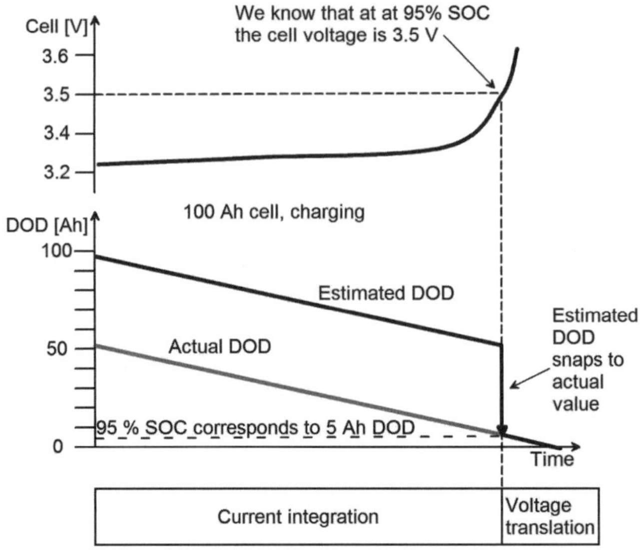

Figure 3.32 The BMS calibrates the estimated DOD when the cell voltage crosses a threshold.

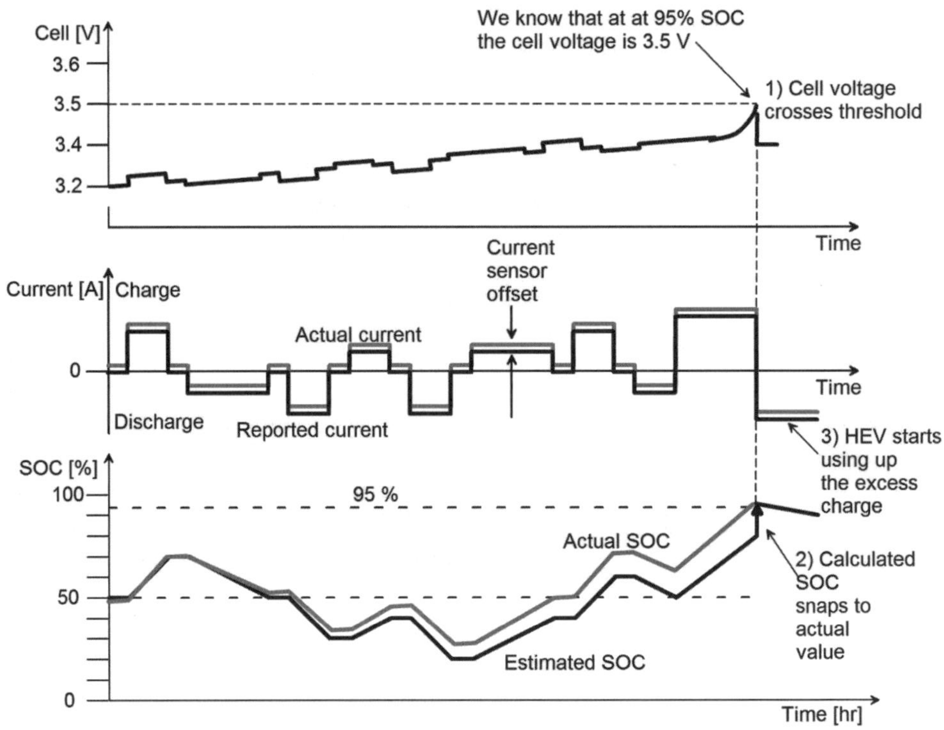

Figure 3.33 Long-term drift in the actual SOC with respect to the estimated SOC in an HEV. Calibration occurs through voltage translation when a cell voltage crosses a threshold

SOC will appear to change too slowly or too quickly, and SOC calibration will be incorrect (Figure 3.34). In an application in which this could present a problem, the battery capacity must be measured (see Section 3.3.2).

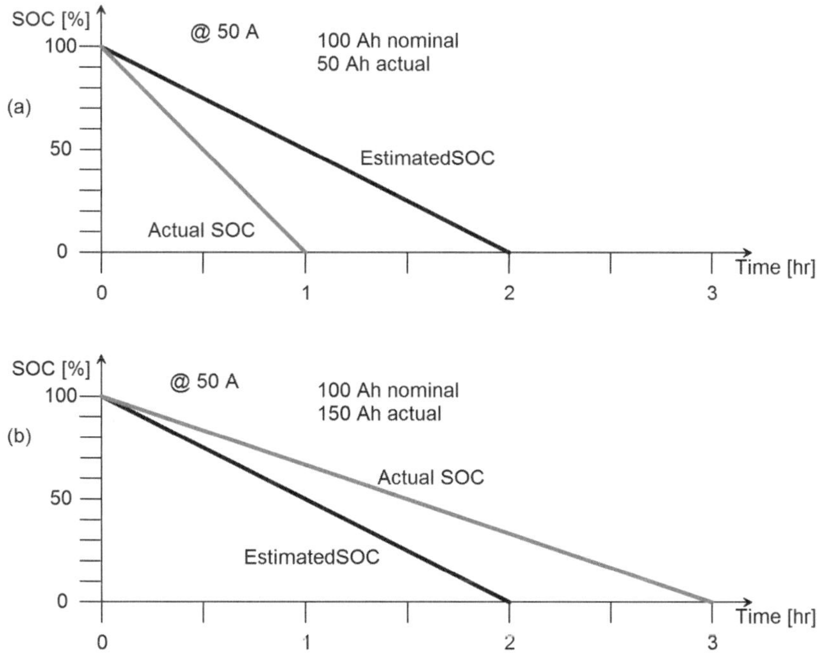

Figure 3.34 A wrong value for pack capacity results in errors in the estimated SOC: (a) low actual capacity and (b) high actual capacity.

3.3.2 Capacity

Measuring the actual capacity of a pack is required for a more accurate SOC estimation. Also, a decrease in actual capacity can be used as one of the parameters considered in SOH evaluation. Capacity is measured by seeing how much charge comes out of a pack from full to empty (Figure 3.35), or the opposite, depending on the application. In general, it is better to do this during a discharge cycle, because the capacity measured by this method (without IR compensation) is a good indication of how much charge can be extracted from the pack under those same conditions each and every time.

An accurate measurement of the actual capacity requires that the pack be truly fully charged and truly fully discharged. The former is, in general, easy to do. The latter may be harder: it is not possible to discharge a pack fully with a load that can only operate at high currents. Unfortunately, in certain applications, is not possible or not acceptable to charge and/or discharge the pack fully:

- In EVs and PHEVs it is never desirable to run a pack all the way down.
- In backup power, in those rare cases the pack is run all the way down, by the time that happens and the capacity can be measured, it's already too late to warn the system that the pack is nearly empty.
- HEVs prefer to keep their traction pack always at mid-SOC levels.

When capacity measurement is possible, it can be rather inaccurate:

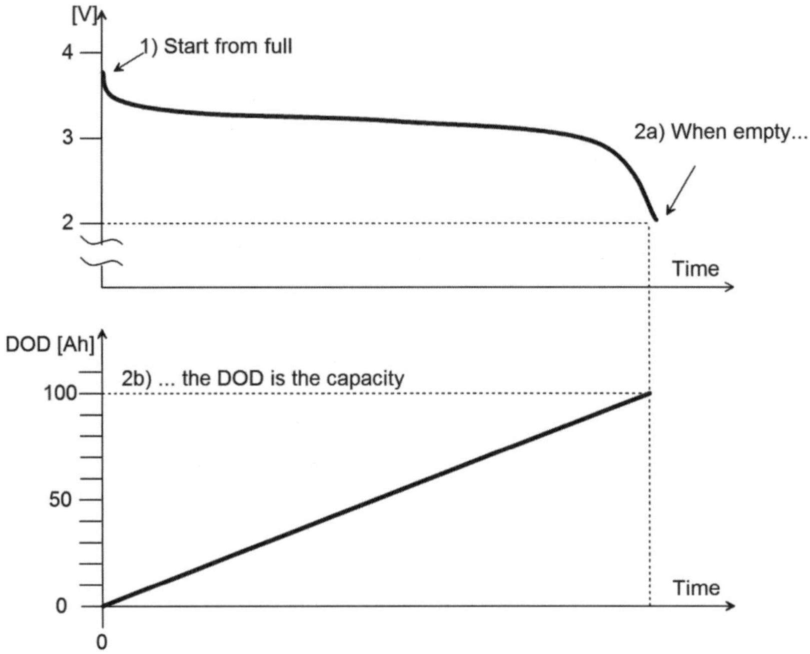

Figure 3.35 Measurement of actual capacity.

- In applications that have a relatively low discharge current, the offset in the current sensor can be so significant that measured capacity may be quite far off: over long periods, the offset in the current sensor can integrate into a large error.
- Systems that do not compensate for cell resistance do not charge and discharge the pack fully.

Having measured capacity once, the BMS cannot rely on its value for long, as capacity changes with pack balance.

The fact is that, in real life applications, the measured capacity will change noticeably each time. Its value is, at best, a good guess, and, at worst, quite misleading.

3.3.3 Resistance

Cell resistance (see Section 1.1.2) by itself is usually of little interest to the user. However, its value is useful in calculating two other parameters that are of interest to the user: SOC and SOH.

- Cell resistance can be used to do IR compensation of the cell terminal voltage in order to estimate the OCV, from which, in turn, the cell SOC can be estimated.
- The increase of cell resistance over time can be one of the parameters used in SOH calculation (see Section 3.3.4).

A cell's resistance varies with SOC, temperature, current direction, and usage. Therefore, a BMS cannot calculate a cell's resistance once and use it from then on. On the contrary, it must try to calculate it whenever possible for the most accurate IR compensation and SOH evaluation. As doing so is often a challenge, some BMS manufacturers will characterize a particular cell in the lab to create an accurate, complex model to estimate the cell resistance under particular conditions.

Cell resistance is a dynamic resistance, which is defined as the ratio of delta voltage over delta current (see Section 1.2.7). Therefore, in order to calculate it, there must be a delta in the pack current (the pack current cannot be constant) resulting in a delta in the cell voltages. In some applications (vehicles during use, large distributed energy sources when providing backup power) the delta occurs naturally. In others (EVs while charging, UPSs in standby), it does not. If the current is constant, and the BMS must measure the resistance, in some cases it may be acceptable to turn the pack current off and on, just long enough to take measurements and calculate the resistance. For example, in a backup power plant with multiple packs in parallel, it may be possible to test one at a time, by have it discharge, stop, then recharge.

3.3.4 State of Health (SOH)

Every BMS that evaluates pack SOH (see Section 1.4.3) defines SOH differently. Two parameters that may be used to calculate an SOH value (Figure 3.36) are:

- Increase of actual resistance with respect to nominal;
- Decrease of actual capacity with respect to nominal.

3.4 External Communications

Whether or not a BMS communicates outside with the external system depends on its type:

- Regulators and meters do not have any external communication by definition.
- Monitors and balancers are required to have at least some form of external communications, to tell the system to reduce the battery current or stop it all together.
- Protectors are self-contained and need no external communications.

For those BMSs that do use external communications, they may carry:

- From the BMS:
  - Request to the system to reduce battery current or stop it;
  - Data on the status of the pack and of the BMS itself.
- To the BMS:
  - System configuration commands;
  - Data from external sensors.

In general communications can be divided into:

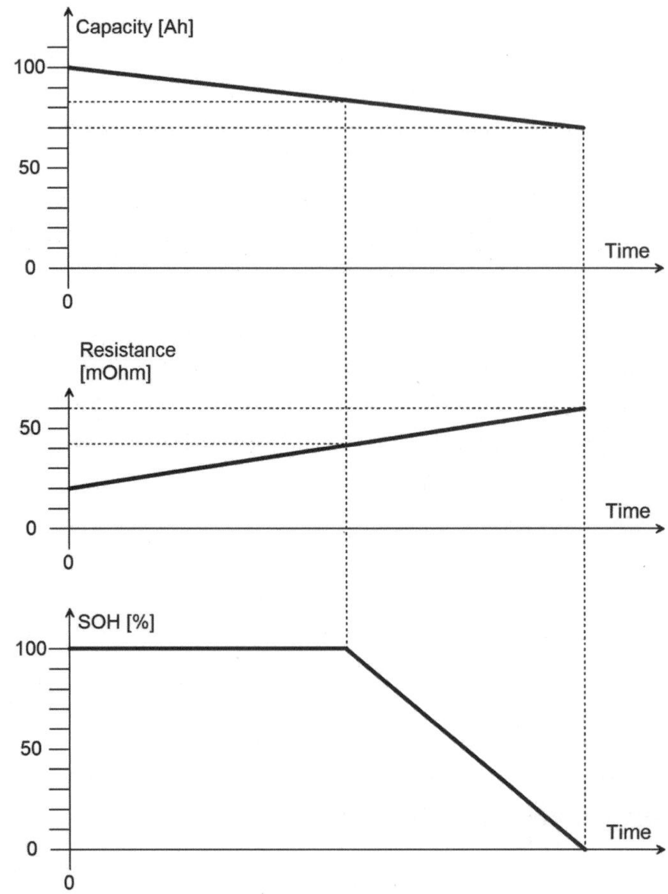

Figure 3.36 Example of SOH determination.

- *Dedicated wire:* a wire that has a specific, dedicated function, which could be:
  - Analog (continuously variable) signal;
  - Digital (on/off) control signal, whether through a solid state device or a mechanical relay.
- *Data link:* a communication link with digital data, which could be:
  - A wired, serial data port (RS232, CAN, Ethernet);
  - A wireless radio link (Wi-Fi, Bluetooth);
  - A light link (fiber optic, Infrared link).

3.4.1 Dedicated Analog Wire

These wires input or output a linear signal, usually in the range of 0 to 5V.

3.4.1.1 Inputs

A BMS may have an analog input to measure the value of sensors outside the pack. An example would be if a charger had an analog output proportional to the output or input current. Other than that, pretty much any other analog signal of interest to the BMS is within the pack, and was covered in Section 3.1.

3.4.1.2 Outputs

One way for a BMS to report status or to request reduction in current is by driving linear outputs with an analog signal that is continuously variable, in proportion to the value of a particular variable. For example:

- *DCL output:* Normally sits at 5V, but as the BMS decides that the discharge current should be limited, it gradually reduces the output voltage, until it brings it down to 0 when no discharge current should be allowed;
- *CCL output:* Similar to the DCL output, but to limit the charging current;
- *SOC output:* Ranges from 0V to 5V as the SOC ranges from 0% or 100%;
- *Fuel gauge output:* Ranges from 0 ohm to 100 ohm, as the SOC ranges from 100% or 0%;
- *Throttle high output:* Just like the DCL output, but with a voltage range suited to feed the high lead of a two-wire or a three-wire throttle pot, to reduce the available range of the throttle in a vehicle as the battery reaches the end of charge;
- *Throttle wiper output:* Same as above, but suited for connection to the wiper lead in a three-wire throttle pot to clamp down its voltage as the battery reaches the end of charge.

Some examples on how such lines may be used are offered in Chapter 6.

3.4.2 Dedicated Digital Wire

The BMS may use on/off lines to communicate with the external system. These are:

- Inputs to allow the external system to control the BMS or report some status to it;
- Outputs to allow the BMS to control the external system or report some status to it.

3.4.2.1 Inputs

Some BMSs are able to drive devices such as contactors, fans and heaters. If so, they would include digital inputs control these devices. The BMS may also have digital inputs to let the external system report status information, such as whether the load is powered (e.g., ignition line or sustain line), whether the source is powered (e,g., a vehicle is plugged in), or to enable charging or discharging into the power grid (e.g., charge control and V2G enabling in a plug-in vehicle or distributed energy source).

Inputs are generally of two types:

- Standard logic level input (e.g., TTL levels); <0.8V is a logic 0, and >2V is a logic 1. These inputs can be interfaced with logic outputs, contacts (relay contacts, switches, interlock, and tamper detectors), open collector/open drain devices (transistors, optoisolators, and solid-state relays), and medium voltage levels (power supplies, 12-V lines) (Figure 3.37).

- Contact closure input compatible with a high-current contact, which include extra capacitive loading to clean the contacts of a high-current relay or switch by running a pulse of high current through its contacts at the time of closure.

3.4.2.2 Outputs

The BMS may use digital (on/off) lines to report to the external system the particular state of the pack or of the BMS, such as:

- *High limit (HLIM)/over voltage limit (OVL):* Asserted when the pack is unable to accept charge, deasserted otherwise;
- *Low limit (LLIM)/under voltage limit (UVL):* Asserted when the pack is unable to deliver discharge;
- *Reserve (or valet, or limp home):* Asserted when the pack is unable to deliver the full discharge;
- *Warning:* Asserted when the BMS has detected a warning condition;
- *Fault:* Asserted when the BMS has detected a fault condition;
- *Charging:* Asserted while current is flowing into the pack;
- *Discharging:* Asserted while current is flowing out of the pack;

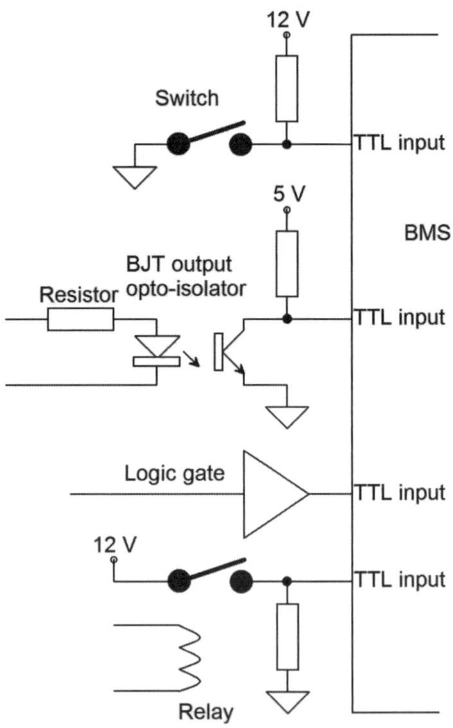

Figure 3.37 Examples of devices wired to a logic input.

- *Idle:* Asserted while current is not flowing in or out of the pack;
- *Charge deplete:* Asserted when the pack is able to deliver substantial charge (may be useful in a PHEV);
- *Charge sustain:* Asserted when the pack is close to empty, and charge needs to flow back and forth to maintain that same general SOC level (such as normally done in an HEV);
- *Battery operating properly (BOP):* A square wave clock whose frequency changes depending on the state of the pack—normal or fault.

Typically, these output lines may be of the following types (Figure 3.38):

- Logic level outputs (e.g., CMOS levels: 0V for a logic 0, 5V for a logic one):
  - Compatible with logic level inputs.
- Low-power open drain/open collector (grounded or open):
  - Capable of carrying small currents when grounded (on the order of 10 to 100 mA);
  - Capable of sustaining small voltages when open (on the order of 20V);

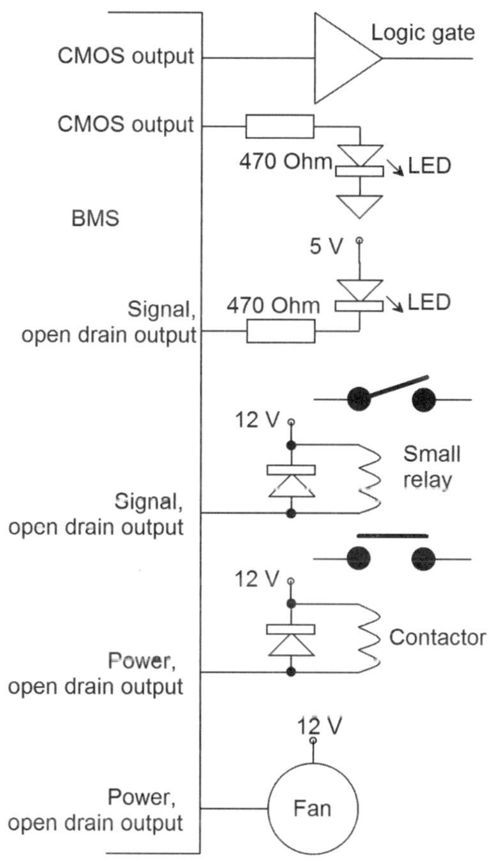

Figure 3.38 Various types of digital output

    - Intended to drive small loads, such as LEDs, small relay coils, and logic inputs.
- High-power open drain/open collector:
    - Capable of carrying large currents when grounded (on the order of 1A to 10A);
    - Capable of sustaining large voltages when open (on the order of 100V);
    - Including protection against inductive kickback;
    - Intended to drive large loads, such as power relays and contactors, fans, blowers, and heaters.
- Relay contacts:
    - Isolated;
    - Normally open, normally closed, or both;
    - May be dry contacts (for low current) or power contacts (for high current);
    - Work with AC or DC.
- PWM outputs (with a variable duty cycle square wave):
    - Intended to drive variable loads, such as variable speed fans and blowers.

3.4.3 Data Link

In professional applications, the BMS uses a minimum of dedicated lines, preferring to rely mostly on a data link to communicate with the external system. This results in a degree of flexibility and in reduced cost, compared to relying mostly on dedicated lines. The data link is usually over wires, though sometime it is done over fiber optics (especially useful if the BMS is electrically connected to the pack, and isolation from the external system is required), radio link, or infrared link (to enable hot swap of batteries requiring only the removal of a power connector for the battery terminals).

The data link can be proprietary or use a standard protocol, but even in the latter case, the coding of the data is usually proprietary. The standards most commonly found in a BMS (Figure 3.39) are:

- *RS232:* Duplex, asynchronous, point-to-point link;
- *RS485:* Balanced, multidrop bus;
- *CAN bus:* Industry and automotive standard bus;
- *Ethernet LAN:* Computer network standard bus;
- *USB:* Computer peripheral, point-to-point link.

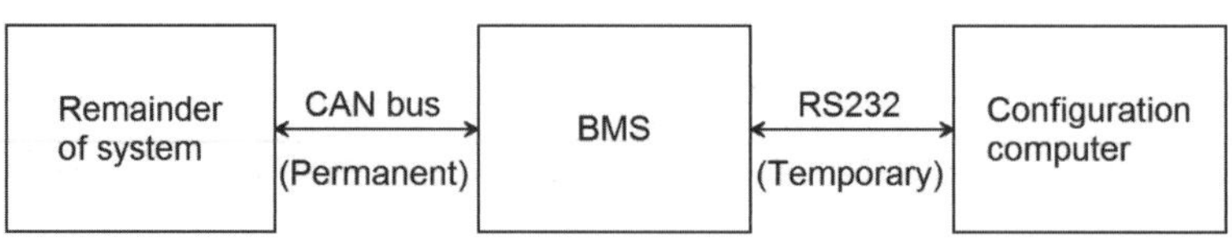

Figure 3.39 BMS with typical communication links: a permanent CAN bus to the rest of the system, and a temporary link to a configuration computer through an RS232 serial port.

To these, we will add one that has gained some popularity in EV enthusiasts' circles:

- *EVil bus:* An isolated, low-speed twisted pair bus for up to 32 nodes.

3.4.3.1 RS232

The RS232 communication standard is a venerable workhorse for point-to-point links. It is still used because RS232 ports are still found in desktop computers (though no longer in laptop computers, which is not a problem as RS232 to USB dongles are readily available). It is not balanced, and therefore not appropriate for permanent use, because the environments in which large packs are used can be electrically very noisy. Yet, it is appropriate for temporary use, during BMS configuration, testing, or troubleshooting.

It is an attractive option for a user interface because terminal emulator applications (such as HyperTerminal, Fetch, and PuTTY) are readily available. That means that there is no need for special software on the host computer; the BMS can communicate with the user, through the terminal emulation application, in a human language. A service technician needs just a laptop or other computer device to service a BMS, without having to locate and install special software to do so. Of course, a dedicated application could be created to provide a more elegant user interface.

Use of RS232 for permanent links is frowned upon in industrial and automotive applications, because of its low speed and poor noise rejection. Still, it is possible for the BMS to use an RS232 link in a noisy environment. In that case, either the device at the other end should be floating (not grounded to the chassis locally) or an RS232 isolator should be placed on the RS232 line. The device at the other end could be, for example, a display that reports the pack status. In that case, the BMS may have to be switched from a human interface mode, to a data dump mode.

3.4.3.2 CAN Bus

The Control Area Network (CAN) bus is the standard in vehicles manufactured today, and a standard in industrial applications. It is not within the scope of this book to describe this bus, but a simplified description may be in order. The standard specifies the middle layers on how to interpret a digital stream (object and transfer layers) but leaves the outer layers undefined: the application layer (at the high level—what the data mean) and the physical layer (at the low level—how the digital stream is transmitted). Nonetheless, industry standard ways of implementing the remaining layers have emerged.

At the hardware level (physical layer), there is no standard for a connector (as there are for other buses), but a de facto standard has emerged for the wiring (balanced, multidrop twisted pair with two 120Ω terminating resistors, one at each end of the bus), the names of the two lines (CANH and CANL), and the signal levels on it (2.5V while idle).

Fundamentally, CAN messages are not addressed to a particular recipient, but are simply broadcast, and any device that needs to can use the data in that message.

If all devices on the CAN bus were to broadcast at all times all the data they have access to, the bus would be too busy. Therefore, only the most essential data are placed in messages that are sent to the bus. To retrieve nonessential data, a device may send a request for it, and the device that has those data responds with a message reporting them. One example of this is PIDs (Chapter 5).

CAN messages may use either an 11-bit or a 29-bit ID field. While both types of messages can coexist on a CAN bus, and newer devices can send and receive both, it is rare for a system to use both types. The priority of a message is set by its ID: the lower the ID the higher the priority.

Industry groups have attempted to define the application layer with a set of standard messages that can be sent on the bus, such as the SAE's J1939 standard and CANopen. Neither of them is focused on the functions of a BMS, and therefore each BMS designer defines CAN differently. In 2007 I proposed a standard set of CAN messages for traction packs, which is now used by a few manufacturers (see Chapter 5).

All messages on a particular CAN bus use the same baud rate (the same speed). Any speed can be used up to 1 MHz, though certain speeds have become standard: 125 kHz, 250 kHz (for example, SAE's J1939 standard), and 500 kHz (common in most passenger vehicles). The CANopen standard can use 10, 20, 50, 125, 250, 500 kHz, and 1 MHz.

3.4.3.3 Ethernet

In large, land-based applications or mobile applications used in research and development, a BMS may include an Ethernet connector for connection to a local area network (LAN). Applications on host computers will be able to monitor, log, or even control the pack's BMS. In mobile applications, a remote communication link (such as through the cellular network) may enable telemetry and remote control.

3.4.3.4 USB

Just as a permanent RS232 link is not appropriate in automotive and industrial applications, nor is Universal Serial Bus (USB). However, USB is ubiquitous in a way that RS232 no longer is. Therefore, USB may be a better choice for a port in a BMS to enable configuration and monitoring through a laptop. Just as with RS232, a standard terminal emulation application could be used to communicate with the BMS in a human language, or a dedicated application could be created to provide a more elegant user interface.

3.4.3.5 Wireless Data Link

Wireless communication may be popular in the consumer world, but not widely accepted in the industrial and automotive environments, when a short cable can do the same job more reliably. Still, there are instances in which a device cannot use a wired connection (such as the Apple iPhone), so a consumer RF link such as Wi-Fi or Bluetooth may be used to communicate with a BMS.

3.5 Logging and Telemetry

While a BMS may store a few records in an error log, expecting it to do more than that may be inappropriate: logging is a job for which a data logger is better suited. A logger can record all of the system data, not just the BMS data. For example, a logger on a vehicle CAN bus can record not only the battery current, but what the engine RPM was at the time. A log may include:

- Pack voltage;
- Pack current;
- Pack SOC and/or SOH;
- Pack resistance;
- Minimum and maximum cell voltage;
- Minimum and maximum temperature;
- CCL and DCL;
- Warnings and errors.

The same data could also be transmitted to a remote location (telemetry). A popular way of doing so is by cellular modem. Other carries include the pager network (slower but much less expensive).

References

[1] Barsukov, Y., *Battery Cell Balancing: What to Balance and How*, Texas Instruments.

[2] Rickard, J., "Get Rid of Those Shunt Balancing Circuits," November 25, 2009, http://jackrickard.blogspot.com/2009/11/get-rid-of-those-shunt-balancing.html.

CHAPTER 4

Off-the-Shelf BMSs

4.1 Introduction

As of this writing, there are just a few dozen companies offering off-the-shelf BMSs, though the landscape is changing rapidly. This chapter is a snapshot of the commercially available BMSs at the time of publication. Some have been announced and will be available by the end of 2010. While this information is likely to change significantly over time, it should useful to give you an understanding of what kinds of products are possible, with what features, and at what prices. For an up-to-date list of commercially available BMSs online, please go to http://book.LiIonBMS.com.

By far the largest proportion of off-the-shelf BMSs consists of simple protectors for small batteries from Chinese companies. As these are not appropriate for large Li-Ion packs, they will not be discussed here in any depth. Very few of the commercially available BMSs are from large, well known, widely trusted companies. Such companies may not even announce their products to avoid inquiries from small customers, preferring instead to offer their products privately to large potential customers. In any case, such BMSs may not really be called off-the-shelf, as much as semicustom. In this chapter we'll explore those few that are publicly announced and truly available.

It appears that the most advanced, most practical, yet most economical BMSs today are offered by tiny U.S. and European companies. These are mom and pop companies that may not receive a second glance from large corporations, yet have on their shelves ready-made, effective, and reliable solutions. This chapter will explore those BMSs in some detail.

4.1.1 Simple

As described in detail in Section 2.2.1, simple BMSs use analog technology and have limited capabilities. Their prices (Figure 4.1) range from \$3 per cell for a protector to \$15 per cell for a distributed BMS. Of the various analog BMS types, regulators, balancers, and protectors, are the ones that may be considered in a large Li-Ion pack.

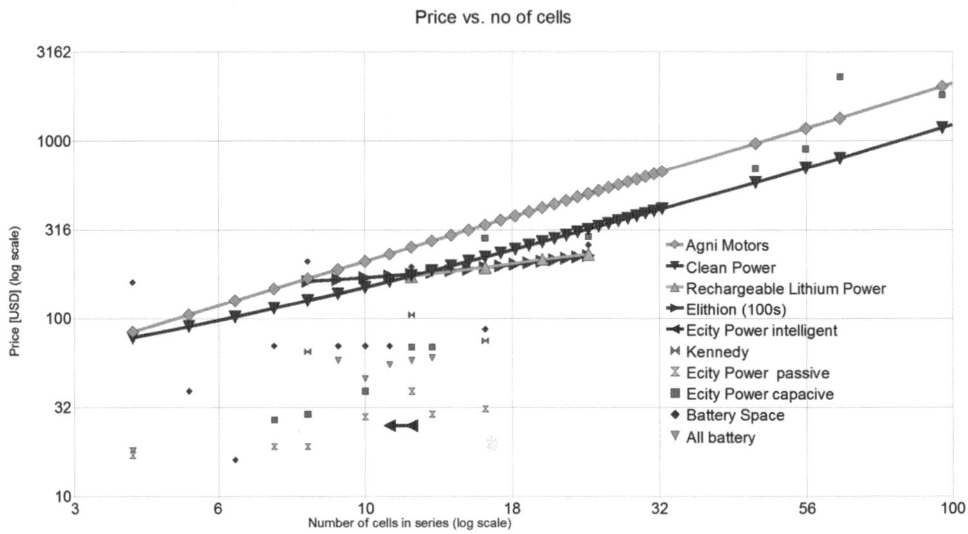

Figure 4.1 Prices of analog BMSs.

4.1.1.1 Regulators

While there are many regulators out there, they tend to be made by hobbyists, and to become unavailable after a little while. One of the two companies that are still making them is LL Labs, with its Bat Guard line of high-power regulators (up to 4A). While shunting, a yellow LED is lit. If the fuse blows, a red LED will light, but the cell will soon be toast, so a human has to watch the LEDs at all times. The fixed 3.67-V knee voltage is a bit too high for $LiFePO_4$ cells, and too low for LiPo, Thundersky, and cobalt cells, though custom threshold voltages can be ordered.

Elite Power sells small regulators for 90-Ah prismatic cells. They bypass less current than the Bat Guard, but they are also less expensive. They too have a fixed voltage, 3.7V, which is a bit too high for $LiFePO_4$ cells, and too low for LiPo, Thundersky, and cobalt cells.

4.1.1.2 Balancers

Analog balancers can have the best value, as they are relatively inexpensive, while fully protecting and optimizing a battery. Most balancers are distributed: a cell board per cell monitor that balances that cell, and a bus between cells reports overvoltage and undervoltage to the system, either directly or through a master controller (which can be as simple as a couple of relays). There are various companies offering such cell boards, including Agni/Stybrook (United Kingdom/India), Black Sheep (United States), Blade (Australia), Belktronix (United States), Clean Power (United States), EV Power (Australia), and Shenzhen Soopower Technology (China). Of those only two will be highlighted in the following sections. I will also discuss the Lithiumind, which, unlike the above products, is a centralized, full featured BMS.

Figure 4.2 Clean Power Auto's MiniBMS distributed or centralized balancer. (*Source*: D. Butvinik, Clean Power Auto, 2010. Reprinted with permission.)

Clean Power Auto

Dimitri Butvinik of Clean Power Auto's (cleanpowerauto.com) in Tampa, Florida, created the MiniBMS (Figure 4.2), which offers the absolute minimum set of functions required by a BMS for a pack of LiFePO4 cells, with a perfect balance between features and simplicity. This no frills, distributed BMS is easy to install and requires hardly any configuration. It is ideal for a DIY hobbyist making their own EV. It handles the basics of $LiFePO_4$ pack protection and enhancement: undervoltage, overvoltage, and balance. It can handle an unlimited number of cells. It can be used in two topologies: distributed or centralized. In the distributed topology it works with a prismatic cell. A cell board is installed on each cell, and a single wire daisy chain loops from cell board to cell board, then to the master controller. In the centralized topology, the cell boards are simply left in a single panel and wired to the cells of any format.

Elithion

As a follow up to its digital balancer, the Lithiumate, in 2010 Elithion (elithion.com) introduced the Lithiumind (Figure 4.3), an analog BMS that sports advanced features only found in digital BMSs; they include current measurement, SOC, and SOH calculation. It is intended for medium-sized batteries (up to 24 cells) in electric bicycles and motorcycles. It is a centralized BMS: 25 wires go to the 24 cells. Four adjustments allow the configuration of threshold voltages: charger turn off, balance, reserve, and motor controller off. A serial output can be sent to an optional display to show the adjustment settings, minimum and maximum cell voltage, battery current, voltage and resistance, capacity, SOC, and SOH. It is bullet proof in that it can withstand all types of miswiring and open faults in the battery.

EV Power

Rod Dilkes, of EV Power (ev-power.com.au) in Western Australia, used a minimalist approach to develop a distributed balancer for any size battery using prismatic Li-Ion cells (Figure 4.4). Each tiny cell module is mounted on a cell, to detect undervoltage or overvoltage. Adjacent cells modules are connected through a sin-

Figure 4.3 Elithion's Lithiumind centralized balancer.

Figure 4.4 EV Power's distributed balancer mounted on a prismatic cell. (*Source*: Rod Dilkes, EV Power, 2010. Reprinted with permission.)

gle-wire, daisy chain loop; two pads on a cell boards are provided for that connection, and they are interchangeable.

Cell modules come in various sizes designed to fit particular terminal spacings. A master controller is simply a box with relays controlled by the bus. A balancing current of 600 mA is sufficient for the majority of applications. The BMS is marketed by EV Works, near Perth, West Australia.

4.1.1.3 Protectors

Various Chinese electronic companies offer a wide selection of analog protectors for small batteries. These include Cyclone, EcityPower, Kennedy Alternative Energy, ONS Power, Rechargeable Lithium Power, Smartec, and Yesa, There is also one company from the United States: Southwest Electronic Energy Group. These protectors are a perfect fit for small batteries, but some try to shoehorn them into large packs. Some Chinese integrators offer a large Li-Ion packs mated to a set of such protectors. To date, every one I have seen simply does not work, and their users are sorely disappointed, toss the protectors, and go searching elsewhere for an effective solution to protecting the pack. That is because those protectors are not designed to be stringed in series, are not designed to handle high currents, are not able to work in high-noise environments, and are nonisolated. In particular, one came with the amusing (but correct) warning: do not communicate with BMS while battery is in use. That warning was due to the fact that each of the six protectors in the pack had its own serial port, each referenced to a different voltage (so connecting them together would short out the battery), and any serial cable connected to them would pick up so much noise that the protector would stop working reliably.

The situation may have changed by the time you read this book, but today I must warn you about anyone who tries to sell you a large Li-Ion pack using a set of inexpensive protectors as a BMS.

4.1.2 Sophisticated

As described in detail (Section 2.2.2), sophisticated BMSs use digital technology and have advanced capabilities. Of the various digital BMS functionalities, meters, monitors, balancers, and protectors are the ones that may be considered in a large Li-Ion pack.

4.1.2.1 Meters and Monitors

The prices of meters and monitors (Figure 4.5) depend on the number of cells in series they support. Today, those price range from \$17 to \$50 per cell (for 100 cells).

Electric Blue Motors

Electric Blue Motors, in Arizona, offers the Blue Window, a dashboard mounted computer/display for custom vehicles. In 2009, it added an extension called Blue

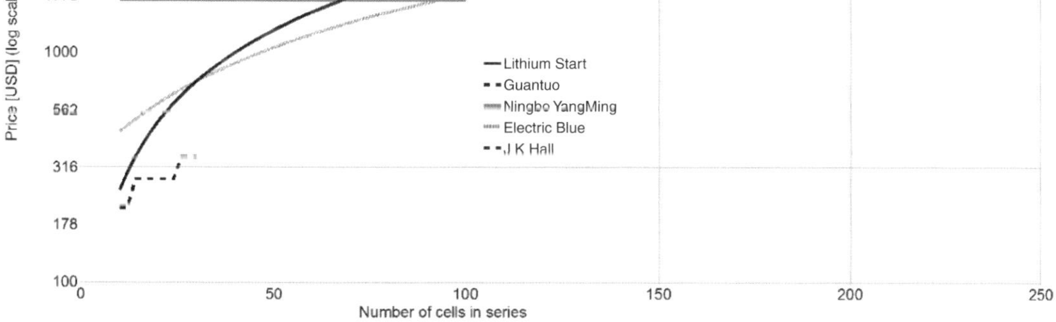

Figure 4.5 Prices of digital meters and monitors.

View, to measure and report the voltage of battery cells, turning their product into a Li-Ion meter, capable of reporting each cell voltage and to calculate SOC, but unable to protect the battery pack.

Guantuo Power

Guantuo Power Equipment Co., Ltd. (http://guantuo.com) specializes in power management system for lithium batteries, using technology developed by major Chinese universities. It offers four versions of its GTBMS005A master/slave monitor, all of which include a central controller, slave modules, a display, a current sensor, and a USB port. They all compute SOC and SOH (Figure 4.6). The versions differ in the display (the MC11 and MC16 have color monitors, while the MC17 and MC8 have monochromatic monitors), the number of cells supported (the MC11 and MC16 support 300 cells, the MC17 100 cells, and the MC8 20 cells), and communications (the MC8 uses relay outputs, and the other three have a CAN bus, which lets them communicate seamlessly with ElCon chargers). Despite the fact that this is just a monitor (it does not have a balancing function), it has been widely used on vehicles. Guantuo is ISO9000 certified; their BMSs are CE certified and come in professional metal cases or hard plastic modules.

JK Hall

Ken Hall, from Windsor, Colorado, developed the PakTrakr (http://paktrakr.com) battery meter for electric vehicles' lead acid batteries. More recently, he made a version of the PakTrakr for Li-Ion and NiCd cells, suitable for medium-sized batteries, such as electric bikes. This meter uses the master/slave topology. Up to six slaves can be used, each measuring up to six cells, for a total of 30 cells. The master is located in an attractive, panel mount, plastic case, and includes a small LCD display and two push buttons. There are no additional connections; the meter is powered by the battery, and it does not control any outside devices.

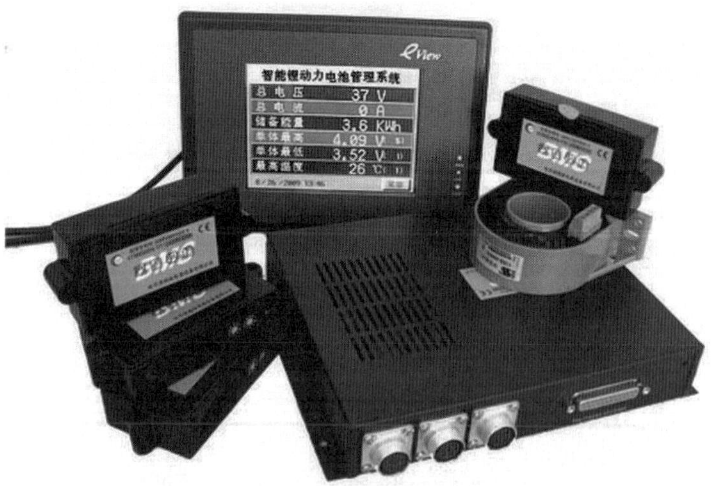

Figure 4.6 Guantuo Power's monitor. (*Source*: Y. Fairy, Guantuo Power, 2010. Reprinted with permission.)

4.1.2.2 Balancers and Protectors

In my research I have found only seven companies that produce balancers or protectors of a quality level appropriate for inclusion in a professional, large Li-Ion pack. Their prices (Figure 4.7) depend on the number of cells in series they support. Today, prices range from \$12 to \$75 per cell, for 100 cells.

Black Sheep Technology

Black Sheep (http://black-sheep.us) is a small company in North Carolina, run by Ron Anderson, with lengthy experience in making electronics for electric vehicles. He is a stickler for quality, offering four different BMS lines, covering a range of needs. They all can handle any cell format.

The Stack V1 line is intended for smaller batteries. It is an open board, modular BMS. One module can handle up to 16 cells in series, and two modules can be joined in a stack to handle up to 32 cells.

The Cage V2 line is designed for medium-sized battery packs for automotive applications. This centralized BMS consists of a master controller in the shape of a motherboard with six slots. Each slot can accept a card, each of which can handle up to eight cells in series, for a total of up to 48 cells. The motherboard uses an automotive grade, sealed connector, and the whole assembly can be housed in a sealed metal housing.

The Mini V3 line is for large packs, as it can handle an unlimited number of cells. It uses a modular topology. Each module is housed in a small, attractive molded plastic case, and can handle 5 cells in series.

The BMS Auto V4 is ideal for large traction packs. It is a modular system, housed in automotive-grade housing. Each module handles up to eight cells in series, and an unlimited number of modules can be used for any size pack.

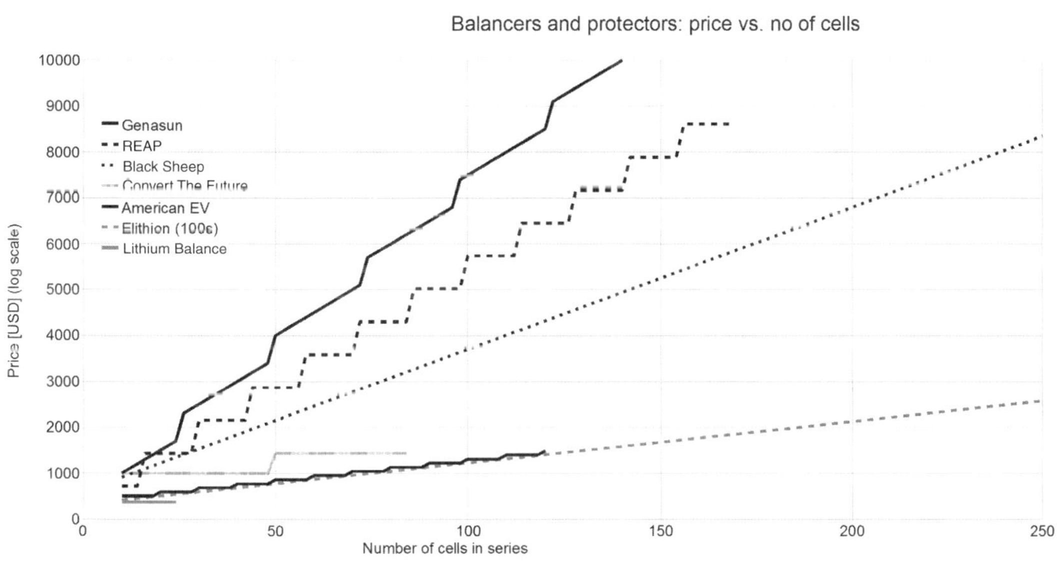

Figure 4.7 Prices of digital balancers and protectors.

Clayton Power

The European company Clayton Power (http://claytonpower.com) offers two off-the-shelf BMSs.

Their protector is a direct knockoff of the Lithium Balance protector (the two companies used to collaborate and then parted ways in 2009). Clayton Power has told me that it will announce a replacement product by the end of 2010.

Clayton Power also offers a master/slave balancer for up to 496 cells. Up to 31 slaves can handle up to 16 cells and four temperature sensors each. The BMS includes an input for a current sensor used to calculate the pack's SOC. The various components are housed in attractive, molded plastic cases.

Elithion

In 2004 I designed a Li-Ion BMS for my Sparrow EV. In 2006 I cofounded Hybrids Plus, a company that converted Toyota Prius and Ford Escape HEVs to PHEVs, and used my BMS technology. A few dozen cars were converted. I left that company in 2008 and started Elithion in order to make my BMS technology available to the industry.

Today, Elithion (http://elithion.com) is a two-person company that offers the Lithiumate (Figure 4.8), a flexible, distributed BMS. Cell boards are available for small cylindrical, pouch, prismatic, and large cylindrical cells, either singly or as a semicustom assembly for multiple cells. Each cell board is powered by its cell, measures the cell voltage and temperature, and provides a medium balancing current. Connection between adjacent cell boards requires a single wire, simplifying installation. From the point of view of the BMS, the pack is divided into banks, up to 16 of them, making it compatible with mod-pack safety disconnects. The BMS controller is housed in a cast aluminum case. It is field programmable through a serial port, using a standard terminal application. Functions include SOC, SOH, resistance and capacity calculation, support of CAN bus, multiple current sensor, cooling and heating, contactor drive, and safety interlock.

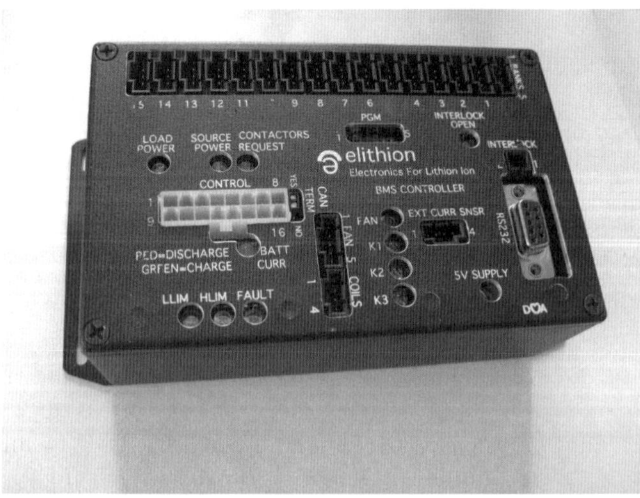

Figure 4.8 Elithion's Lithiumate, a distributed balancer.

An optional high-voltage front end detects pack isolation, measures pack current and voltage, and drives a set of contactors. An optional, simple LED display shows the SOC. Third-party vendors offer more advanced displays compatible with the Lithiumate. Current sensors are available between 5A and 600A, either cable mounted or inline.

Elithion uses a large per-order fee as a way to encourage individuals and small companies to buy products and get support from an international network of resellers, allowing it to focus on larger, direct customers.

EVPST

From Guangzhou, China, comes EVPST (http://evpst.com), a company that presents itself as strongly environmental. It offers batteries and related electronics for solar and marine applications. Their EVPST-BMS-4 is a master/slave Li-Ion balancer that includes a display and a current sensor (it calculates SOC). It communicates with EVPST chargers through an RS-485 bus and with the rest of the system through a CAN bus. Confusingly, its products are also sold by a neighboring company, ECityPower (a company that also offers a large array of protectors for small batteries), though EVPST insists that there is no relationship between the two companies.

Genasun

Alex MeVay formed Genasun (http://genasun.com) in Boston to produce electronics used in solar systems such as charge controllers. When he began switching from lead acid to Li-Ion batteries, he wasn't satisfied with the choices for a BMS, so he decided to design his own. The result was the GLD series protectors (with a high-power contactor) or balancers (without). It is a distributed BMS: cell boards are mounted directly on the cells. The cell boards are designed to fit just one particular size of prismatic cells, though they could also be shoehorned into working with small cylindrical cells. Communications with the master controller is through modular (telephone-type) cables.

Lithium Balance

Lithium Balance is a Danish-based company specializing in Li-Ion BMSs and also manufactures complete turnkey Li-Ion packs. They offer two BMS types: a line of centralized protectors and a master/slave balancer.

The i-BMS series of centralized protectors (Figure 4.9) is designed for quick installations and conversions from existing 48, 60, or 72 volts lead acid packs to Li-Ion packs, using cells of any format. It can interrupt a rather high current, making it ideal for medium-sized battery packs, such as for NEVs, though it could also be used for smaller batteries such as for electric wheelchairs and e-bikes. An RS232 port allows a connection to a computer running their PC Diagnostic software for diagnostics and status, including SOC and stored event and error logs. It is compatible with various intelligent chargers on the market, is very nicely packaged in an extruded aluminum case, and is CE certified by TÜV to meet international standards. It is available in various versions, as listed in Table 4.1.

The Scalable Battery Management System (s-BMS) is a full featured, master/slave balancer for large packs up to 600V (Figure 4.10). Each slave (an LMU)

Figure 4.9 Lithium Balance's protector. (*Source*: Adetunji Adebusuyi, Lithium Balance, 2010. Reprinted with permission.)

Table 4.1 Lithium Balance's Protector Line's Specifications

Model	*Nominal Voltage*	*Cells in Series*	*Continuous Current*
1101-0053	48V	15	140A
1101-0037	60V	19	250A
1101-0058	72V	23	250A

can handle eight cells in series, and the master (a BMCU) can handle eight slaves, for up to 256 cells in series. A daisy-chained, four-wire bus connects the various components. This BMS is available either as a collection of open assemblies (available for installation in your custom enclosure), or encased in sturdy, sealed metal enclosures using sealed circular plastic connectors. The BMS is configurable for all common cell chemistries, reports pack data and computed parameters (SOC, SOH), and features a CAN bus and an RS232 port (for connection to a PC for diagnostics and configuration), charger control (analog, PWM, or CAN), an interface to the current

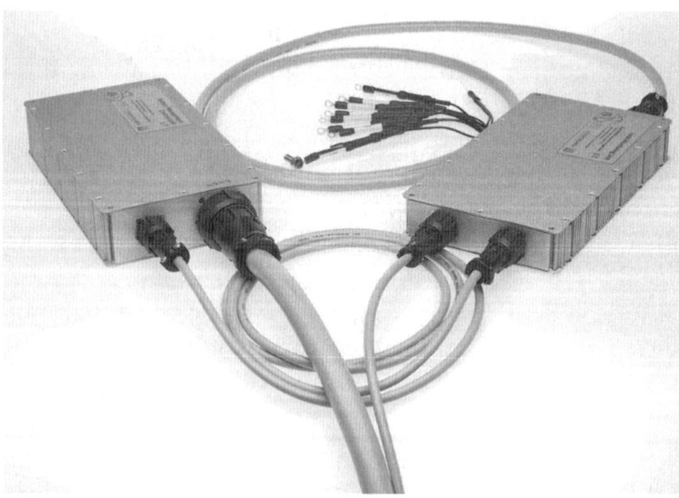

Figure 4.10 Lithium Balance's balancer. (*Source*: Adetunji Adebusuyi, Lithium Balance, 2010. Reprinted with permission.)

sensor (Hall effect or shunt), and isolation fault detection. Balancing is done with a 0.5A current, which can be increased to 4A with external resistors.

REAP

As a researcher at the University of Southampton, Dennis Doerffel became an expert in Li-Ion cells and used that knowledge to develop a BMS technology that was both cost effective and very functional. He and Stephanie Pielot created REAP (http://reapsystems.co.uk) to bring that technology to the market in 2003, which was greeted with good response. They produce a very nice, modular, balancer BMS (Figure 4.11). Each module can handle 14 cells in series, and up to 12 modules can be used to handle up to 168 cells in series (one of the modules is configured as a master). The BMS is full featured, with extensive communications (CAN bus, RS232, RS485/RS422), current sensor input, and fuel gauge function. It is housed in an attractive molded plastic case, and has an optional display.

4.1.3 Cell Manufacturers' BMSs

Certain cell manufacturers sell a BMS together with their cells. As these BMSs are not available by themselves, we do not consider them commercially available, in the way the BMSs described previously are.

4.1.3.1 Elite Power

Elite Power (http://elitepowersolutions.com) sells blocks of 4 prismatic cells (from Thundersky/Zhejiang GBS Energy) already bound in metal plates, and with BMS cell boards already riveted to the cells. The BMS is a digital balancer that can handle at least 24 cells in series. Communication is thorough a two-wire daisy chain loop.

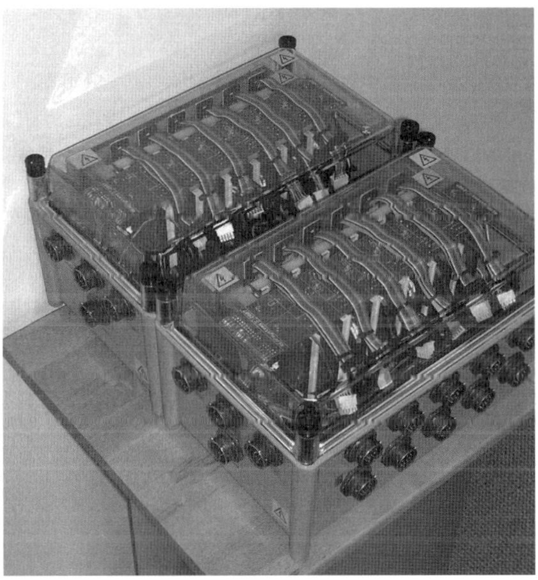

Figure 4.11 REAP's BMS for 168 cells. (*Source*: Stephanie Pielot, REAP Systems, 2010. Reprinted with permission.)

Unusually, the BMS controller is incorporated into the body of the current sensor. An LCD display as well as an LED SOC display are included. The BMS interfaces seamlessly with the company's chargers.

4.1.3.2 Kokam/N Tech

Kokam (http://kokam.com) is the long-time provider of pouch LiPo cells for small applications (such as model airplanes). Recently, its focus has grown to include large traction packs, and, with that, the need for a BMS has appeared. In 2010, Kokam started offering a BMS from N-Tech. This modular, digital balancer includes a CAN bus and computes SOC, measuring current with either a Hall effect current sensor or a current shunt. Up to eight modules can be used, and each module can handle up to 14 cells in series, for a maximum total of 112 cells in series. A balance current of 65 mA may be a bit on the low side.

4.1.3.3 Thundersky/Ningbo Yangming

Ningbo Yangming's BMS40, a master/slave, digital monitor (no balancing), has been offered with the Thundersky cell (http://thunder-sky.com). It can handle up to 10 slaves, and each slave handles up to 10 cells, for a total of 100 cells in series. It includes a display and current sensor and performs SOC calculations. This BMS is more suited for a laboratory environment than for commercial use.

4.1.3.4 Valence

Cell manufacturer Valence (http://valence.com) fully understands the need for a good BMS to ensure the reliability of Li-Ion cells. For that reason, since its inception it has offered its cells only when bought together with its U-BMS, a full-featured, modular, digital balancer. Each BMS module handles a U-Charge battery module, which contains four or six cells in series, and as many as 128 modules can be used in a system, for a maximum total of 768 cells (in a system with multiple strings in parallel), though the maximum number of cells for any series string is 220. The modules communicate through a CAN bus, which can be the same CAN bus as the rest of the system (its speed can be set to be the same as the system's CAN bus speed).

4.1.4 Comparison

Table 4.2 compares the BMS described in this chapter.

Table 4.2 Off-the-Shelf BMS Comparison

	Technology	*Topology*	*Function*	*Cells*	*Features*
Black Sheep, V1	Digital	Modular	Balancer	32	SOC, RS232, contactor
Black Sheep, V2	Digital	Centralized	Balancer	48	SOC, RS232, contactor
Black Sheep, V3	Digital	Master/slave	Balancer	Any	SOC, RS232, contactor
Black Sheep, V4	Digital	Master/slave	Balancer	Any	SOC, RS232, contactor
Clayton Power	Digital	Centralized	Protector	4, ... 23	SOC
Clayton Power	Digital	Master/slave	Balancer	496	SOC
Clean Power Auto	Analog	Distributed	Balancer	Any	
EVPST	Digital	Master/slave	Balancer	?	SOC, CAN, RS485, optional display
Electric Blue Motors	Digital	Distributed	Monitor	255	SOC
Elite	Digital	Distributed	Balancer	24	SOC
Elithion Lithiumate	Digital	Distributed	Balancer	256	SOC, SOC, CAN, RS232, contact
Elithion Lithiumind	Analog	Centralized	Balancer	24	Current sense, SOC, SOH, optional display
EV Power	Analog	Distributed	Balancer	Any	
Genasun	Digital	Distributed	Balancer	Any	RS232
Guantuo	Digital	Master/slave	Monitor	100	SOC, CAN, USB, display
JK Hall	Digital	Master/slave	Meter	30	SOC, display
Lithium Balance	Digital	Centralized	Protector	4, ... 23	SOC
N-Tech	Digital	Distributed	Balancer	112	SOC, CAN, display
Ningbo Yangming	Digital	Master/slave	Monitor	100	SOC
REAP	Digital	Modular	Balancer	168	SOC, CAN, RS232, optional display
Valence	Digital	Modular	Balancer	220	SOC, CAN

CHAPTER 5

Custom BMS Design

If you chose to use an off-the-shelf BMS, you may safely skip this chapter and all its technical details, and proceed to Chapter 6.

So you decided to build your own BMS. Let's try to expedite that process, by looking at some sample circuits, common pitfalls and their solutions. This chapter assumes that you have an electrical engineering background. It revisits the same topics as Chapter 3, but while Chapter 3 discusses each topic in general terms from the point of view of a BMS user, this chapter discusses them in technical detail, from the point of view of a BMS designer.

Please start from Section 5.1, with advice on choosing a BMS application-specific integrated circuit (ASIC). Then if you plan to design an analog BMS, read Section 5.2. If you plan to develop a ready-made design for a digital BMS, read Section 5.3. If you plan to design a digital BMS from scratch, read Section 5.4. Then, regardless of your choice, go on to Section 5.5, which talks about mechanical installation techniques, which are applicable to all designs. Section 5.6 is not strictly about BMS design, but about distributed charger design, which some may wish to implement instead of a balancing BMS.

5.1 Using BMS ASICs

This chapter explores BMS ASICs available in 2010, or about to be announced. As this is a rapidly changing field, by the time you read this, some of these ASICs may no longer be available, and others will certainly be added (I will keep updating an online list of BMS ASIC at http://book.LiIonBMS.com). Therefore, in addition to mentioning these specific BMS ASICs, I would like to offer some general tips on selecting a BMS ASIC for a large battery pack BMS.

5.1.1 BMS ASIC Selection

ASICs that implement a complete BMS for small Li-Ion batteries for consumer products (such as laptop computers) are readily available.

Some common highlights include:

- Ideally suited for mounting directly on the cells, for cleanest design and minimal risk of short circuits;
- Low cost (less than one dollar per cell) and readily available.

Unfortunately, there are two limitations to these chips that limit their usefulness in large Li-Ion packs. The first one is that they are designed for low battery currents sensed by a PCB mounted resistor, and not for the high currents typically present in large packs and sensed with a Hall effect sensor or current shunt (see Section 3.1.3). Without knowledge of the battery current, any powerful intelligence in the chips goes to waste. The second limitation is that they handle at most a few cells in series.

To create a BMS for larger packs using such ASICs presents multiple challenges:

- At first glance, you may think that multiple BMS ASICs could be placed on the same bus (such as I2C or ISP). However, because their ID is fixed, there is no way to address a particular IC directly. Therefore, a multiplexer must be used to address individual ICs, or a small processor must be added to each BMS ASIC between it and the bus to respond to a particular ID.
- As each chip has a mind of its own, it is hard to coordinate them, and to make use of the parameters that each ASIC calculates.
- In large packs, the master controller reads the pack current from its own current sensor. BMS ASICs do not expect a master, and therefore do not have provisions to read the battery current from the master. Without knowledge of the current, the BMS ASICs are not able to perform their more advanced functions.
- There may be no way for the master controller to control the loads on each chip. Therefore the ASIC's balancing function may not be used, and instead extra circuitry must be a added to perform cell balancing.

Despite this, some may try to shoehorn small battery BMS ASICs into a large battery pack design, with disappointing results (see Section 5.3.5). I urge you to avoid this route, and instead start from ASICs that are intended for large packs (or from generic ICs). When selecting an ASIC for a BMS for a large Li-Ion pack, look for the following characteristics, and avoid the following pitfalls. Look for a BMS ASIC that:

- Has an accuracy of better than 25 mV, for effective balancing and protection;
- Has at least one temperature sensor for every six cells (though you can always add sensors, separately);
- Has the ability to handle the number of cells in your pack;
- Has separate outputs for charging and discharging (if it has limit outputs);
- Has current source outputs, and carries a continuous data stream or at least a clock (if it includes ports for cascading ICS);
- Enables you to program the ID of each IC (if it is to be placed on a common bus);
- Has an individual data acquisition circuit per cell, for excellent common noise rejection;

- Is from a manufacturer who has years of BMS experience and is likely to continue offering that IC for many years to come;
- Implements a complete BMS, if you do not have the know how and resources to design a BMS past the front end;
- Implements balancing and uses dedicated lines for balancing, separate from the lines used for measuring;
- Is configurable for your cell's voltage.

Avoid protector ICs for small batteries because:

- They include terminals for an onboard current sensing resistor.
- They include terminals to drive MOSFETs.
- Their ID (for a communication port) is not programmable.
- While they may include sophisticated BMS functions, they do not have provisions for receiving a measurement of the pack current from a master, and therefore cannot use those features.
- They are not able to be controlled by a master, such as to turn on balancing loads.
- They may not have a digital communication port.

Avoid ICs that can be cascaded but use inappropriate methods of doing so because:

- The cascading port uses voltage level shifters (poor noise immunity).
- The cascading port requires opto-isolators (extra cost and high current draw).
- The cascading port uses more than three or four wires (a problem if using physically separate modules).
- The cascading port uses steady logic levels (which cannot be distinguished from a shorted or open wire).

Avoid ICs that do not implement balancing correctly because they:

- Share the same pin for voltage measurement and for balancing (cannot balance two adjacent cells at the same time, cannot measure while balancing);
- Cannot support balancing with external components (balancing current done internally to the IC is too low for a large pack);
- Only balance among cells handled by a single IC if they do active balancing (cannot balance across adjacent blocks of cells);
- Implement all-the-time balancing or bottom balancing (only top balancing and SOC history balancing are effective algorithms);
- Do not have any provision for balancing (you will have to add balancing separately).

Avoid ICs that cannot easily be configured to work with your cells because they:

- Work with LiPo voltages (3.7V) when you use $LiFePO_4$ cells (3.3V), or vice versa;
- Must be configured for a particular voltage by the factory;
- Require a digital programmer to be configured, yet you do not plan to include a processor in your BMS able to do that programming.

Avoid ICs with poor specs because:

- The current drain from the cells is above 30 μA in stand by and above 300 μA during operation.
- They have poor cell voltage accuracy: more than 25 mV.
- They use a single out-of-bounds limit output for both charging and discharging.

Be aware of ICs that require complex circuits with many external components, whose cost eclipses the cost of the IC itself. While the specs may show direct connection to the cells, in reality most BMS ASICs require five or more extra components per cell for filtering and protection.

Avoid vapor ware, ephemeral ICs, and paranoid manufacturers because:

- Johnny-come-lately manufacturers who may be half-serious about BMS and may stop offering their ICs after you designed them into your BMS;
- ICs that are manufactured to order: there is no stock anywhere of those ICs, and even their distributors do not list them;
- Some manufacturers are obsessed with secrecy, do not publish their IC's spec sheet, and may even have you sign an NDA before they show you the spec sheet.

5.1.2 BMS ASIC Comparison

The applicability of ASICs available in 2010 (or just announced) may be analyzed by viewing them within a graph of number of cells versus BMS sophistication (Figure 5.1). BMS ASICs can be divided into these groups:

- Front ends for large packs (left edge of Figure 5.1): They are intended for large pack BMS, though you need to design the remainder of the BMS yourself (for which you can use this chapter as guidance).
- Complete BMS for small batteries (bottom edge of Figure 5.1): They cannot easily be integrated in a large pack BMS, so, in general they should be avoided.
- Front ends for small batteries (bottom right corner of Figure 5.1): They are both inconvenient and incomplete for a large pack, so avoid them at all costs.
- Complete BMS for large packs: They offer good performance and ease of development, so they are ideal, though today the choice is still very limited.

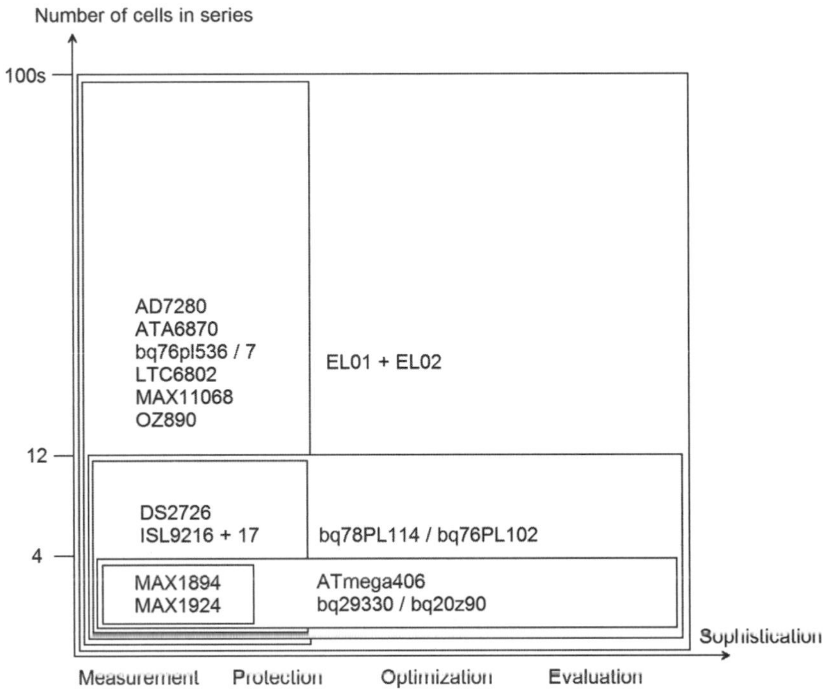

Figure 5.1 Number of cells versus sophistication for various BMS ASICs.

5.2 Analog BMS Design

Most Li-Ion analog BMSs are based on some common designs. Let us take a look at them.

5.2.1 Analog Regulator

A regulator is placed across a Li-Ion cell [Figure 5.2(a)] and bypasses part or all of the charging current when the cell is full. The regulator is a voltage clamp, and the breakdown voltage at which it starts conducting is also called the knee voltage (because, in its current versus voltage characteristic curve, that section of the curve looks like a sharp knee). Below its breakdown voltage (selected to be somewhat less than the maximum cell voltage), a regulator should draw negligible current. Above the breakdown voltage, the regulator starts drawing current. Some regulators are very stiff, meaning that past the breakdown voltage they conduct fully [Figure 5.2(b)], while others are softer, meaning that the current starts at 0 at the breakdown voltage, and increases linearly as the cell voltage increases [Figure 5.2(c)].

5.2.1.1 Zener Diode

At first glance, one may think that a Zener diode could work well as a regulator. In reality, at the low voltages of Li-Ion cells, Zener diodes have a very soft knee. They conduct so much current that they would soon discharge the cells. For example, a

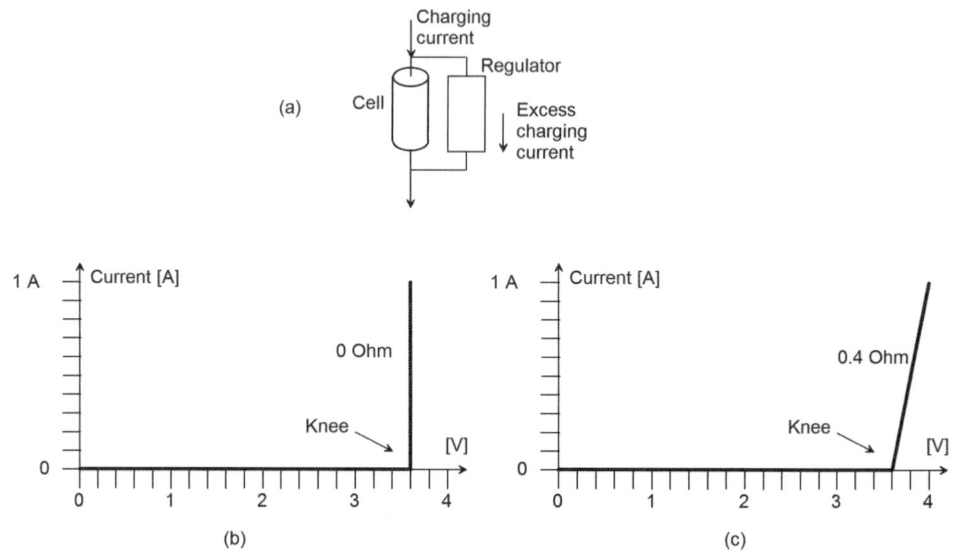

Figure 5.2 (a) Regulator across a cell. Characteristic curve for an ideal regulator: (b) stiff and (c) soft.

3.6-V Zener diode (appropriate for use across a $LiFePO_4$ cell), will leak about 0.8 mA at the nominal cell voltage of 3.2V [Figure 5.3(a)], which is unacceptable. The characteristics of Zener diodes become closer to ideal at 6.2V and above. For example, a 7.5-V Zener will conduct negligible current below its breakdown voltage [Figure 5.3(b)].

For a very inexpensive regulator, one might use a Zener diode every two cells, because, at the voltage of two cells in series, Zener diodes work well. A resistor should be added in series to soften the knee and protect the Zener diode from excessive currents (Figure 5.4). The two cells could still be unbalanced with respect to each other, but not nearly as much as cells in a long string.

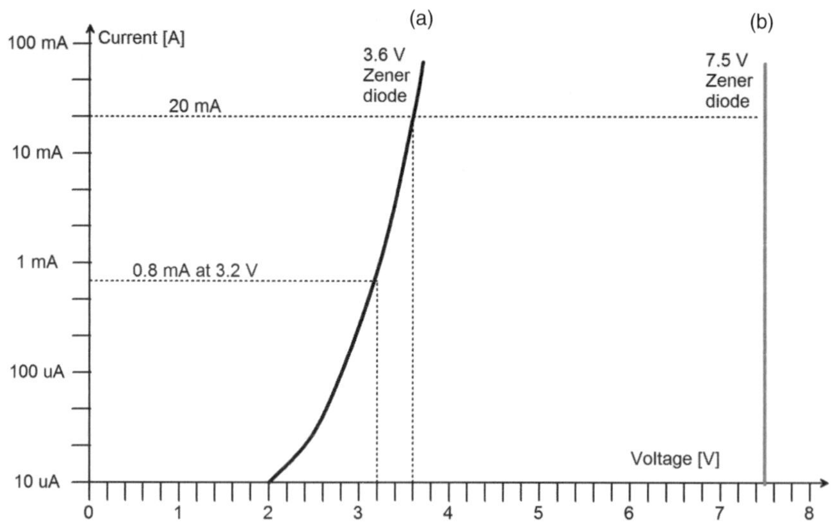

Figure 5.3 Characteristic curve for two Zener diodes: (a) 3.6V and (b) 7.5V.

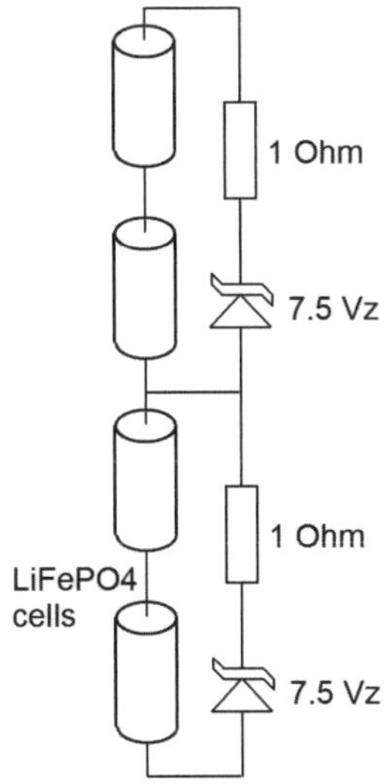

Figure 5.4 Using a Zener diode as a regulator for two cells.

5.2.1.2 Integrated Circuit

In practice, instead of a Zener diode, regulators use a generic analog IC and some form of a balancing shunt (Figure 5.5). The IC might be a voltage regulator, microprocessor supervisor, power supply supervisor, or voltage detector. Regardless, the IC includes two elements: a voltage reference and a comparator (or differential amplifier). The comparator compares the cell voltage (divided down by a resistive voltage divider) and the reference voltage so that its output flips state as the cell voltage crosses the threshold, turning the balancing shunt on or off. Figure 5.6 shows a few sample circuits, using various ICs.

5.2.2 Analog Monitor

The main function of an analog monitor is to send a request to the external system to stop the battery current whenever any cell voltage is too low or too high. It does so through limit outputs, which may be called various names, such as low voltage

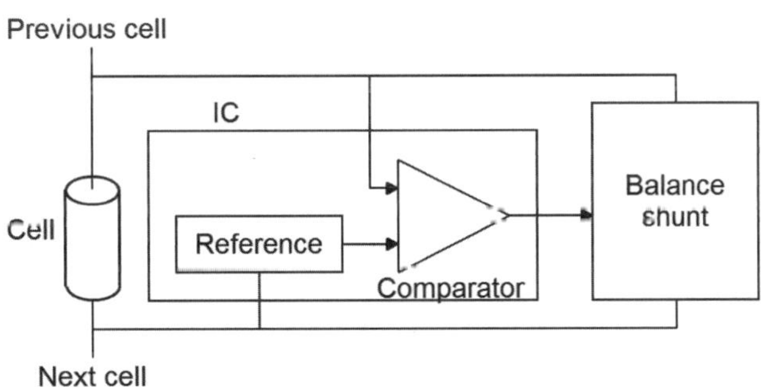

Figure 5.5 Basic block diagram of a regulator.

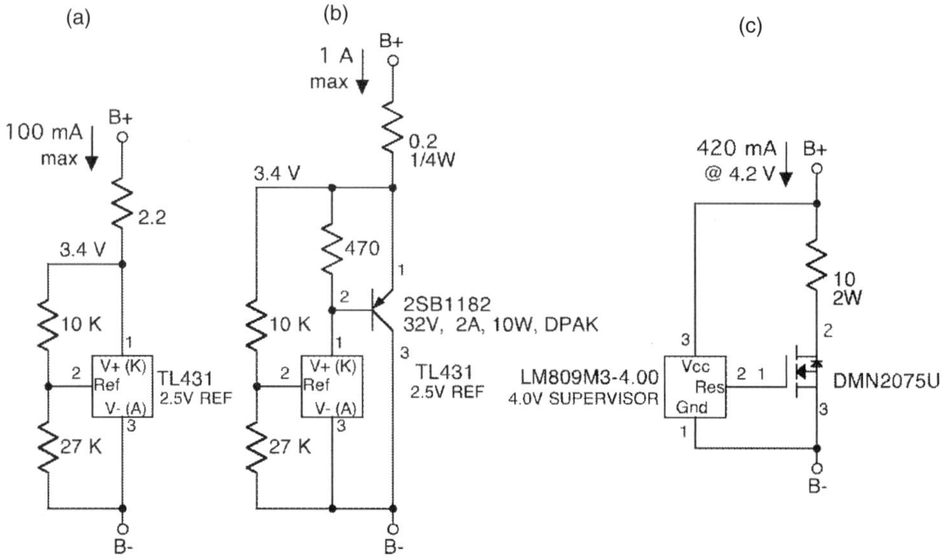

Figure 5.6 Practical regulator circuits: (a) linear, low current; (b) linear, high current; and (c) digital, medium current.

limit (LVL) and high voltage limit (HVL). A monitor cannot tell the system which cell is low or high, nor what its actual voltage is (though it may tell a user, through LEDs).

An analog monitor can be distributed or localized. We will look at the distributed topology first.

5.2.2.1 Distributed BMS

A distributed BMS consists of a cell circuit across each cell, plus a master controller. The cell circuit's outputs are interconnected to report to the master controller if any cell is out of bounds. As each cell circuit is referenced to its cell, their outputs are all at different potentials, so they cannot be connected directly together. Instead, either each output must be isolated, or there must be a voltage stepping circuit between adjacent cells, as described in the next subsections.

Whether the outputs should be normally open (NO) and then closed if there is a problem, or normally closed (NC) and then open if there is a problem, has significant consequences on reliability. Therefore, let us make this design choice first.

NO—Parallel Bus

When using a cell circuit with NO outputs, all the outputs must be connected in parallel. If any cell voltage is out of bounds, the output of its cell circuit closes, shorting the common bus. The master controller will detect this, and interrupt the battery current (Figure 5.7).

There are several problems with a parallel bus:

1. If the bus is accidentally disconnected, the master controller will have no way of knowing if a cell is out of bounds.

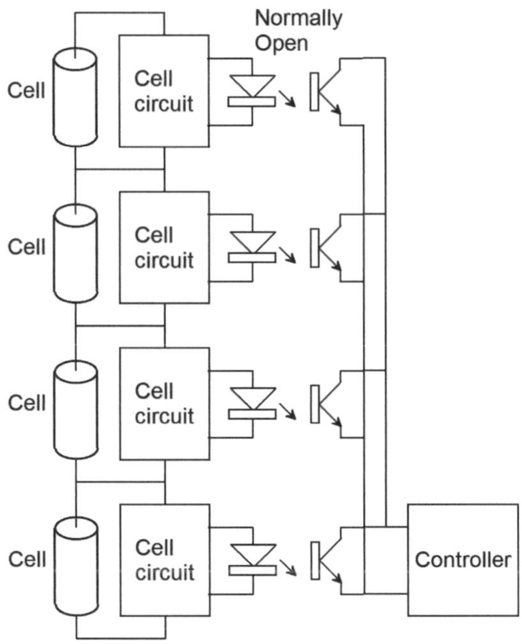

Figure 5.7 Parallel bus with NO opto-isolators.

2. A cell circuit on a discharged cell finds itself driving the LED in its LVL opto-isolator, discharging the cell even more rapidly.
3. When a cell voltage gets so low that it can no longer power its cell circuit, that LVL opto-isolator is no longer powered, making the master controller think that that cell is OK, allowing the system to attempt to draw even more current from that empty cell.

NC—Series Bus

When using a cell circuit with NC outputs, all outputs must be connected in series, forming a daisy chain. If any cell voltage is out of bounds, the output of its cell circuit opens, opening the daisy chain. A controller will detect this, and interrupt the battery current (Figure 5.8).

Two advantages of the NC bus are immediately apparent.

1. The series bus uses two wires between cell boards (LVL and HVL), making it more attractive than the parallel bus, which uses a common bus of three of four wires (LVL, HVL, and one or two commons).
2. When the cell voltage is too low to power its cell circuit, both limit outputs open, giving a clear indication of the cell's state.

A truth table of the four possible states of the two limit outputs (Table 5.1) reveals that there are no ambiguous conditions that would result in a low cell being discharged further, or in a high cell being charger further. For proper operation, the master controller simply needs to interpret the state of the two limit lines as follows:

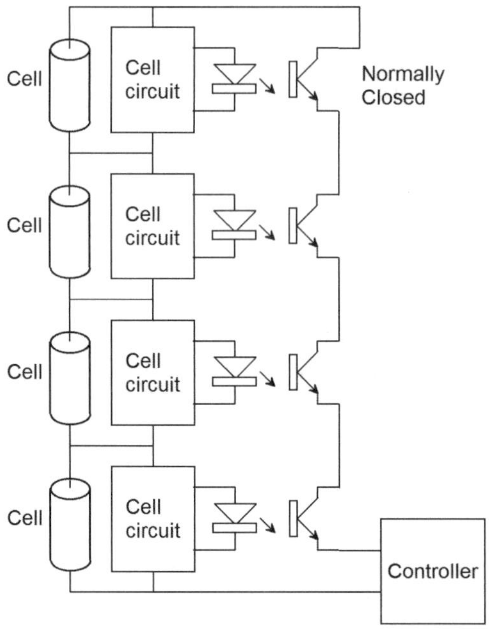

Figure 5.8 Series bus with NC opto-isolators.

Table 5.1 Truth Table for the LVL and HVL Limits in an NC Series Bus

Cell Voltage	*$LiFePO_4$ Example*	*LVL*	*HVL*	*Enable*
Very low	< 2.0V	Open	Open	Charging
Too low	2.0 ... 2.5V	Open	Closed	Charging
OK	2.5 ... 3.6V	Closed	Closed	Both
Too high	> 3.6V	Closed	Open	Discharging

- Discharging is allowed if the LVL line is closed.
- Charging is allowed if the HVL is closed or if both lines are open.

However, there are several problems with a series bus:

1. An opto-isolator with a BJT output, when on, has a voltage drop of a few tenths of a volt. In a long string of opto-isolators in series, these voltages quickly add up to a few volts and even tens of volts, and can easily exceed the master controller's supply voltage (e.g., 12V).
2. Both opto-isolator LEDs are powered whenever the cell is within bounds, wasting charge all the time even if the pack is not in use.
3. An accidental short across any cell board output, across the entire bus, or from the bus output to a supply rail will prevent the master controller from seeing that a cell is out of bounds.

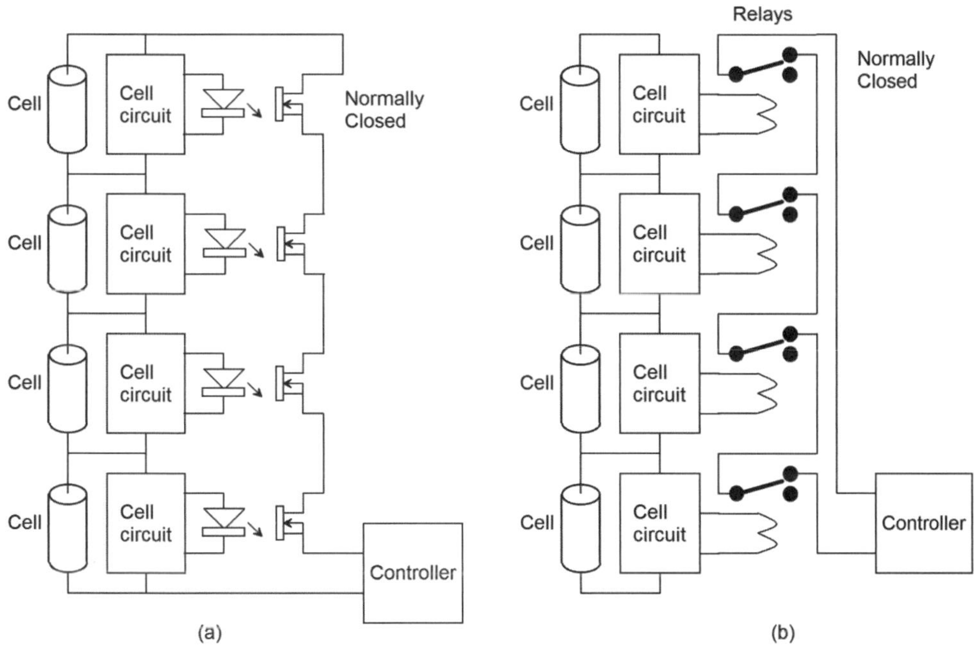

Figure 5.9 Series bus with low voltage drop: (a) MOSFET opto-isolators and (b) relays.

We saw that both the NO, parallel bus and the NC, series bus have limitations. Yet, I would argue that the third item in the list of problems with the NO, parallel bus is so significant to make that solution unacceptable. Therefore, let us choose to implement the NC, series bus, and explore how we can reduce its limitations.

The first problem in the NC, series bus is addressed by using a sufficiently high supply voltage to feed the positive end of the string, higher than the total voltage drop across all the opto-isolator outputs when they are all turned on. One readily available source of high voltage is the battery voltage itself, though that defeats the isolation that the opto-isolators provide. Another solution is to use opto-isolators with MOSFET outputs, and use a very low bus current, so that the voltage drop per isolator is negligible [Figure 5.9(a)]. Yet another solution is to use relays (whose contacts have effectively 0 voltage drop), instead of opto-isolators [Figure 5.9(b)]. Still, relays are expensive, and their coils take even more power than opto-isolators, worsening the second problem in the list above.

A very different approach is to use current sources to step the voltage between cell circuits (Figure 5.10). The current sources are normally on, and are turned off (opened) if the voltage of their cell is out of bounds. The current sources can be designed to carry a relatively low current, lower than the current required to drive opto-isolator LEDs, reducing the charge taken from the cells. Even better, the current sources can be turned on only when the BMS is in use (or even only for a few ms every second), reducing the drain on the cell significantly when the battery is not in use. The easiest way to do this is with an opto-isolator. By using an opto-isolator at the bottom of the bus as well, isolation is achieved between the battery and the master controller.

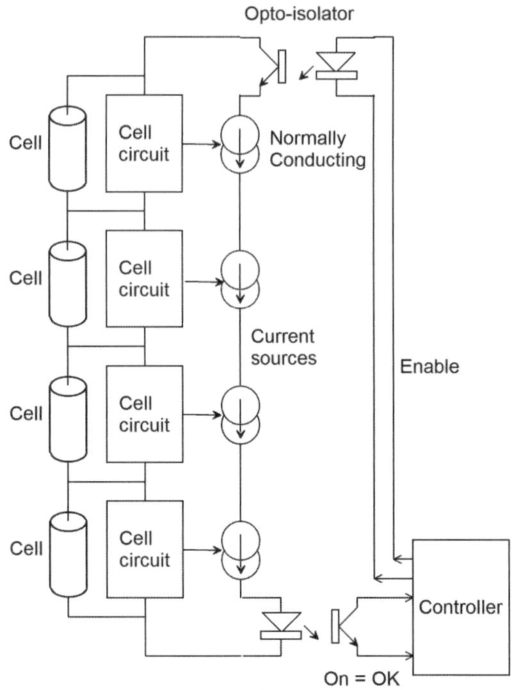

Figure 5.10 Series bus with current sources and isolation, and shut down when not in use.

In practice, the third problem (a short on a series bus) is not nearly as likely as experiencing an open in a parallel bus, making the series bus more reliable. To reduce that risk even further, modulating the input of the series bus allows the master controller to detect if the bus has been accidentally shorted to a supply rail (Figure 5.11). The controller drives the bus with pulses, and looks for the same pulses coming back. If it does not see them, either a cell circuit detected that its cell is out of bounds, the daisy chain is open, or the bus is shorted to a rail. In any of these cases, the BMS shuts down the battery current.

Single Bus

Ideally, the two limit outputs (HVL and LVL) are routed separately to turn off the charger and the load, respectively [Figure 5.12(a)]. In practice, it is possible to simplify the circuit by combining the HVL and LVL outputs into a single out of bounds (OOB) output, which is used to turn off both the charger and the load [Figure 5.12(b)]. Doing so is advantageous because only one bus is required instead of two, saving components and reducing the number of wires in the bus.

However an impasse may result if a cell voltage is low and therefore the OOB line is asserted, which disables the charger; the battery can never be recharged. When using a high current load, this is not a problem, as the cell voltage will drop under load and shut down the load, and then recover after a while, re-enabling the charger. For the same reason, this is not a problem after charging, as a cell whose voltage first becomes high enough to stop the charger will then settle down, enabling the load. The impasse occurs when a battery sits for a long time, and a cell voltage becomes low enough to assert the OOB line. Then, recharging is no longer possible,

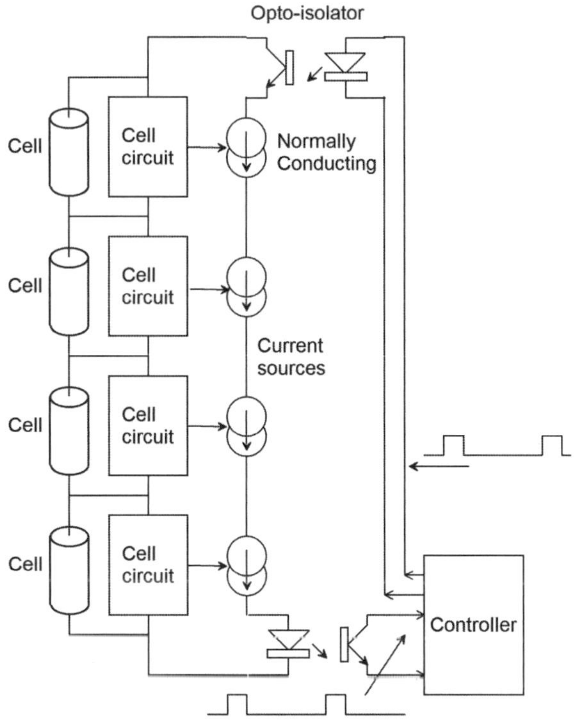

Figure 5.11 Modulating a series bus to detect problems on the bus.

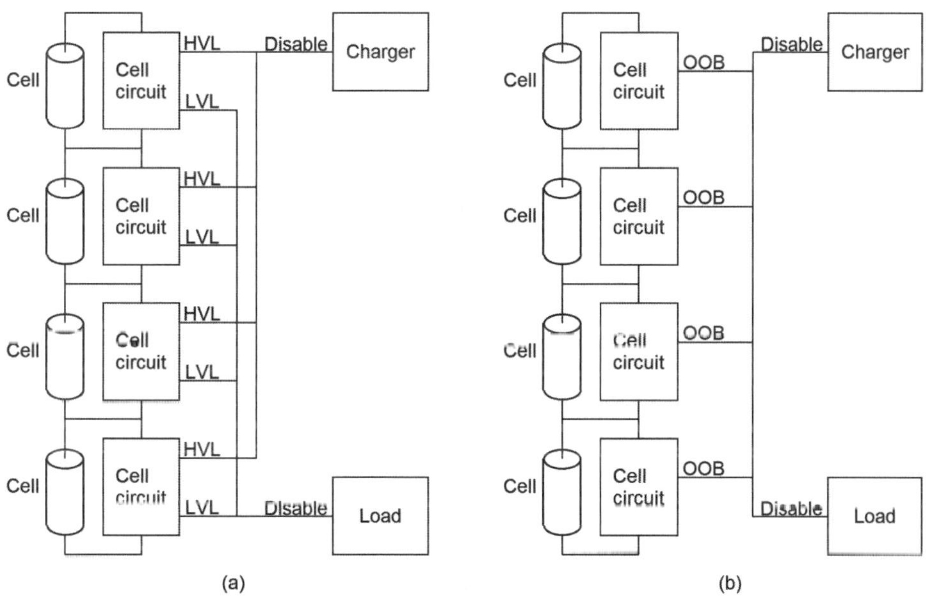

Figure 5.12 (a) Two separate buses: HVL and LVL. (b) A single OOB bus.

and user intervention is required to override the OOB line and enable the charger (Figure 5.13).

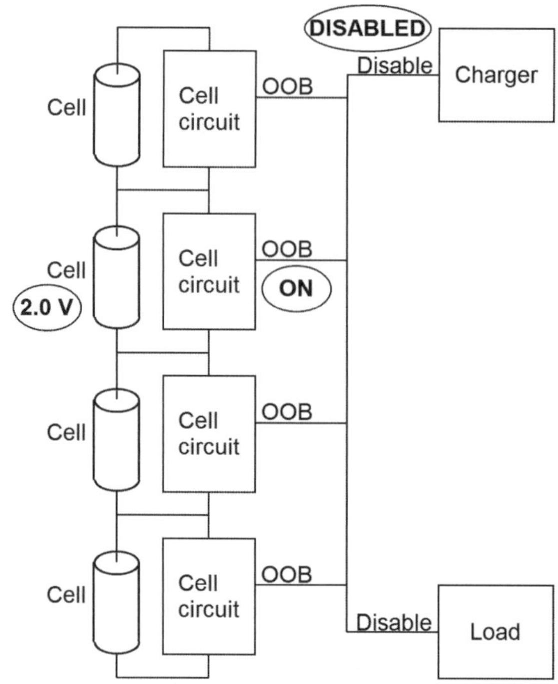

Figure 5.13 Impasse when using a single bus. A low cell voltage prevents charging.

A solution may involve a timer to ignore the OOB line for a while when the charger is first powered up. This is risky, as it would be possible to overcharge the battery by applying and removing power to the charger repeatedly, which would restart the timer each time. Another solution may be to look for transitions in the OOB line in order to turn off the charger when the OOB line becomes asserted. This is also risky, as a noise transition would interrupt charging. Also, if for some reason the transition is missed, charging would continue, damaging the battery. Therefore, I strongly recommend that two separate limits be used.

Cell Circuit

The typical analog cell circuit includes a pair of detectors: one to detect the low voltage and output the LVL signal and one to detect the high voltage and output the HVL signal. There are many ways of implementing such a cell circuit. The following example addresses the issues raised in the previous section (Figure 5.14).

This cell circuit includes the following: a resistor divider to sample the cell voltage using high resistance, precision resistors; a series voltage reference that works down to 1.8V and produces a precise 1.25V reference; a dual comparator IC that works down to 2.0V and draws little supply current; and a dual opto-isolator with a Darligton output able to operate at particularly low LED current, to couple the LVL and HVL signals to the rest of the system.

For compatibility with a series bus, the opto-isolators are NC; they are both on when the cell voltage is within bounds, and one of them is off when the cell is out of bounds.

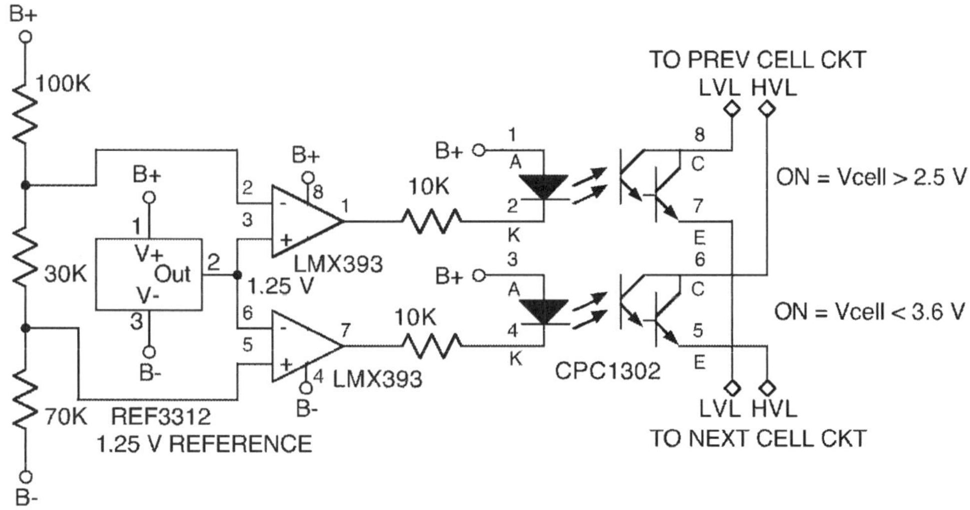

Figure 5.14 Analog monitor cell circuit.

When the cell is within bounds, the circuit draws 300 μA nominally from the cell (100 μA for each opto-isolator that is on, plus 100 μA for the remainder of the circuit), which would drain a 10-Ah cell in about 3 years. It draws less current (about 200 μA) if the cell voltage is out of bounds. That current will drain a discharged 10-Ah cell in a few weeks, down to 1.8V, at which point the ICs will shut down and save the cell from discharging completely. At that point, as we saw (Table 5.1), the master controller will notice that both the LVL and HVL buses are open, meaning that a cell is too low to power its circuit, and it will enable charging.

5.2.2.2 Localized BMS

In the localized topology, a single circuit handles all the cells in a small battery by selecting a cell voltage at a time, and then using just one pair of comparators to analyze that voltage. This reduces the cost and parts count. This solution is limited to a small number of cells in series (typically of 4 to 12 cells) to keep the voltage from becoming too high for the electronics.

Multiplexed Circuit with Generic ICs

To select a particular cell voltage one could use a pair of 4:1 analog multiplexers followed by a differential amplifier to take the difference in the sampled voltages, whose output voltage will be proportional to that cell's voltage (Figure 5.15). The differential amplifier must be able to handle the full battery voltage (which is done with resistive voltage dividers, whose tolerance unfortunately attenuates the sampled voltage and adds errors to it). A clock and a counter select each cell in turn.

A similar circuit that doesn't suffer from the same limitations uses a "flying capacitor" to sample a particular cell's voltage (Figure 5.16). The flying capacitor is first connected across a cell to sample the voltage by the switches on the left. Then it is placed across the sampling capacitor to transfer the cell voltage to it, by the switches on the right. This is repeated a few times for each cell, to make sure that the

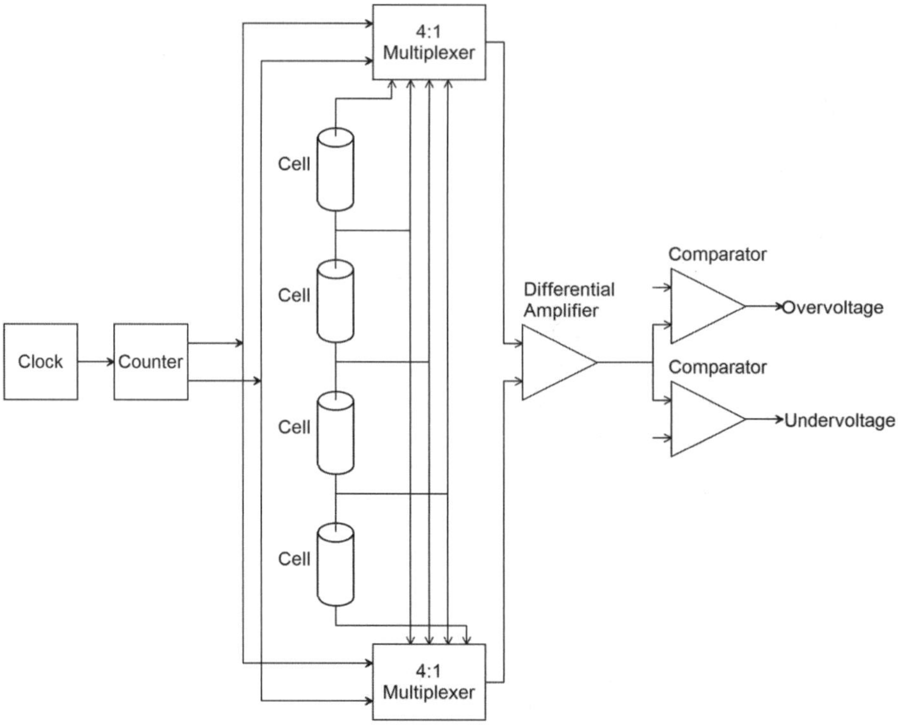

Figure 5.15 Analog monitor using a multiplexer and a differential amplifier.

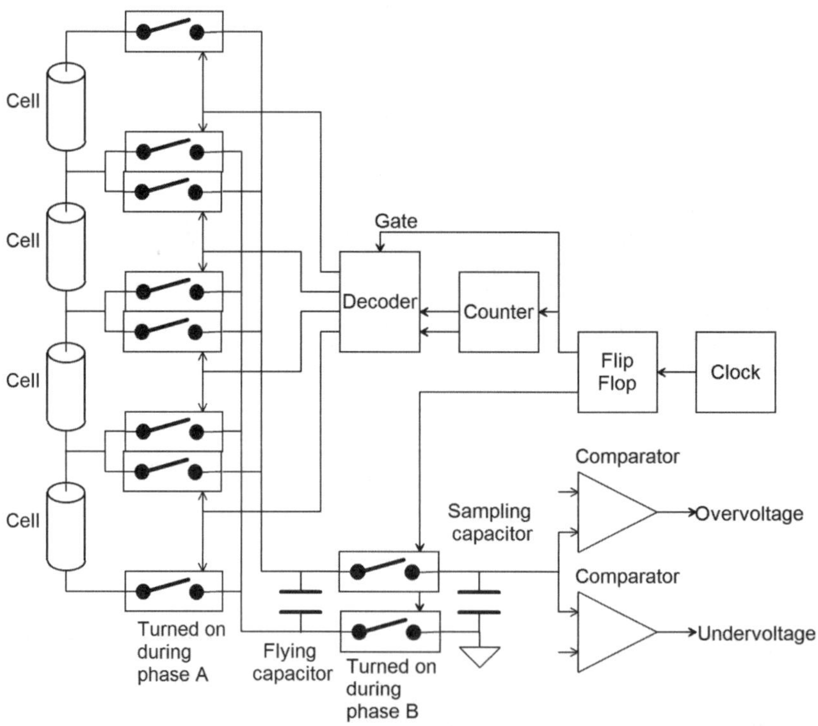

Figure 5.16 Analog monitor using a multiplexer to select each cell in turn and a flying capacitor.

sampling capacitor voltage reaches the same voltage as the cell. Then the comparison is made, and the next cell is sampled, and so forth. The advantage of this circuit is that there are no resistors introducing errors. The disadvantage is that it may take a bit longer to sample each cell's voltage.

LTC6801 Fault Detector

Linear Technology's LTC6801 fault detector is designed to add an independent, redundant test to a digital BMS, to ensure the cells are within their SOA. However, this IC could be used by itself, as the core of an analog monitor. Even though, internally, this IC is a digital BMS, I have included it here with the analog BMSs as it can be viewed as a black box that does not report the cell voltages, and so in that sense it is no more functional than any analog BMS.

The circuit (Figure 5.17) consists of a string of LTC6801 ICs, one for every 4 to 12 cells in series. The master controller feeds a clock to the string, which is returned

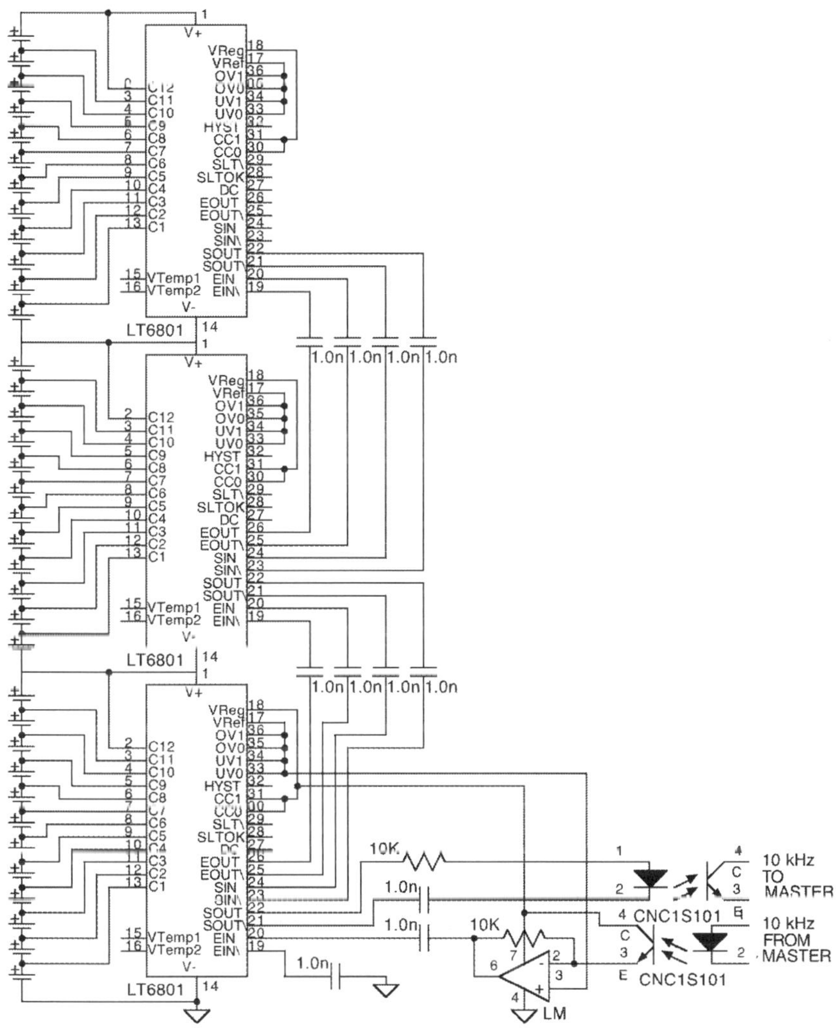

Figure 5.17 LTC6801 as an analog monitor for 36 cells.

back to the master. Should any IC see that one of its cells is outside the SOA, it gates off that clock. Seeing that the clock is gone tells the master controller that one of the cells is not OK (though it does not know which one or if it is too high, too low, or too hot). The thresholds and delays are programmed by strapping certain programming input pins. There is no need for a programming tool, because the PCB can be used for that purpose.

MAX1894/MAX1924 Protector

Maxim's MAX1894/MAX1924 series of ICs implement a three- or four-cell protector for standard Li-Ion cells or LiPo cells (with a top cutoff voltage around 4.2V). The cutoff voltages are factory programmable, which may be fine for high-volume production, but otherwise quite inconvenient. The difference between the various parts in the series has to do with how many cells they handle (three or four) and whether there is hysteresis in the cutoff voltages. A typical protector circuit (Figure 5.18) uses a MAX1894 and a few additional parts. Although intended for up to four cells, this IC can be shoehorned into a large battery pack by having independent circuits for four cells, each using one IC, and combining the charge and discharge outputs of all the ICs into a single set of limit lines (Figure 5.19).

MAX11080 Fault Detector

Just like the LTC6801 fault monitor can be used as the core of a simple monitor, so can Maxim's MAX11080 fault monitor (Figure 5.20). The Maxim part uses fewer lines between adjacent sections.

5.2.3 Analog Balancer

An analog balancer performs all the essential functions of a BMS. It is similar to an analog monitor (which protects against low and high cell voltage), except that it also handles balancing (see Section 5.4.5.1). Compared to a digital balancer, an analog one cannot report which cell is low or high, or what its voltage is. You may design an analog balancer using generic ICs or BMS ASICs.

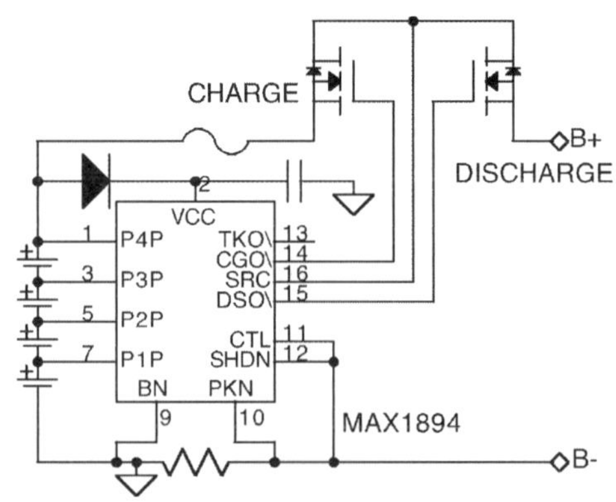

Figure 5.18 MAX1894 ICs in a protector for four cells.

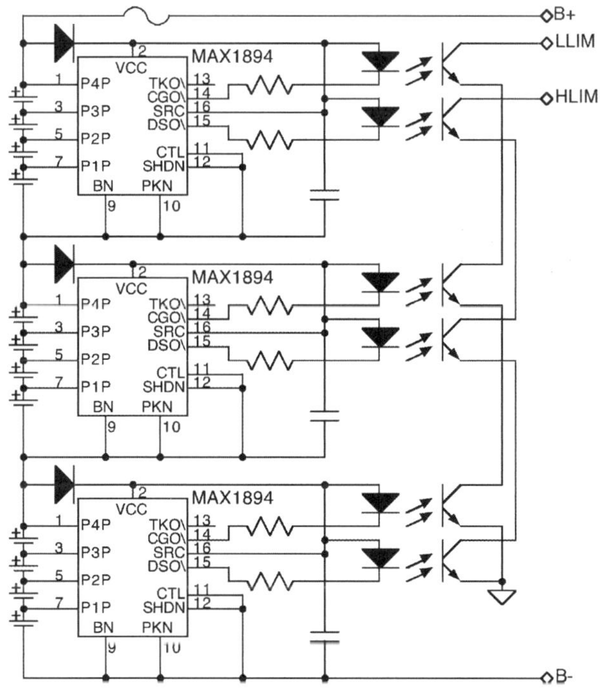

Figure 5.19 MAX1894 ICs shoehorned into a monitor for 12 cells.

5.2.3.1 Generic ICs

The circuitin Figure 5.21 is an example of a distributed analog balancer using generic ICs. This circuit is identical to the analog monitor circuit (Figure 5.14), with the addition of a third channel to detect when the cell voltage exceeds the balance voltage, and to turn on a load in that case.

The balance bus is only required if you intend to use a charger that can reduce its current down to the balancing current, so that the cell circuit bypasses the charging current around the cell. Without the balance bus, the charger current is not reduced; instead, the BMS will naturally fall into a pattern of turning the charger off and on every few minutes, with a duty cycle such that the average charger current will equal the balance current (the time constant and the duty cycle are set by the cell's relaxation effect).

5.2.3.2 DS Series, DS2726

Maxim became a major player in small Li-Ion batteries when it acquired the DS series of ICs from Dallas Semiconductors. It carries protectors for small batteries with anywhere between 5 and 10 cells in series. The part in the DS series that handles the highest number of cells is the DS2726. It is a standalone, analog balancer, with a respectable level of balance current (Figure 5.22).

This part is intended only for standard Li-Ion (Cobalt) cells or LiPo cells, as the top cutoff voltage is on the order of 4.2V. The part threshold voltages and delay times are configured by connecting select pins through PCB traces, so a programming device is not needed. Top balancing uses external resistors, with a maximum

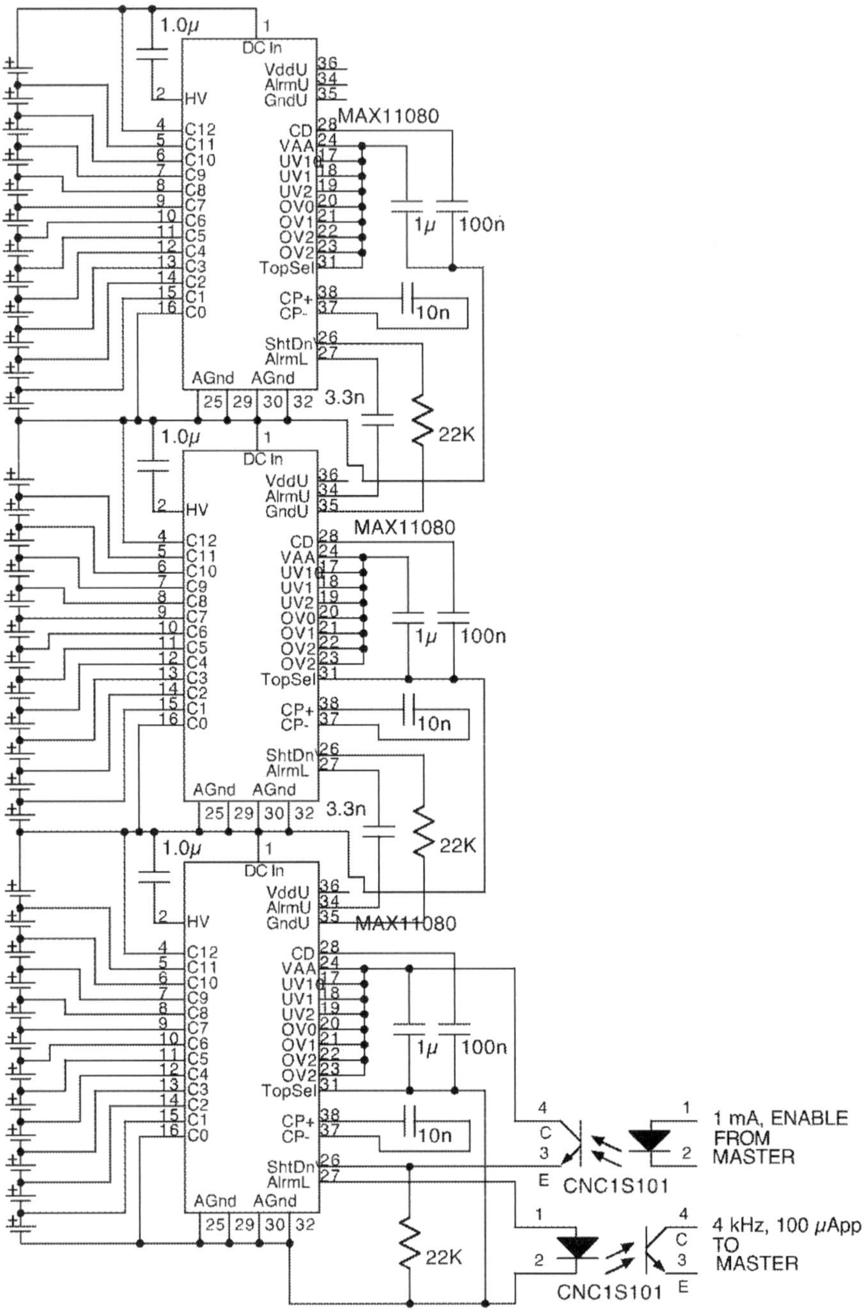

Figure 5.20 MAX11080 ICs in an analog monitor for 36 cells.

balancing current of 300 mA. Cell voltage measurement and the cell balancing share the same pins, which prevents the cell from balancing and measuring at the same time. To use this part in a large pack, each IC can monitor and balance 10 cells, and the charge and discharge outputs of each IC must somehow be combined to derive overall limit outputs (Figure 5.23).

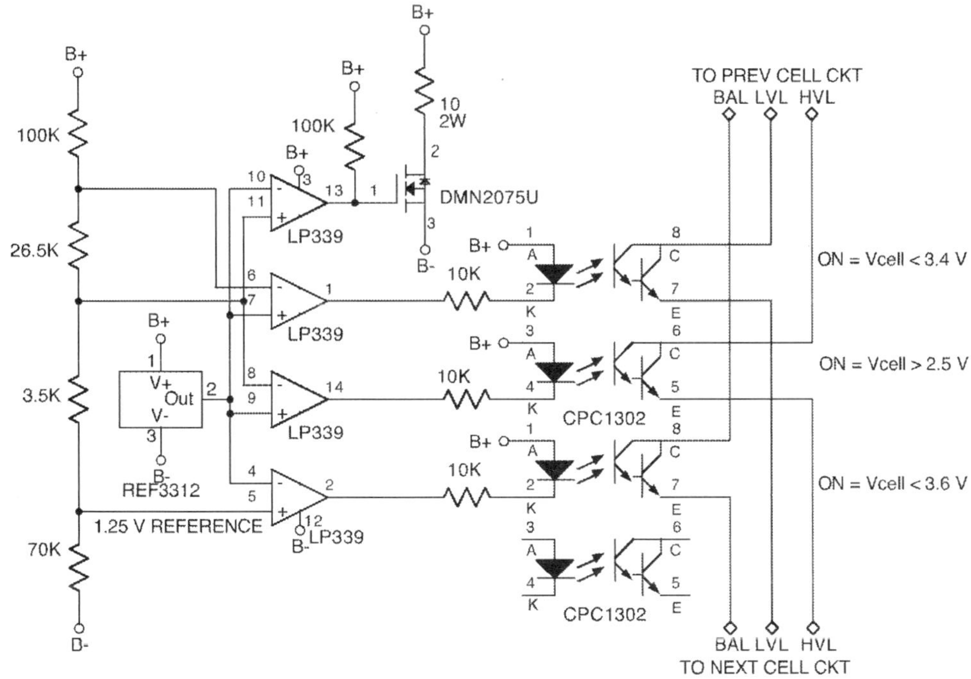

Figure 5.21 Analog balancer circuit using generic ICs.

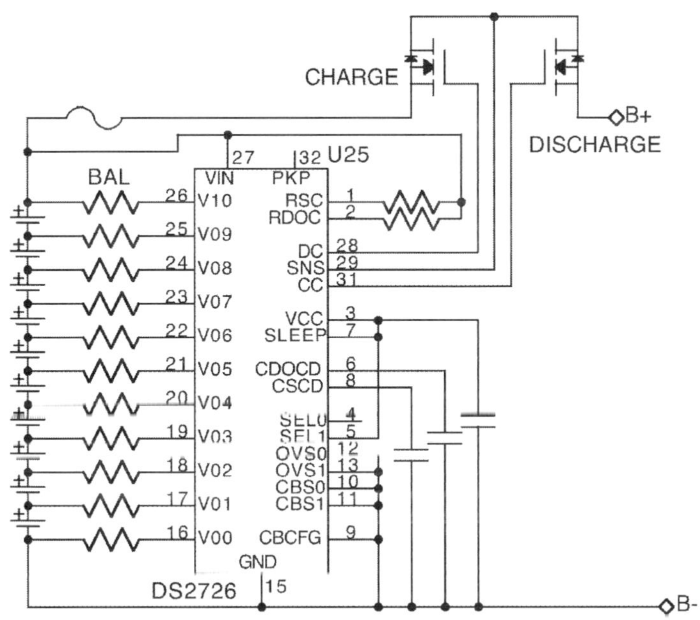

Figure 5.22 Maxim's DS2726 in a 10-cell analog protector.

5.2.3.3 bq76pl536 BMS Front End

Earlier in this chapter we saw how the Linear Technology LTC6801 and the Maxim's MAX11080 are digital monitor ICs that do not report cell voltages, and

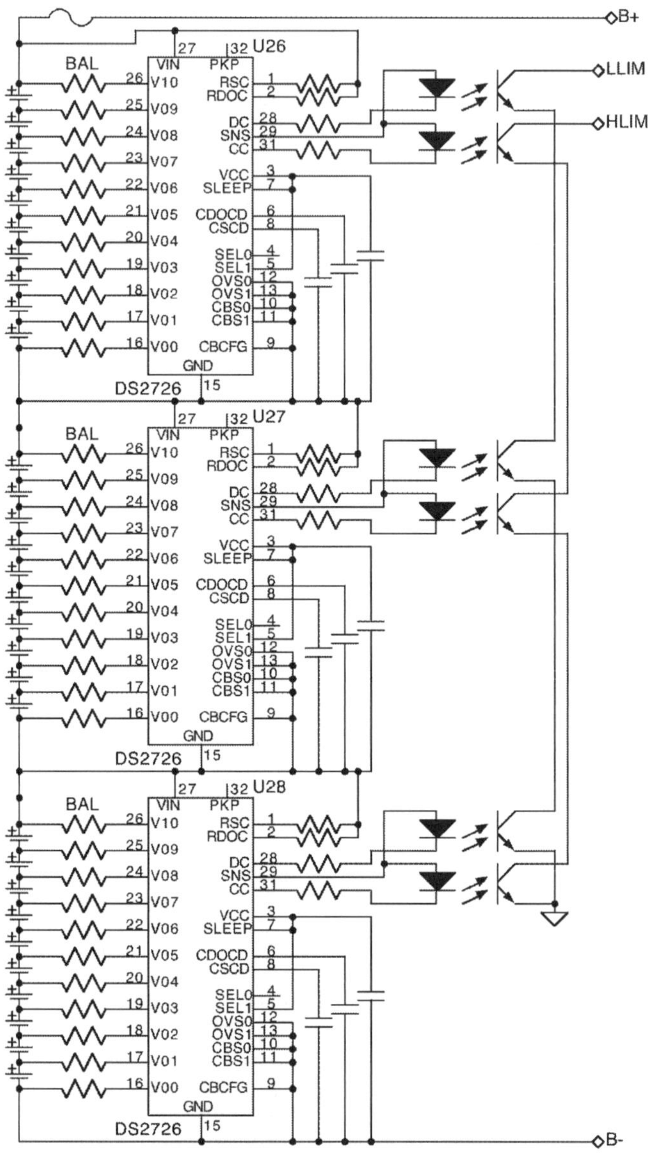

Figure 5.23 Maxim's DS2726 in a 30-cell analog balancer.

hence can be used as part of an analog monitor. Similarly, Texas Instruments' bq76pl536 is a BMS front ends for digital balancers that, when used in a standalone mode, can form the core of an analog balancer. This IC can handle three to six cells in series, and up to 32 ICs can be cascaded together to handle up to 192 cells in series. It includes three SPI bus ports: one for a host computer (which is not used in a standalone application such as the one described here), and the other two for cascading to the adjacent ICs. The IC uses separate pins for cell voltage sensing and for balancing (using external components), for best performance.

For example, three ICs can be used in an 18-cell balancer (Figure 5.24). Internally, the ICs work digitally; externally, there are only two dedicated lines: an input

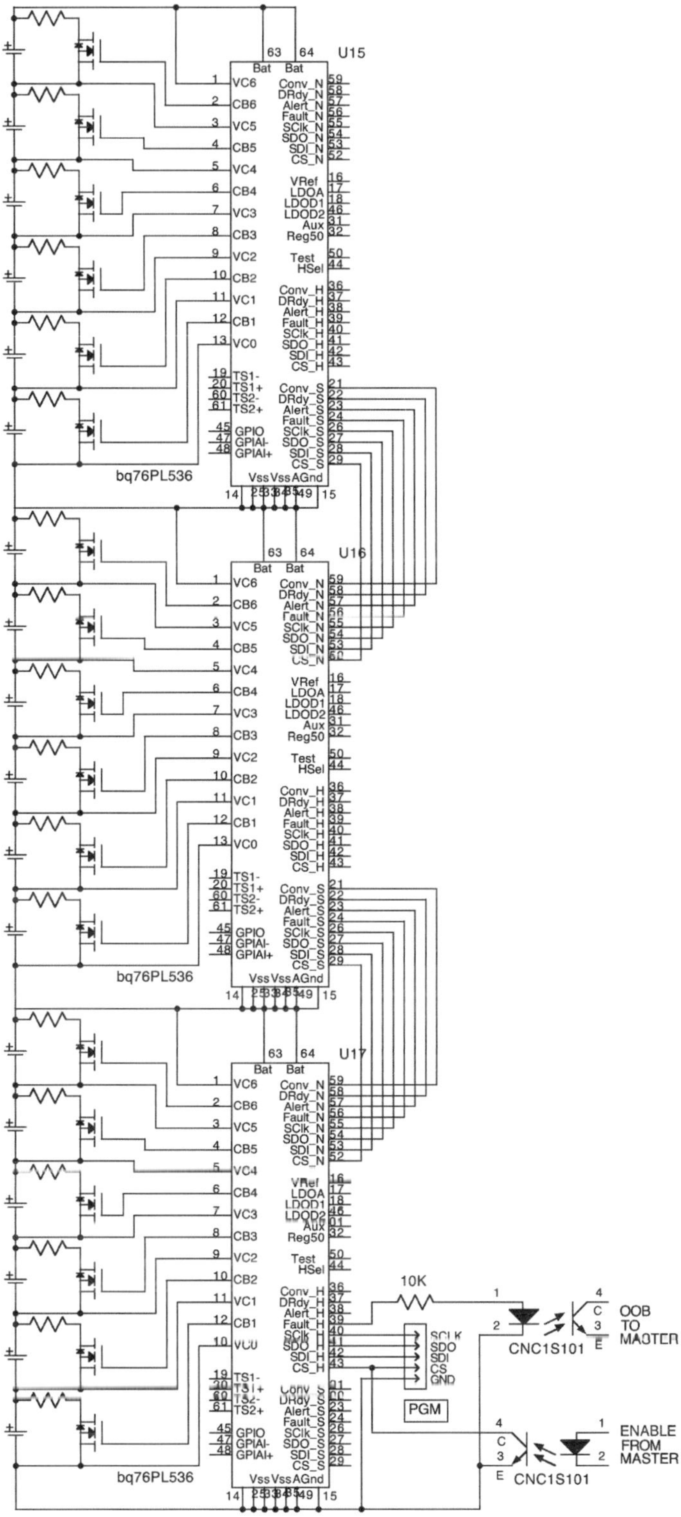

Figure 5.24 Texas Instruments' bq76pl536 in standalone mode in an 18-cell analog balancer.

to enable the circuit (if disabled, it draws little current), and an output to report whether everything is OK or any parameter is out of bounds. A digital programmer is connected to a configuring port when first setting up the BMS.

Later (Section 5.4.1.2) we will see how this IC can be used in a digital BMS. This IC is scheduled for release by the end of 2010.

5.2.3.4 bq76pl537 BMS Front End

Texas Instruments' bq76pl537 is identical to the bq76pl536 (passive balance), except that it uses active balancing. At first glance, active balancing is better, but that may not be true in reality because of how inefficient small active balancers are (Section 3.2.3.3). In particular, active balancing, as implemented here, is not helpful, because an IC only balances its own cells with respect to each other, and cannot balance them with respect to the cells in the rest of the pack.

5.2.4 Analog Protector

An analog protector is like a balancer, but also has a power switch to cut off the battery current. This switch is the same whether the protector is analog or digital. Section 5.4.6 will go into more detail.

5.3 Ready-Made, Digital BMS Designs

The fastest way to make your own BMS, with a minimum amount of risk, is to start from an existing design. Various BMS designs are available to implement, adapt, and expand according to particular needs. Some are open source projects using standard ICs, while others are complete BMS chip sets.

Table 5.2 compares the designs available at the time of publication.

5.3.1 ATMEL's BMS Processor

Atmel's ATmega406 is actually a full microprocessor that includes the peripherals to manage a small Li-Ion battery. Its features are comparable to the TI chips, but it is also programmable by the customer. Some highlights are:

- One chip handles up to four Li-Ion cells in series.
- A chip can include the smart algorithms available to manage Li-Ion batteries and is field programmable.
- The ATmega406 requires few additional parts for protection.
- It includes cell balancing, using on-board MOSFETs, and external resistors.
- It performs fuel gauge (SOC and DOD calculation) functions.
- It includes SMB (I2C) serial interface to the outside world.

Its limitations are:

- It is not appropriate for large packs, as it can only handle four cells.

Table 5.2 Comparison of Ready-Made BMS Designs

	Atmel	*Elithion*	*Perkins*	*Texas Instruments*	
Part number	ATmega406	E01/EL02	V series	bq29330 + bq20z90	bq78PL114 + 4 × bq76PL102
Type	Programmable IC	Chip set	Generic ICs, open source	Chip set	Chip set
Max. number of cells/ battery	4	255	255	4	12
Number of ICs/number of cells	1/4	1/1 + 1 master	1/1 + 1 master	2/4	2/12
Balancing[a]	Passive Shared pins, external resistor 2 mA	External, dedicated pin. Passive (resistor) or active (DC-DC) Any current	External, dedicated pin. Passive (resistor) or active (DC-DC) Any current	Passive Internal 10 mA	Active Shared pins External LC tank
Accuracy at 3.6V	58 [mV]	15 [mV]			15 [mV]
Temperature sense	1/cell	1/cell	1/cell	One total	1/cell
Communications	I2C/SMB	CAN, RS232	RS232	I2C/SMB	I2C/SMB
Cost of ICs only[b]	\$0.9/cell	\$3/cell + \$20 master	\$1/cell + \$10 master	\$1.4/cell	\$1.25/cell
Cost, total[c]	\$1.10/cell	\$4.50/cell + \$50/system	\$5/cell + \$20/ system	\$1.90/cell	\$3/cell
Availability[d]	Excellent	Good	Excellent	Excellent	Poor

[a]Using shared pins to drive balancing loads reduces pin count, but there are severe limitations on balancing. For example, adjacent balancing loads cannot be on simultaneously.
[b]An IC with a lower cost per cell is not necessarily the least expensive solution, as some ICs require many more external components than others (costs are in large quantity).
[c]Rough estimate of costs of all components, in large quantity.
[d]As of time of publication.

- An accuracy of 58 mV may be a bit too low for accurate SOC estimation from OCV.
- A 2-mA balance current is insufficient for large battery packs.

5.3.2 Elithion's BMS Chip Set

Elithion makes available the chip set at the core of its Lithiumate BMS (Section 4.3.2.2) for incorporation into a BMS that implements only the functions needed in a given application.

The procedure to get started with this chip set is:

1. The client installs an off-the-shelf Lithiumate BMS to confirm that it meets the application's requirements.

2. The client enters in a contract with Elithion to protects its intellectual property.
3. Elithion gives the client the design files for the Lithiumate, and, if requested, will adapt the Lithiumate design to the application's needs.
4. The client sets up its own BMS manufacturing line, using Lithiumate ICs that it buys directly from Elithion.

The advantage of using this chip set is the off-the-shelf availability of the Lithiumate BMS (Section 4.1.2.2), a well documented, sophisticated BMS with a proven track record. A Lithiumate BMS can be purchased from stock, installed, configured, and tested by a skilled technician in the span of a week. This allows the customer to determine whether it meets the application's needs in a far shorter time than other ready-made designs, even those that offer evaluation modules. Soon after that, and with relatively little effort, a Lithiumate BMS using this chip set can be repackaged into a BMS that the customer can call its own, putting its name on the box (Figure 5.25). Some highlights are:

- It can handle from 1 to 255 cells in series (~1 kV max).
- There is no practical limitation on battery capacity and battery current.
- Cell-mounted cell-boards with only one daisy-chain wire result in a clean layout with little space requirements and a simpler installation.
- The chip set includes two ICs: an EL01 per cell in series and an EL02 in the BMS controller.
- The BMS controller is fully configurable to match the needs of the system.
- The BMS reports the voltage and temperature of each cell.
- The BMS offers fuel gauge (SOC and DOD) and SOH evaluation.
- The BMS can perform cell balancing using external components. (Active balancing is possible.)

The Elithion solution does have some disadvantages as well:

- The algorithms are not as evolved as in the Texas Instrument chips.
- The BMS is not fully integrated. Fourteen additional parts are required to complete a cell board.
- It is more costly than other solutions. In high quantities, the chip cost is $3/cell in series, while the complete cost is about $5/cell in series, plus the cost of the controller (approximately $50).

5.3.3 National Semiconductors' Complete BMS

National Semiconductors is about to announce a chip set to do a complete BMS solution for large Li-Ion battery packs. As a latecomer into the Li-Ion management field, National enters it at full steam by offering more than just an analog front end (like Lithium Technology) or a BMS for small batteries (like Texas Instruments). They are the first and only semiconductor company to offer a complete BMS for a large Li-Ion packs.

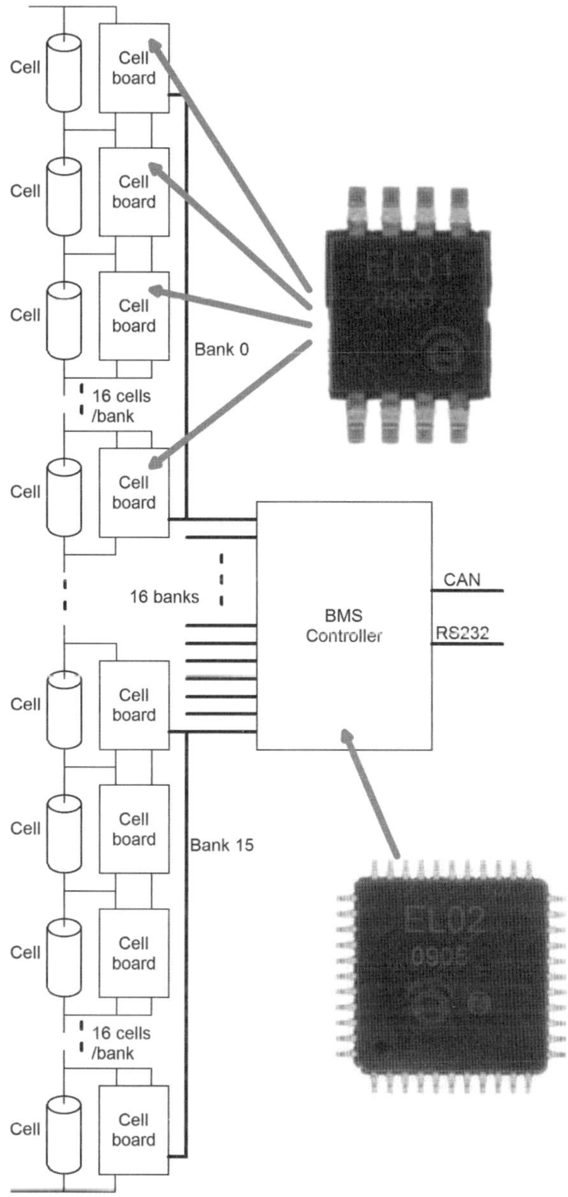

Figure 5.25 Block diagram of a BMS front end using the Elithion EL01 and EL02 ICs.

5.3.4 Peter Perkins' Open Source BMS

While many groups are more or less actively pursuing open source BMS projects, I know of only one that was completed: Peter Perkins' BMS. Peter Perkins is not an engineer, he's a police officer; he is a 48-year-old police sergeant in North Yorkshire, England, with no formal qualifications in anything electronic or related to programming. However, he is assiduous and clever, and succeeded where more qualified people got bogged down. Having converted a Bedford Rascal van to be solar and wind assisted (the "Solarvan"), he needed a BMS for its Li-Ion battery at a time when no decent BMS was available. So he designed his own distributed balancer, and then he made the design available to anyone who wants it.

The BMS is simply a home-built project, which started in 2003 when he was one of the group in the United Kingdom who bought some of the first Thundersky cells. That group included Dennis Doerffel (who went on to found REAP—see Section 4.1.2.2) and Cedric Lynch (who went on to create the Agni analog balancer). Peter was in very clever company and used his connections to his advantage. His first BMS used relays and an old hacked 386 laptop, which left a lot to be desired. Peter developed his next BMS, based on the Picaxe microcontrollers. He did so with the help of members of the Battery Vehicle Society forum, specifically Greg Fordyce.

Peter's present BMS is a sophisticated (digital) balancer for Li-Ion packs of up to 256 prismatic cells (Figure 5.26). It can be made as a distributed BMS or a centralized one. Each cell board circuit is powered by its cell, which measures its voltage and can balance it with about 300 mA of current, but does not measure its temperature. A two-wire daisy-chain loop goes from a master controller through each cell board, and back to the controller. The system includes a current sensor (for SOC calculations), an RS232 serial port, and a display. Its cost is quite low: a BMS for 16 cells costs about $100 in parts. It can be built by anyone who can do basic soldering. It is fairly flexible, as the software can all be modified to suit. Peter and Greg have built two samples of this BMS, and others are starting to do so as well. You can find the BMS design information at http://batteryvehiclesociety.org.uk/forums/viewtopic.php?t=1245.

Peter has since sold his Solarvan, and is working on his Honda Insight converted to PHEV and getting a whopping 150 mpg.

Figure 5.26 Peter Perkins' BMS. (*Source*: P. Perkins, 2010. Reprinted with permission.)

5.3.5 Texas Instruments' bq29330/bq20z90

Texas Instruments is the de facto leader in ICs used in small Li-Ion batteries, such as cell phones and laptops. It entered this field in 1999 when it acquired Unitrode, which had just acquired BenchMarq, at the time the leader in the battery management IC field. Hence, the ICs' prefix is still bq, as in BenchMarq. Most of its parts are for up to four cells in series.

A typical solution uses the bq29330 and bq20z90 ICs, which together form a complete BMS (Figures 5.27and 5.28). Highlights include:

- Implements the smartest algorithms commercially available today to manage Li-Ion batteries;
- Ancillary ICs available to provide redundant protection for bullet-proof reliability;
- Very few additional parts required;
- Includes cell balancing, using on-board components (external loads can be used as well);
- Fuel gauge (SOC and DOD calculation) functions with better than 1% accuracy;
- Includes a SMB serial interface.

Limitations include:

- It has a very low balancing current. While it is possible to increase the current using external MOSFETs, the fact that there are no dedicated pins to drive the MOSFETs poses severe limitations to balancing.

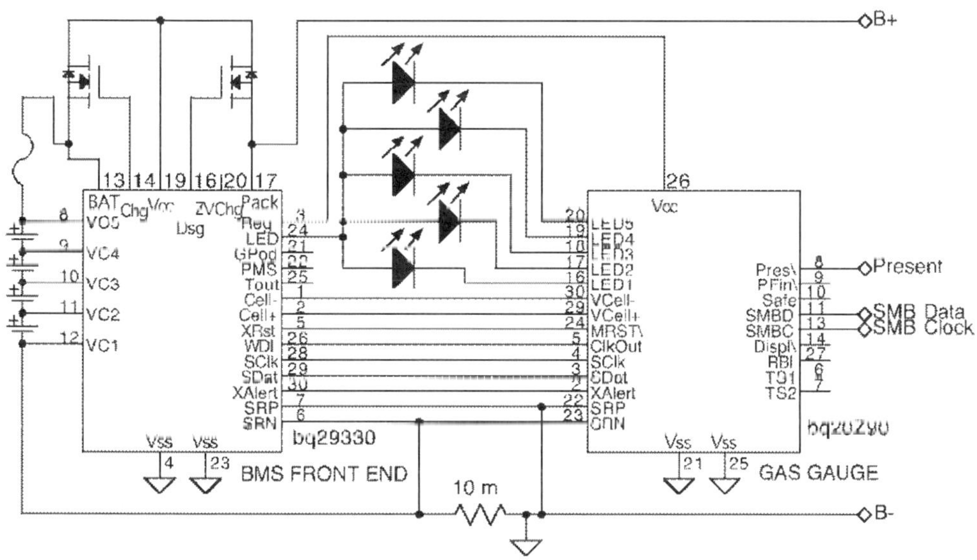

Figure 5.27 A four-cell protector using Texas Instruments' bq29330 and bq20z90 ICs.

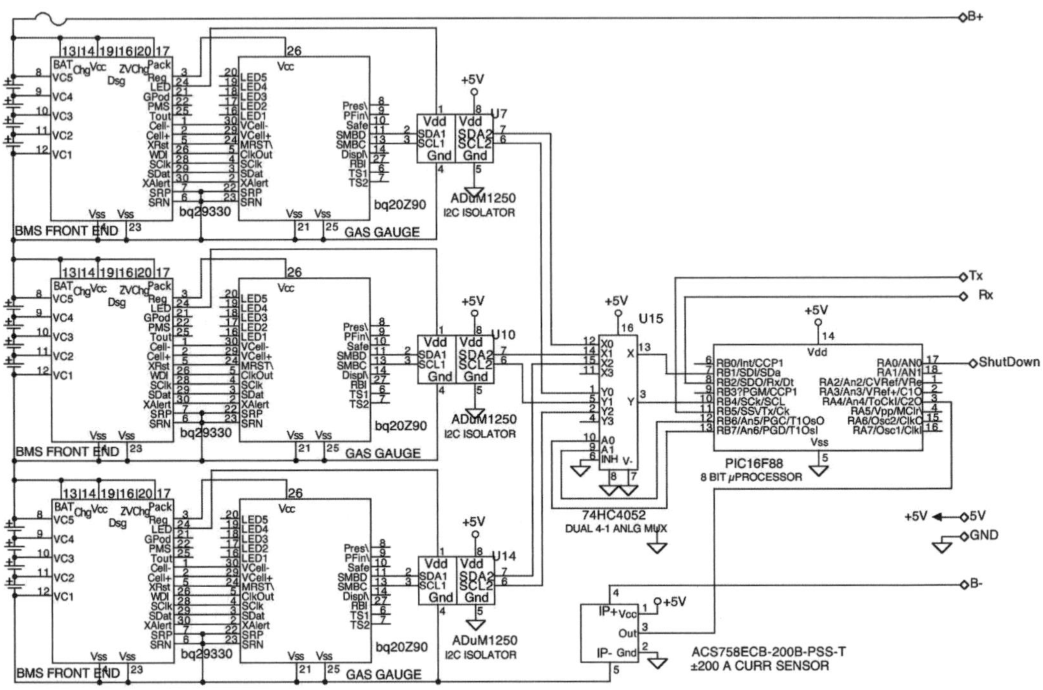

Figure 5.28 Texas Instruments' bq29330 and bq20z90 ICs shoehorned into a 12-cell battery pack.

Unfortunately, as recently as 2008, TI application engineers inappropriately encouraged clients to shoehorn this chip set into BMS for large battery packs, with rather unsatisfactory results. (Vectrix motorcycle BMS, using 26650 $LiFePO_4$ cells, each group of four cells in series fitted with a small BMS using TI ICs, plus a regulator on each cell to make up for inability to balance the cells with the TI chips, plus a master controller to try to herd all the cats, with rather unsatisfactory results.) This chip set works well for what it was designed for: a small, four-cell battery, but it is quite unsuited for a large pack.

While theoretically you could make slaves based on this chip set, each able to handle four cells in series, you would encounter these limitations:

- Because all the TI chips use the same I2C ID, they cannot be placed on the same bus. Therefore, a multiplexer is required to allow a master controller to communicate individually with each slave, one at a time.
- It is meant to be used with a small current sense resistor for each IC. You would have to jump through hoops to make it work with a single, high current sensor.
- In a large pack, the master controller measures the current with a single, large current sensor, so it knows the current. This chip set has no provision to be told what the current is, so this chip set's sophisticated algorithms to evaluate battery status and performance are not put to use.
- Even if one were to add a current sensor in each and every slave to enable those algorithms, these chips are meant to be used with a small resistor as a current

shunt and are unsuited for use with a large shunt. Additionally, a current shunt per slave would result in unacceptable losses.

- Each slave would have a mind of its own, and the master controller would not able to make them work together. For example, the master is not able to control the balancing loads directly because the chip set does not have provision for external control of the loads. Therefore, this chip set is not recommended for BMSs for large packs.

5.3.6 Texas Instruments' bq78PL114/bq76PL102

As of this writing, Texas Instruments' bq78PL114/bq76PL102 is the only commercially available BMS chip set that includes active balancing. The bq78PL114 IC controls up to four cells. By adding one or more bq76PL102 ICs (each of which handles two cells), the set can handle up to 12 cells in series.

For example, an eight-cell protector uses one bq78PL114 and two bq76PL102 ICs (Figure 5.29). Seven DC-DC converters transfer energy between the eight cells, controlled by the ICs. A small board-mounted resistor senses the battery current. Two MOSFETs (one for the charging direction and one for the discharging direction) act as protection switches. Highlights of this chip set include:

- Active balancing using two MOSFETs and an LC tank to dump energy from one cell to one of its two adjacent cells;
- Smartest algorithms commercially available today to manage Li-Ion batteries;
- Fuel gauge (SOC and DOD evaluation), with better than 1% accuracy;
- SMB serial interface (an extension of the I2C standard, used in laptop batteries).

Again, the TI application engineers will have you believe that this chip set is ideal for large battery packs.[1] However, this is not the case. This chip set should work well for what it was designed to do: a small, 12-cell battery with active balancing. It is quite unsuited for a large pack.

While theoretically you could make slaves based on this chip set, each able to handle 12 cells in series, you would encounter all the limitations we saw in Section 5.3.5 with the bq29330/bq20z90 chip set, plus this one:

- The active balance circuit could transfer charge within the 12 cells handled by one slave, but would not be able to transfer charge between adjacent slaves, resulting in groups of 12 cells that cannot be balanced with respect to other groups of 12 cells.

There is a workaround to this limitation: you could keep on unbalancing and rebalancing the most charged set of 12 cells, to dissipate energy through the inefficiency of the active balance circuit. Of course, that would waste as much energy as a passive balance circuit (defeating the purpose of active balance) and would take longer than passive balance would. So, it would be pretty pointless.

1. Late in 2009, TI presented a Web seminar touting this chip set as ideal for HEV traction packs.

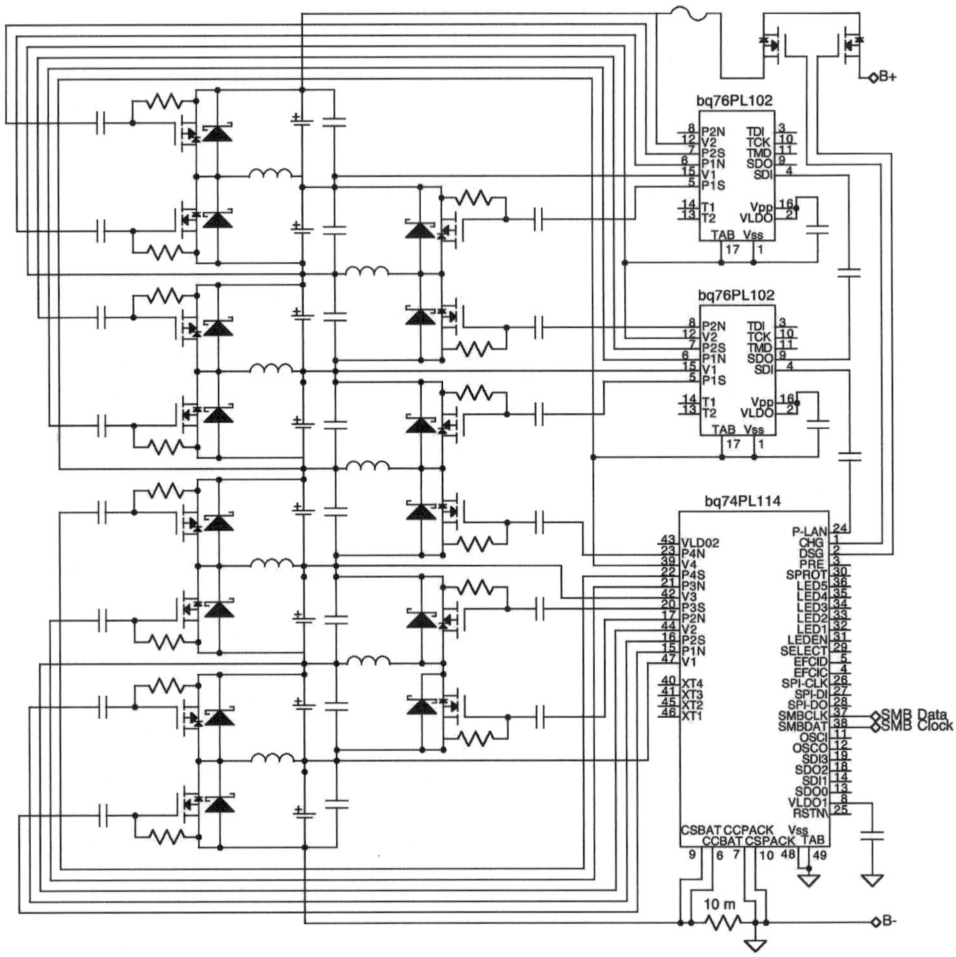

Figure 5.29 An eight-cell protector using a Texas Instruments' bq78PL114 and two each bq76PL102 ICs (simplified).

5.4 Custom Digital BMS Design

This section is dedicated to the many facets of the design of a digital BMS. Not all of its subsections apply to all types of BMS given that some BMS only implement some functions. Table 5.3 specifies which functions are implemented in various types of BMS, ranging from a plain digital meter (which implements only the measurement function) to a digital protector (which implements all functions).

5.4.1 Voltage and Temperature Measurement

The first BMS function is measurement. Every type of digital BMS implements this function. There are two ways of measuring cell voltage (and temperature): distributed (one circuit per cell) or localized (a circuit shared among various cells).

Table 5.3 List of Functions Implemented by Various Types of BMS

Function	*Section*	*Meter*	*Meter with SOC*	*Monitor*	*Balancer*	*Protector*
Measurement	5.4.1.2	✔	✔	✔	✔	✔
Evaluation	5.4.3	✔	✔	✔	✔	✔
Communications	5.4.4			✔	✔	✔
Optimization	5.4.5				✔	✔
Switching	5.4.6					✔

5.4.1.1 Distributed Measurement

Discrete measuring, one device per cell, allows simultaneous measurement of all the cell voltages. Doing so is ideal, because each cell voltage can be related to the battery current in that same instant, which is important when calculating cell resistance. (Even though the measurements are taken simultaneously, the readings can be streamed serially a bit later.)

Discrete measurement is the obvious solution in distributed systems (with one cell board per cell). Having the measuring device mounted to the cell has advantages and disadvantages. On one side, the measuring device is immersed in the electrical noise typically present in a high-power environment, which can reduce the measurement accuracy, and may affect data communications. On the other side, the measurement can be more accurate, as the cell voltage is measured locally, so there are no long wires bringing the voltage to the measuring device and acting as antennas picking up electrical noise. Also, the measurement is inherently immune to common mode noise, as the measuring device is referenced to the cell being measured. All things considered, a distributed discrete measurement done by a cell board tends to be more reliable and accurate than a remote, multiplexed measurement.

Typically, a cell board uses a small microprocessor with an A/D input to measure the cell voltage, cell temperature, control a balancing load, and communicate with a BMS controller (Figure 5.30). The microprocessor must have a decent A/D converter (10 or 12 bits) to get a good voltage measurement. It must have a peripheral for communications (such as a UART) for synchronous or asynchronous serial communications (such as an I2C or SPI bus), or a CAN bus machine. The following example uses simplex asynchronous communications. A description of this circuit may be appropriate.

Protection

The fuse opens if the cell board is misconnected or exposed to the pack voltage (if the power connection to its cells is suddenly opened under load). A TVS clamps the input voltage, to blow the fuse in case of reverse connection or excessive voltage. The LC filter removes some of the high frequency noise present on the cell.

Power Supply

The voltage regulator generates a fixed 2.0-V supply to power the circuit. It also functions as the first line of defense if the cell board is misconnected.

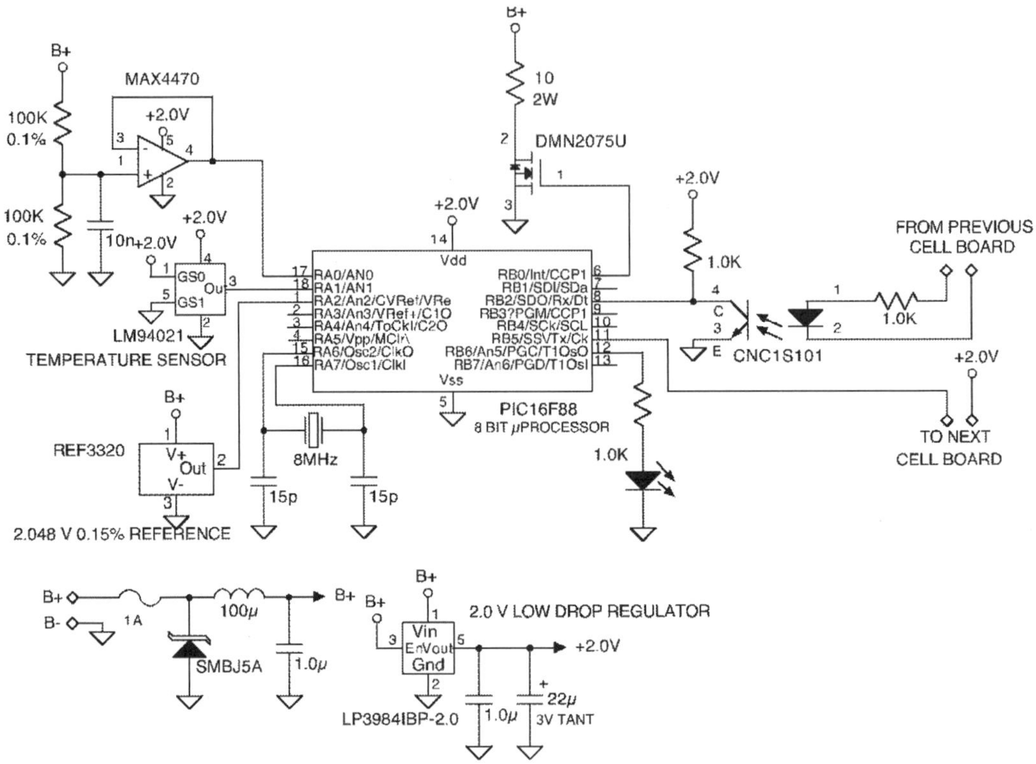

Figure 5.30 Digital cell board circuit.

Microprocessor

A MicroChip PIC16F88 processor is the heart of the cell board: it measures the cell voltage and the temperature, it drives the balancing load, and it communicates with the BMS controller through the previous and the next cell board. It works down to 2.0V, drawing less than 1 μA when asleep, has a 10 bit A/D converter, a PWM generator, and a UART for communications. It uses a crystal to generate an 8-MHz clock.

Analog/Digital Converter Reference

The band-gap voltage reference provides a precise 2.048-V reference to the A/D converter. This voltage is slightly higher than the supply voltage, but the micro can handle it. The reference's voltage drop, unloaded, is just a few millivolts, so it will work down to a cell voltage of 2.050V.

Cell Voltage Sensor

Two precision resistors form a voltage divider to halve the cell voltage down to a range of 1.0V to 2.1V, which is below the reference voltage. Their resistance must be really high to avoid draining the cell. The micro requires a low source drive to its A/D input, so a voltage buffer is required (an op-amp as a unity-gain voltage-follower). The op-amp is able to operate down to 2V, and draws less than 1 μA from the supply; its output feeds an A/D input of the micro.

Temperature Sensor
A sensor IC measures the cell board temperature (which is close to its cell's temperature as long as the cell board is mounted on the cell and the balancing load is off). It generates a voltage that goes lower as the temperature increases. It is pin-strapped for a low gain, to generate 1.88V at −40°C, down to 0.86V at 85°C.

Balancing Load
A MOSFET controls the high-power resistor used as a passive load.

Status Display
An LED provides visual feedback on the activity and status of the cell board.

Communications
Two opto-isolators provide isolation between the cell board (which is referenced to a high voltage inside the battery) and the bus (which is referenced to ground). One opto-isolator is between the previous cell board and the RX input of the micro's UART, to receive data from the master controller; the other one is between the TX output of the micro's UART and the next cell board, to send data to the master controller.

The circuit is powered by the cell itself, so it must draw low power (it must draw less current than the leakage of the cells being measured), especially during standby, in order not to discharge the cell; a typical standby current of 10 μA is good, 100 μA is acceptable, and 1 mA is unacceptable. Many micros have a sleep mode, which shuts down peripherals and clocks to reduce the battery consumption drastically. They can be woken up from sleep either regularly (by a watchdog) or in case of interrupt (such as from receiving a message in a communication port). The reference and the temperature sensors do not draw too much power, and the bias current of the regulator and the supply current of the cell voltage buffer can be very low.

This example circuit implements the desired functions by the cookbook. In reality, much simpler circuits have been developed to accomplish the same functionality. For example, an Elithion midbank cell boards use only 15 parts, including just two ICs and no opto-isolators. The software in the micro could operate as follows:

1. At power-up, set up the ports and the peripherals.
2. Turn off the balancing load.
3. Go to sleep, leaving the UART on.
4. Wake up when a message is received.
5. Receive a message.
6. Interpret and obey the command from the master controller.
   - *Read voltage:* Measure the cell voltage divider and convert to volts.
   - *Read temperature:* Read the temperature sensor voltage and convert the reading to degrees Celsius.
   - *Report voltage:* If the master requested this cell's voltage, transfer that reading.
   - *Report temperature:* If the master requested this cell's temperature, transfer that reading.

  - *Load on:* If the master requested that this cell's load be on, do so.
  - *Load off:* If the master requested that this cell's load be off, do so.
7. Go get the next message (item 5).
8. If no commands for 10 seconds, go back to sleep (item 3).

5.4.1.2 Localized Measurement

We saw how, in a distributed BMS, each cell circuit measures the voltage of its cell. In a localized BMS, a shared circuit measures the voltages of multiple cells. Usually, this results in a reduction in the number of components. Localized measurement can be done with general purpose ICs or with one of a few commercially available BMS ASICs.

Multiplexed Measurement

Typically, localized measurement of cell voltages uses multiplexing, so that measurement components may be shared among a few cells. Many designs share an A/D converter among 4 to 12 cells, by using an analog multiplexer to select which cells is measured. For example, a dual 4:1 analog multiplexer can be used to sample the voltage of one of four cells in turn and feed that voltage to a single, shared A/D converter [Figure 5.31(a)]. This circuit is similar to its analog equivalent (Section

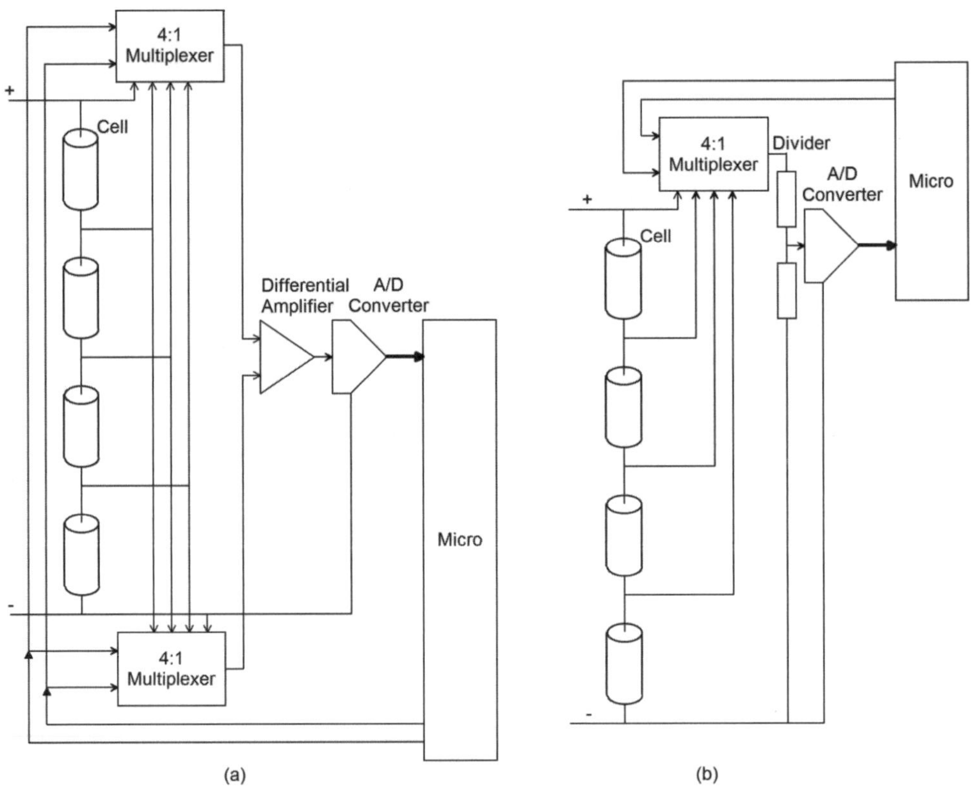

Figure 5.31 Multiplexed measurement of a four-cell battery using generic components: (a) two multiplexers and a differential amplifier and (b) one multiplexer.

5.2.2.2). Or, as single multiplexer could be used [Figure 5.31(b)] to take measurements of the tap voltages (instead of the cell voltages), and leaving it up to the micro to take the difference between two adjacent tap voltages to calculate a cell voltage.

There are some minor limitations with these approaches. With both circuits, measurement of the various cell voltages, which is done consecutively, must be done rapidly to avoid introducing errors in pack voltage and cell resistance calculation due to the lack of simultaneity in the measurements. Also, due to limited common-mode rejection, the noise level with respect to the negative reference becomes higher and higher as you go up the chain of cells in series (Figure 5.32).

There are problems with each of these circuits. With the two-multiplexer circuit, the accuracy of the cell voltage measurement get worse as you go from the most negative to the most positive cell, due to tolerances in resistor dividers and limited common-mode rejection in the differential amplifier.

With the single multiplexer circuit, except for the most negative cell, the cell voltage is not measured directly, but is calculated later by subtracting the measurements of the taps on either side of the cell. As the two measurements are not done at the same time, and given that the cell voltage is usually not constant, an error is introduced.

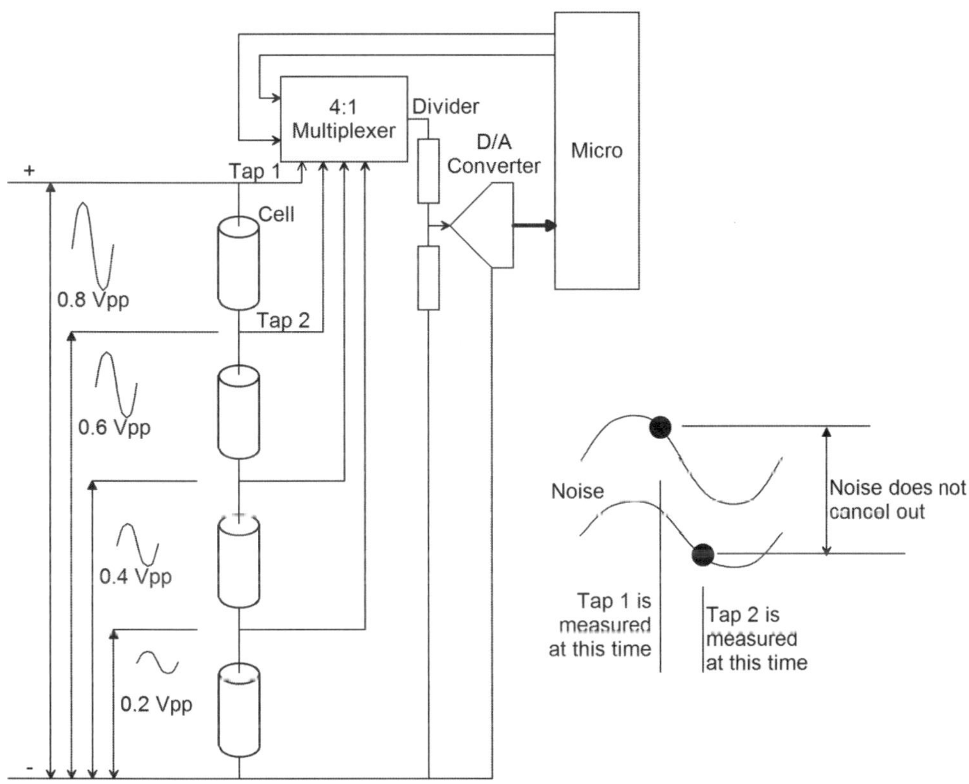

Figure 5.32 Noise error increases as you go up the string of cells.

General Purpose ICs Versus ASICs

A multiplexed circuit to measure cell voltages may use either generic IC or one of the specialized BMS front-end ASICs available on the market today. These ASICs can be divided into two groups: those intended for large packs, and those intended for small batteries (laptop, cell phone) that some try to shoehorn into a large pack. These ICs may include just a multiplexer, or they may also include an A/D converter. Some may also include the logic to drive balancing loads. They all include some intelligence to perform measurements and to communicate with a master controller. Some of these are not actually true ASICs, but are microprocessors or Field Programmable Gate Arrays (FPGAs) that are preprogrammed or preconfigured for BMS operation. Table 5.4 compares some of the effects of using generic ICs, or BMS ASICs (small battery or large pack) in the design of the front end of a BMS for a large pack. Some highlights from this table are:

- A circuit using generic ICs requires more design effort, compared to a circuit using ASICs (as their manufacturer provides sample circuits). This is balanced by the effort required to learn how to use the ASICs, which tend to be more complex than a circuit designed with generic ICs.
- The effort required to adapt an ASICs for small batteries to use in a BMS for large packs can be very high.
- It is unlikely that a design using generic ICs can approach the performance (measurement accuracy, flexibility, self testing, low current consumption) achieved by BMS ASICs, given their manufacturers' experience in this area. However, while ASICs perform beautifully in the lab, their performance may suffer when subject to real-world interference.
- Using ASICs locks the design at the mercy of a particular manufacturer. Today, the availability of some of these ASICs is problematic.
- While the IC cost and real estate is lower for ASICs, both approaches require similar quantities of external components (external filtering, protection and balancing components, communication buffers and isolators, master controller), making the total cost and real estate of both approaches quite similar.

Digital BMS Using General-Purpose ICs

Various generic ICs can be used in a BMS front end:

- Analog ICs: pp-amps, comparators, multiplexers, voltage regulators, voltage references.

Table 5.4 Comparison of Cell Voltage Measurement Technologies in a Large Pack: Generic ICs Versus BMS ASICs Intended for Small Batteries or for Large Packs

	Generic ICs	*Small Battery ASICs*	*Large Pack ASICs*
Circuit Complexity	High	Medium	Medium
Design Effort	Medium	High	Medium
Performance	Good	Excellent	Excellent
Cascading	Good	Very poor	Very good
Second Source	Very good	None	None
Cost	Same	Same	Same

- Interface ICs: A/D converters, line drivers.
- Data processing ICs: small microprocessors with one or a few A/D converter channels, powerful microprocessors with multiple A/D conveters and communication peripherals, digital signal processors (DSPs).

The electronic circuit that is connected to the cells must be able to operate at the maximum voltage achieved by that number of cells in series (for example, 51V for 12 standard Li-Ion cells in series, which is a tall order for most ICs). To handle high voltages, discrete transistors can be used instead of ICs. The following example uses a dual 4:1 CMOS multiplexer IC that can handle 18V, which is higher than the 16.8V for four standard Li-Ion cells in series.

Many circuits for a BMS front end using generic ICs are possible, including the following one (Figure 5.33). This slave handles four cells in series (any Li-Ion chemistry), includes balancing, and communicates through a daisy-chain digital bus (from the master, through the previous slaves, through this slave, through the following slaves, and back to the master). A description of this circuit follows.

Supply The four cells in series power the slave (from 8V to 16.8V, depending on the cell chemistry and their SOC). The bottom end is connected to the circuit's local ground. The top end is sent through a protection resistor and then to the internal circuits. This supply powers an LP2980 voltage regulator to generate a 5V supply.

Multiplexer Each tap voltage between cells goes through a current limiting resistor and a filter capacitor to ground, then to a 4,053, dual 4:1 analog multiplexer. A 2-bit address selects one of the two cells at a time. The top section of the multiplexer selects one of the four taps connected to the "−" cell terminals; the bottom section of the multiplexer selects one of the four taps connected to the "+" cell terminals. The entire cell voltage of the selected cell appears across the two outputs of the multiplexer (not referenced to ground). A capacitor across the outputs of the multiplexer is charged to the voltage of the selected cell.

Switched Capacitor Block An LTC6943 switched capacitor block changes the cell voltage reference to ground. It operates in two phases: in the first phase, it connects a floating capacitor to its input, charging it to the cell voltage; in the second phase it disconnects the floating capacitor from the input and connects it to the output, to a capacitor across its output. The output capacitor is referenced to ground, and, after a few cycles, its voltage is equal to the selected cell voltage. That voltage is measured by the micro through one of its A/D inputs. The micro can shut down the switched capacitor's clock by grounding its oscillator capacitor to reduce power while the battery is not in use.

Balancing Across each cell, a MOSFET switches a power resistor to remove charge from that cell. The micro turns on a MOSFETs through an opto-isolator (not to provide isolation, but to step the voltage—the micro's output are referenced to ground and go up to 5V, while each MOSFET gate is referenced to the "−" terminal of its cells, and needs a few volts of gate voltage to be turned on). Resistors between the gate the source turn off a MOSFET when the opto-isolators are off.

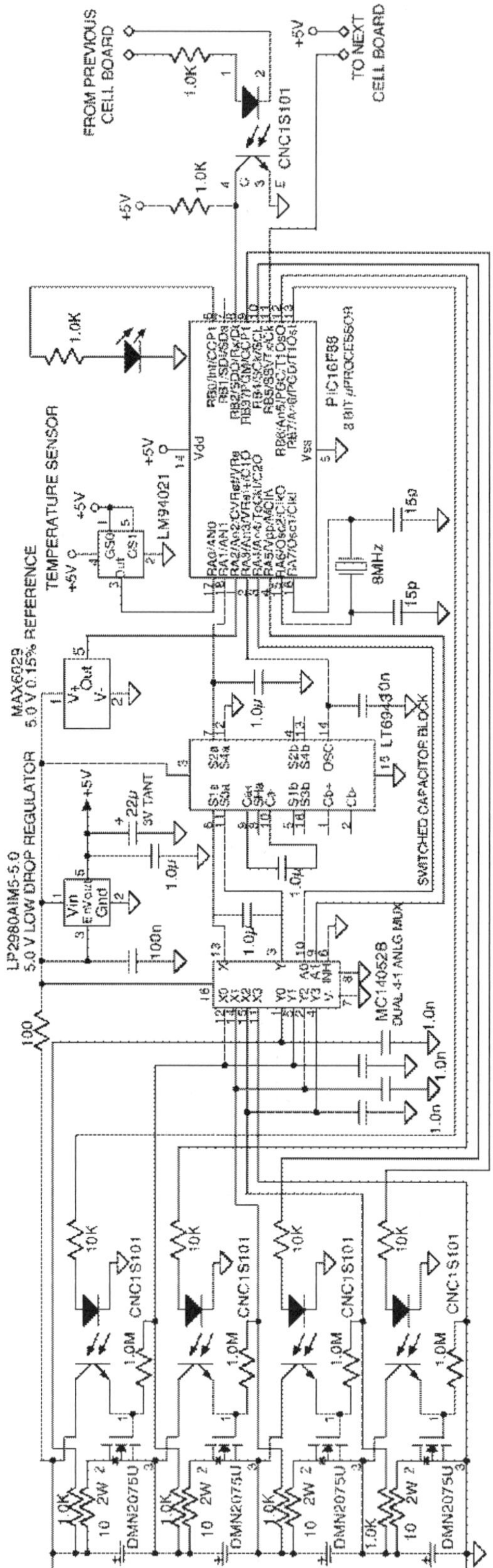

Figure 5.33 Four-cell BMS slave board using generic ICs.

Temperature Sensor An LM94021 sensor IC measures the temperature of the assembly (which is close to the cells' temperature as long as the slave is mounted on the cell and the balancing loads are off), generating a voltage that goes lower as the temperature increases. It is pin strapped for the highest gain, to generate 3.16V at −40°C, down to 1.47V at 85°C. That voltage is sent to an A/D input on the micro to be read.

Voltage Reference Since the 5-V regulator voltage is not sufficiently accurate, a MAX6029 precision reference is used instead to give the micro's A/D converter a precision 5-V reference.

Status Display An LED provides visual feedback on the activity and status of the slave.

Communications Two opto-isolators provide isolation between this slave (which is referenced to the negative terminal of its most negative cell) and the adjacent slaves. One opto-isolator is between the previous slave and the RX input of the micro's UART, to receive data from the master controller. The other one is between the TX output of the micro's UART and the next slave, to send data back to the master controller.

Microprocessor A MicroChip PIC16F88 processor is the heart of the slave: it measures the cell voltages and the temperature, it drives the balancing load, and it communicates with the BMS controller through the previous and the next slaves. It draws less than 1 μA when asleep, has a 10 bit A/D converter, a PWM generator, and a UART for communications. It uses a crystal to generate an 8-MHz clock.

The circuit is powered by the cells themselves, so it must draw low power (it must draw less current than the leakage of the cells being measured), especially during standby, in order not to discharge the cells. A standby current of 10 μA is good, 100 μA is acceptable, and 1 mA is unacceptable. The micro has a sleep mode, which shuts down peripherals and clocks, to reduce the battery consumption drastically. The micro is woken up from sleep regularly (by a watchdog) or in case of interrupt (from receiving a message to its communication port). The reference and the temperature sensors do not draw too much power (hen not scanned, the multiplexer draws negligible current); the micro can turn off the oscillator in the switched capacitor block to minimize its current consumption.

Digital BMS ASICs

As of this writing, a few ASICs are available to form the front end of a digital BMS. Most of them can be cascaded directly to handle many cells in series in a single bank (without isolators), while the rest require isolators and additional components to communicate independently with a master controller through a shared bus.

These ICs integrate functions required to interface to a number of cells, in order to simplify the design of the BMS front end. In general, they contain (Figure 5.34):

- A multiplexer for reading the voltage of 2 to 12 cells, and a few temperature sensors;

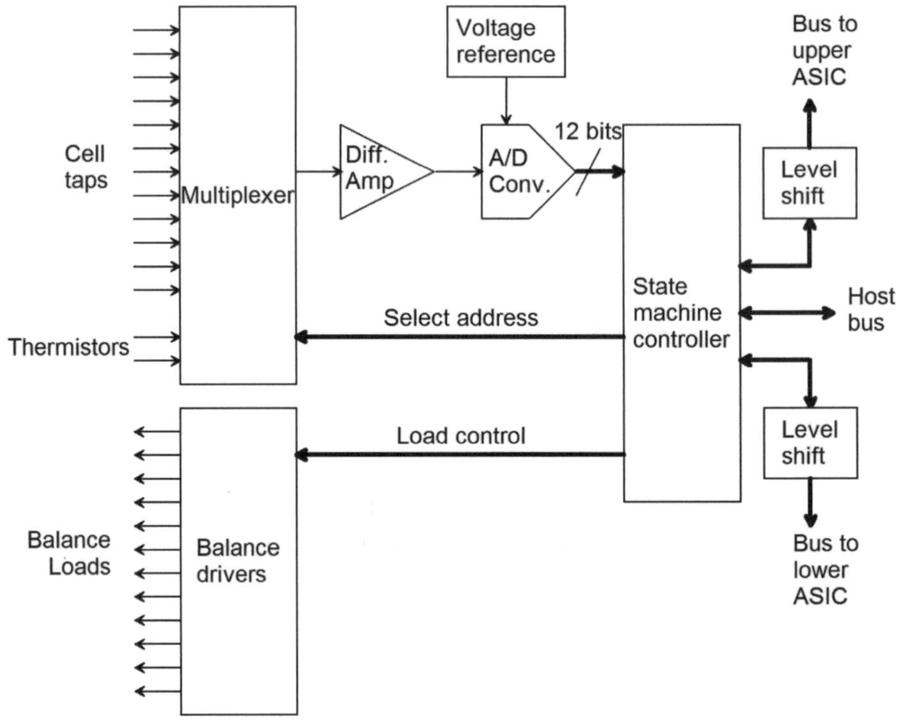

Figure 5.34 Block diagram of the internal functions of a typical BMS front-end ASIC.

- A precision A/D converter or one converter per cell;
- A voltage reference;
- Drivers for balance loads for those cells;
- Digital interface to a micro;
- Daisy-chain link between adjacent ICs to expand the system by cascading multiple chips.

Using one of these BMS front-end ASICs has certain implications:

- It is your responsibility to design a BMS controller (hardware and software), just as it is if using generic ICs.
- They are ideally suited for centralized and master/slave topologies. Therefore, "spaghetti" wiring through the battery poses a risk of short circuits and plasma events.
- If used in a master/slave topology, direct communication between adjacent slaves may work well on the bench, but may fail due to electrical noise in real-world applications (from devices such as chargers and motor controllers); there may nothing you can do about it (with generic IC you may be able to change the design to solve such problems).
- If the first or the last tap from the cells (the ones that provide power to the chip) open up, the chip may be destroyed.

- Even if the manufacturer implies that only a few external components are required, in reality many additional components are needed for protection, noise filtering, and balancing at practical levels, reducing significantly the perceived cost advantage of ASICs over generic ICs.

Table 5.5 compares the ICs described in this section. Note that:

- Recommendations are based on parts' performance in a large battery pack and availability at time of publication.
- Using shared pins to drive balancing reduces the pin count, but there are severe limitations on balancing. For example, adjacent loads cannot be turned on simultaneously.
- Better measurement accuracy improves the ability to determine cell SOC from its open circuit voltage. A lower number of wires in the link between adjacent ICs lowers system cost and improves reliability. Capacitive coupling in the link is not reliable in the presence of noise, and opto-isolators increase cost, leaving current sources as the best solution.
- Costs are in large quantity. An IC with a lower cost per cell is not necessarily the least expensive solution, as some ICs require many more external components than others.
- Not shown is ACTEL's A40MX02-PL537: an FPGA customized for BMS operation.

A detailed look at each BMS front-end ASIC follows.

Analog Devices' AD7280 Analog Devices' AD7280 is a BMS front end for four to six cells in series. Up to 50 ICs can be connected directly, for up to 300 cells in series. Communication uses a 7-wire daisy chain between adjacent cells, using current sources. Cell balancing, using external components, is driven by dedicated pins for maximum versatility. Each IC can read six temperature sensors, one per cell. A disadvantage is that a relatively high number of daisy-chain wires are required between adjacent ICs.

An 18-cell BMS (Figure 5.35) uses three AD7280 ICs. A diode on each communication line offers protection against damage caused by an unexpected open in the power wiring between cells while under load. This IC is not recommended for large pack designs due to very poor availability.

Atmel's ATA6870 The latest entry in the field is from Atmel. Its ATA6870 handles up to six Li-Ion cells in series. Up to 16 ICs can be connected directly, for up to 96 cells in series. Communication and control uses an eight-wire daisy chain between adjacent cells. Cell balancing, using external components, is driven by dedicated pins for maximum versatility. This ASIC is unique in that it uses a discrete topology: it has six independent acquisition blocks, one for each cell, which allows simultaneous readings of all the cells with excellent noise immunity and common mode noise rejection. Other than that, this IC is very similar to the Analog Devices' AD7280, described above.

Table 5.5 Comparison of BMS Front-End ASICs and Their Applicability to Large Packs

Brand	*Analog Devices*	*Atmel*	*Intersil*	*Linear Tech*	*Maxim*	*O2Micro*	*Texas Instruments*		
Model	AD7280	ATA6870	ISL9216+ ISL9217	LTC6802-1	LTC6802-2	MAX11068	OZ890	bq76pl536	bq76pl537
Recommended		✔		✔✔	✔✔			✔	
Cells:IC	6:1	6:1	12:2	12:1	12:1	13:1	6:1		
Cells/Bank	300	96	12	192	12	208	372	192	
Passive Balancing	Dedicated pins	Dedicated pins	Dedicated pins External resistor 200 mA	Dedicated pins Internal balancing	Shared pins External resistor	Dedicated pins	Dedicated pins External transistor and resistor	Dedicated pins Active External MOSFETsLC tank	
Accuracy @ 3.6 V	10 [mV]	7 [mV]	58 [mV]	8 [mV]	15 [mV]	—	3 [mV]		
Temperature Sensors:Cells	1:1	2:6	1:12	2:12	2:12	—	2:6		
Communication Between Adjacent Boards	7 wires Current sources	8 wires Current sources	N/A	3 wires Current sources	N/A	4 wires Voltage source	3 wires Opto-isolators	8 wires Current sources	
Communication to Master	SPI	SPI	I2C	SPI	I2C/SMB	I2C/SMB	SPI		
Cost/Cell, IC Only	—	$0.70	$0.60	$0.80	—	—	—	—	
Cost/Cell, Total	—	$1.30	$1.20	$2.50	$2.80	—	—	—	—
Availability	Dismal	Announced	Good	Very good	Very bad	Very bad	Announced		

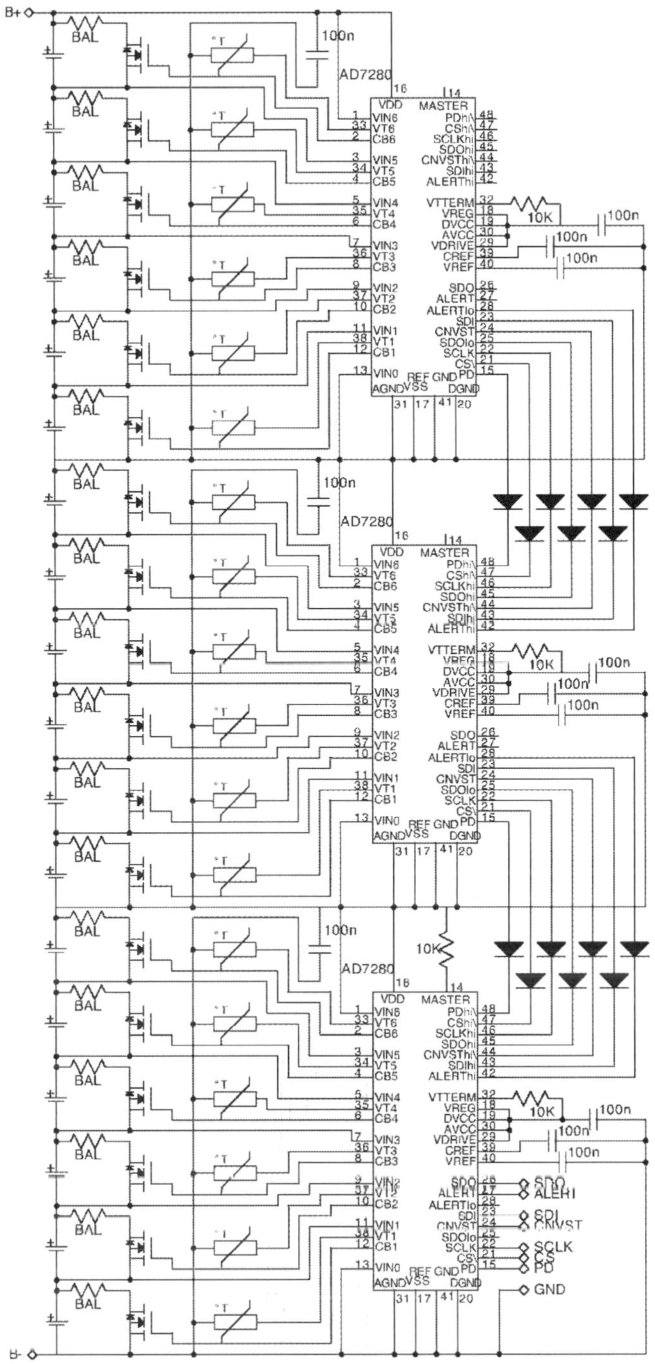

Figure 5.35 Analog Devices' AD7280 BMS ASIC in an 18 cell BMS.

An 18-cell BMS uses three ATA6870 ICs (Figure 5.36). The master controller enables the ICs through an opto-isolator to the top IC, and communicates to them through a serial port to the bottom IC. This IC is particularly recommended for

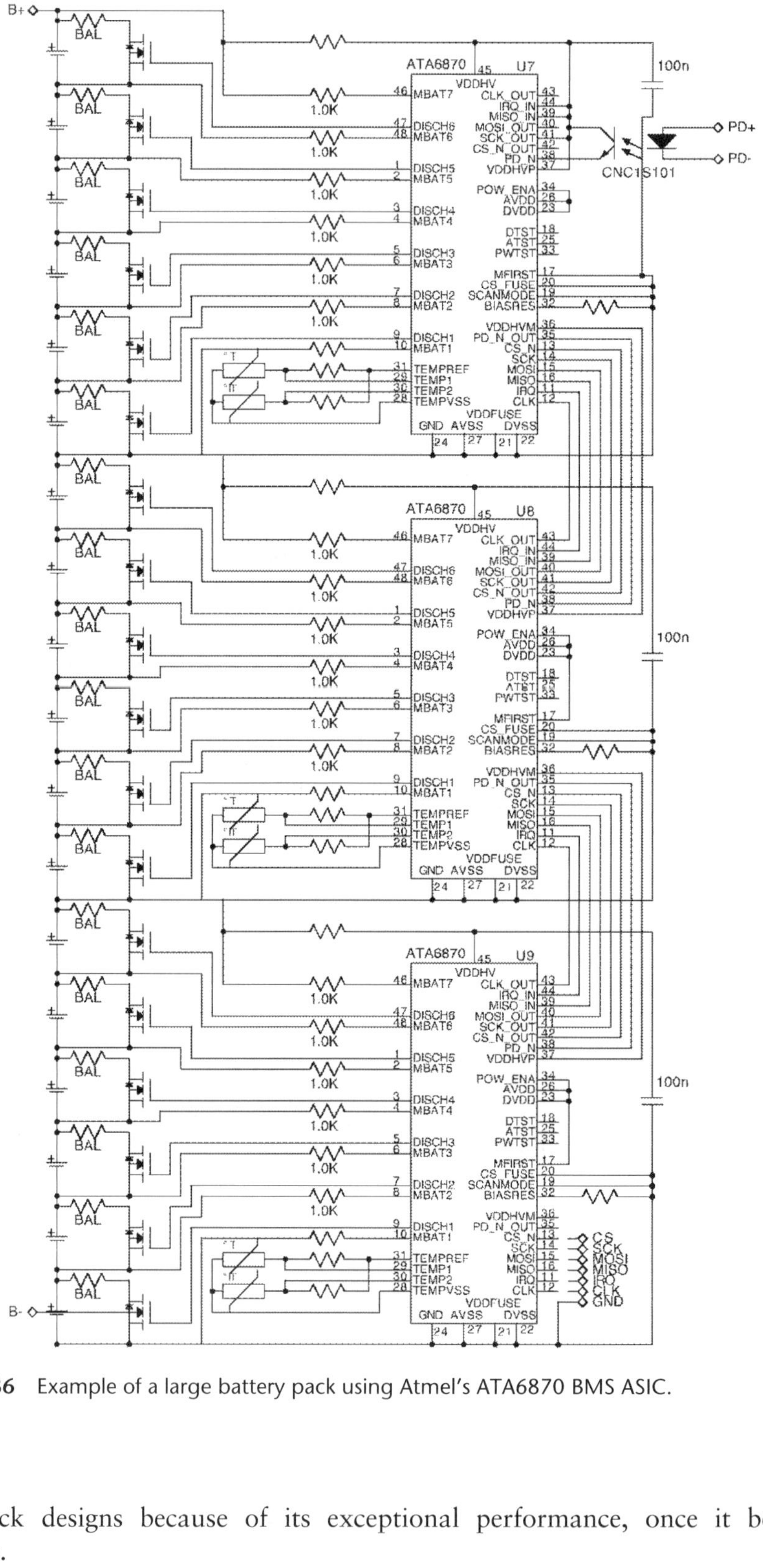

Figure 5.36 Example of a large battery pack using Atmel's ATA6870 BMS ASIC.

large pack designs because of its exceptional performance, once it becomes available.

Intersil's ISL9216/17 Intersil offers a few BMS analog front-end ASICs without A/D converters, including the ISL9208, ISL94200, and ISL94201, each of which can handle up to seven cells. The ISL9216 plus ISL9217 chip set can handle up to 12 cells in series, The ISL9216 handles the first five cells, and the ISL9217 handles an additional seven cells. This chip set features cell balancing, using external resistors (internal MOSFETs), and an SMB (I2C) serial interface. An accuracy of 58 mV is too poor for accurate SOC determination. A 12-cell protector based on these ICs uses an onboard current sense resistor and two MOSFETs, one for the charging current and one for the discharging current (Figure 5.37).

This chip set could be cascaded for use in a large battery pack, though the circuitry required would be extreme, because the analog voltage to be measured would have to be somehow level shifted between each set of ICs. It would make much more sense to make separate slave boards, each handling 12 cells, and each with its own microprocessor with an A/D converter and communication link to talk to a master controller (Figure 5.38).

This chip set is not recommended for large pack designs due to limited functionality and excessive number of external components.

Linear Technology's LTC6802-1 Linear Technology designed the LTC6802-1 specifically for HEV traction packs. It is identical to the LTC6802-2 (next section), except that it includes provisions for cascading multiple chips directly to each other.

Each chip handles up to 12 Li-Ion cells in series. Up to 16 ICs can be connected directly, for up to 192 cells in series. Communication uses a three—wire daisy chain between adjacent cells. This IC has dedicated pins for cell balancing for maximum versatility. Cell balancing uses external components (internal balancing is available as well, but of limited use due to heat dissipation limitations). Mixed internal/exter-

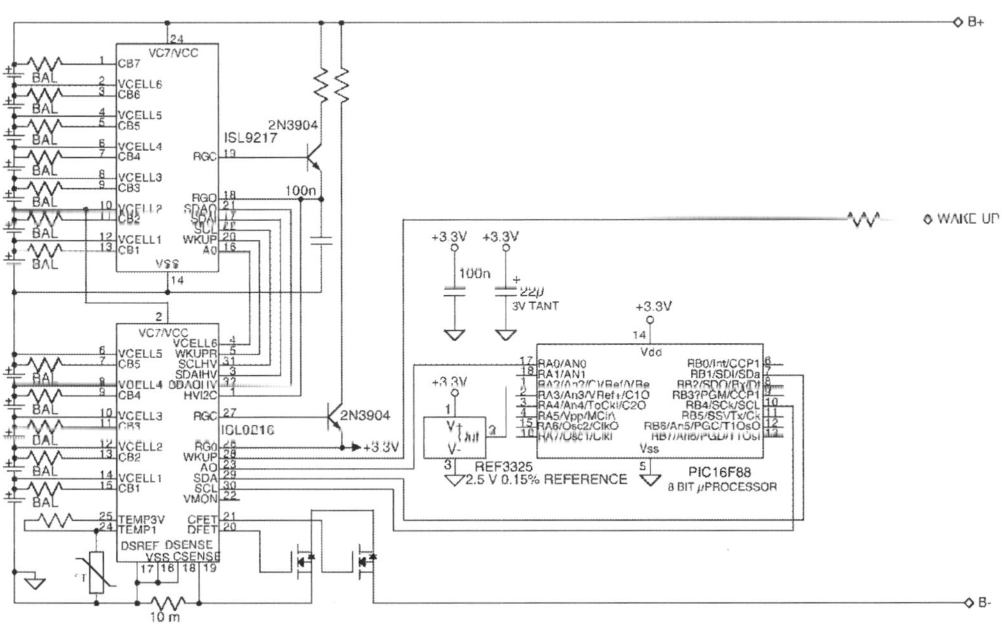

Figure 5.37 A 12-cell protector using Intersil's ISL9216/17 ICs.

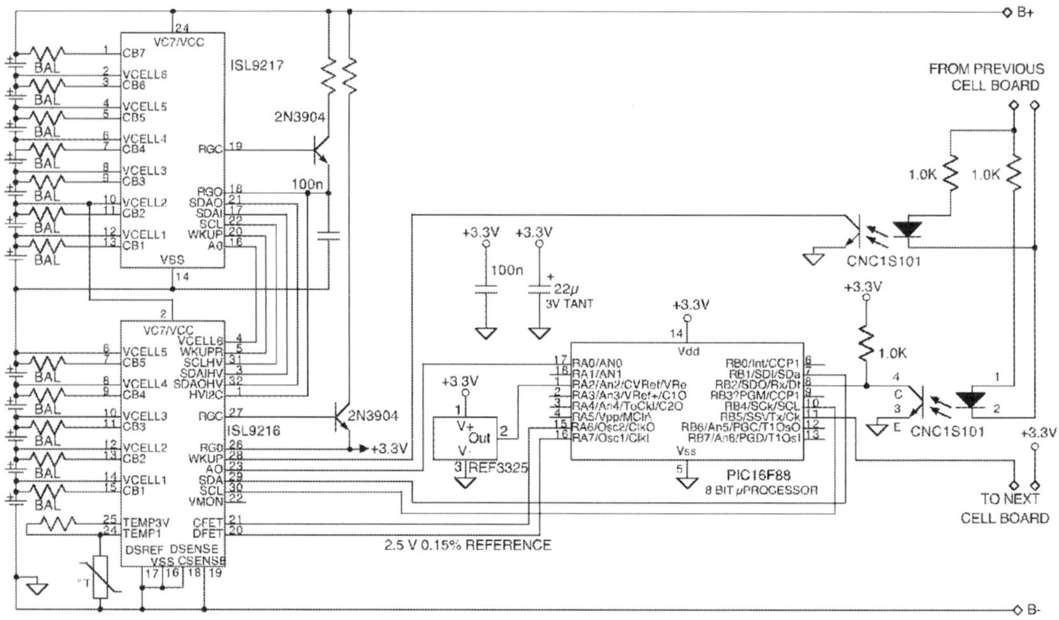

Figure 5.38 A 12-cell slave using Intersil's ISL9216/17 ICs.

nal balancing is possible as well, using just external resistors, for slight improvement on chip temperature.

A BMS for 36 cells (Figure 5.39) uses three LTC6802-1 ICs cascaded together. Three digital communication lines between adjacent ICs use current sources (as programmed by the VMODE pin). Diodes in these lines protect the IC against damage should a power connection between cells be opened under load. Balancing uses external resistor and the internal MOSFETs. Even though the spec sheet implies that very few additional parts are required, in reality six additional protection and balance components are required per cell. These components are not shown in this simplified schematic, but are shown in the schematic for the LTC6802-2, in the next section. Two thermistors measure the temperature at two points in each group of 12 cells. The bottom IC is programmed to use a voltage mode for its communication port, connected to a microprocessor that handles the overall BMS functionality and drives two optoisolated limit outputs. The micro requires its own supply, as the LTC6802-1 IC cannot provide enough current to power it.

This IC is recommended for large pack designs, especially because of its performance and availability. A companion IC, the LTC6801, is available to add an independent, redundant test to ensure the cells are within their SOA. This IC can be also be used as the core of an analog monitor (Section 5.2.2.2).

Linear Technology's LTC6802-2 Linear Technology's LTC6802-2 is identical to the LTC6802-1 (previous section), except that it doesn't have provisions to cascade multiple chips directly. Instead, it has an SPI port for direct connection to a micro. This IC is better suited for BMSs that are exposed to levels of EMI that the

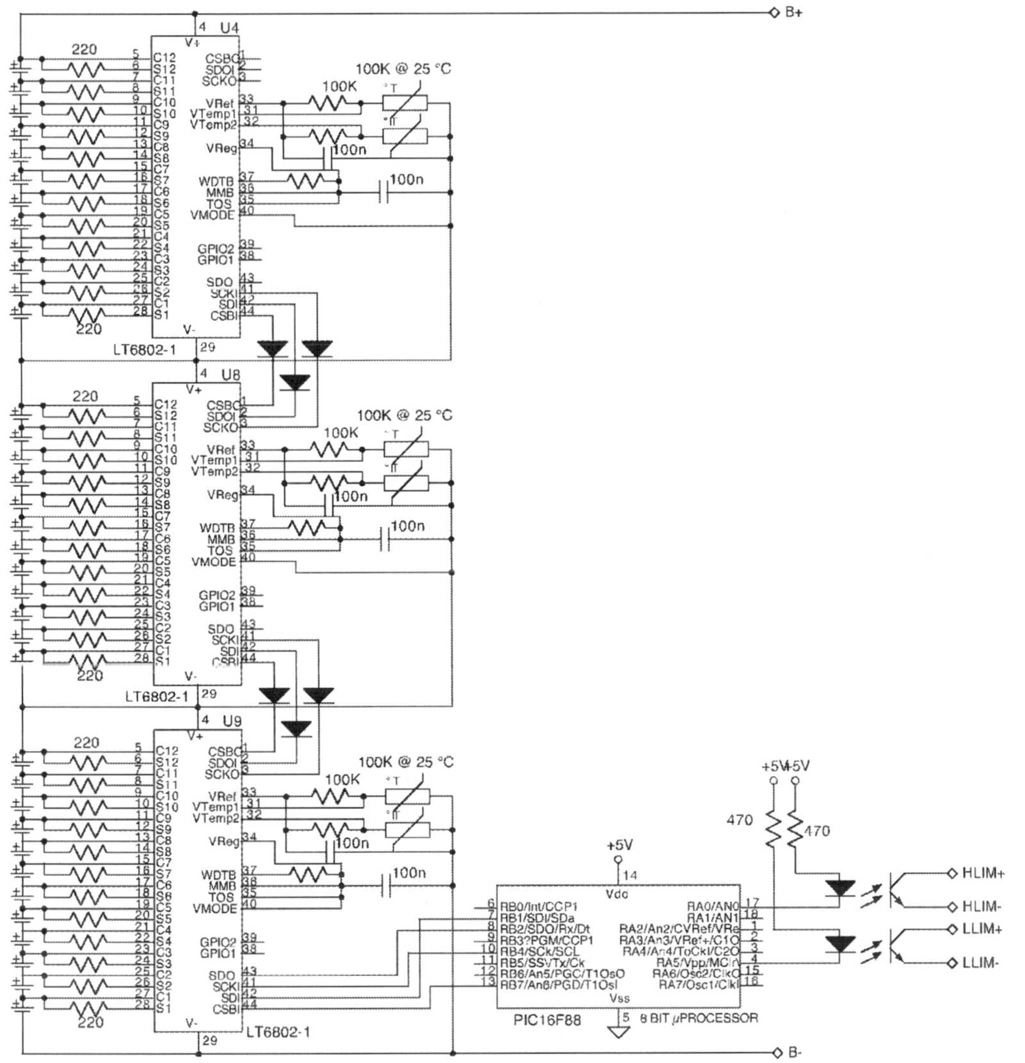

Figure 5.39 Example of a 36-cell BMS using Linear Technology's LTC6802-1 ICs (simplified schematic).

LTC6802-1 may not be able to handle, because there is only one reference: the most negative cell. It can still be used in large battery packs:

- Without a micro: each IC needs its own SPI isolator, and a common SPI bus is wired to all the 12-cell circuits; separate lines from the master controller to each slave enable communications to that particular slave.
- With micros: each IC has its own micro to translate between the IC's SPI bus and an external bus common to all 12-cell circuits.

This IC may be used in a 12-cell slave (Figure 5.40). This schematic does show all the external protection and balance components (unlike all the other examples in this section, which show simplified schematics): ferrite impedances, capacitors,

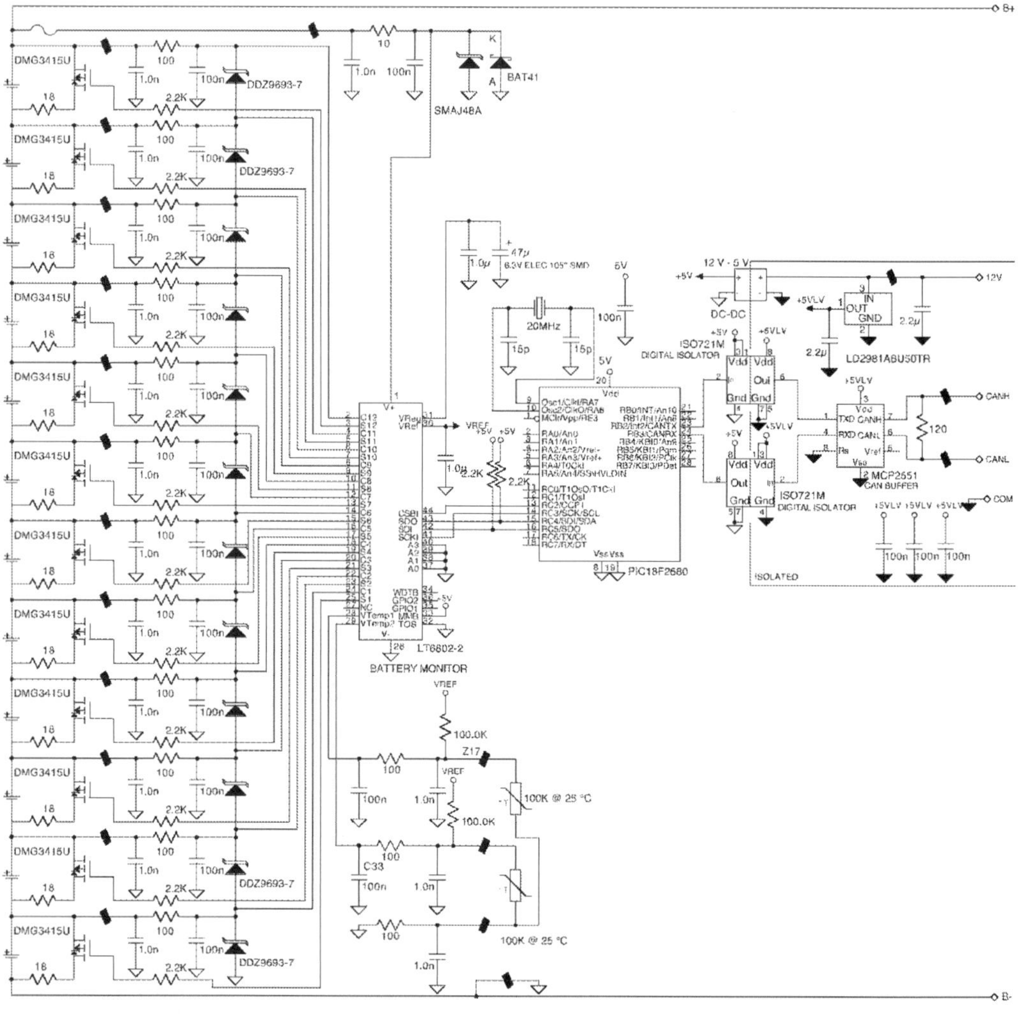

Figure 5.40 Example of a 12-cell slave with CAN bus interface and all external components, using Linear Technology's LTC6802-2.

resistors, and TVS diodes are used to add protection (supplementing the protection built into the IC) and noise filtering. Even though the IC could balance the cells on its own, it could only do so at low power to avoid overheating; instead, external MOSFETs and power resistors are used for balancing at a higher current. A microprocessor interfaces to the LTC6802-2 SPI bus and to a CAN bus through digital isolators and a CAN buffer IC. A small DC-DC converter provides an isolated 5-V supply to the micro. The bus has four lines: two CAN lines, a ground, and a 12-V supply.

A master controller and a set of slaves can then be connected together through the CAN bus (Figure 5.41). Each slave must be somehow programmed with its own ID, to differentiate the various slaves. This IC is recommended for large pack designs, especially because of its performance and availability.

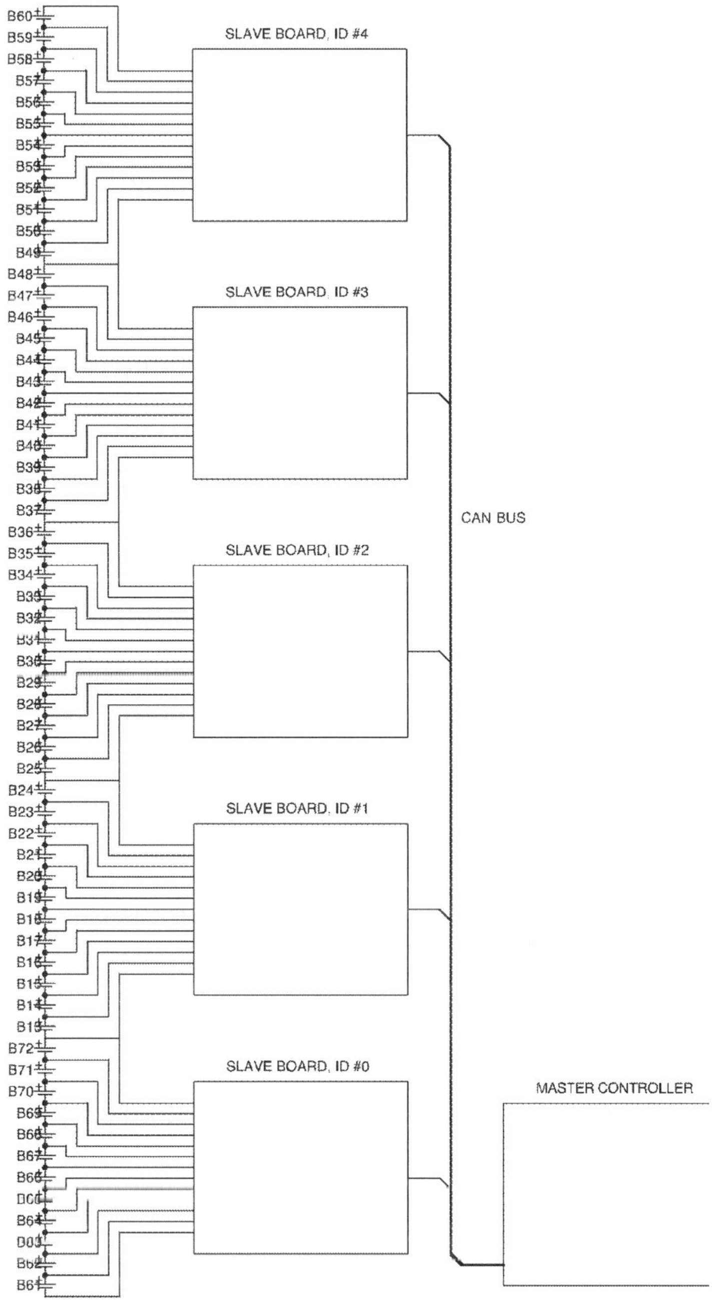

Figure 5.41 Block diagram of a BMS for a large pack with slaves connected through the CAN bus.

Maxim's MAX11068 Even though Maxim announced its MAX11068 ASIC in 2008, 2 years later there is no stock available (though DigiKey lists it at $204 each), and it is quite difficult to get a data sheet. While comparable to Linear Technology's 6802-1, the Maxim part uses shared pins for measurement and balancing (which limits the flexibility in balancing), and its communication ports to the next cascaded IC are level-shifted voltages (which works in the lab, but is susceptible to electrical noise in real world applications). Both are inferior solutions that limit this IC's performance.

A 36-cell BMS (Figure 5.42) uses three cascaded ICs, each handling 12 cells. Resistors in series to each cell tap are used for balancing as well as for filtering. Capacitors between adjacent ICs level-shift the communication lines and offer a bit of protection in case of an unexpected open in the power connection between groups of cells. The bottom IC communicates with a microprocessor that controls the three ASICs, handles the overall BMS functionality, and drives two opto-isolated limit

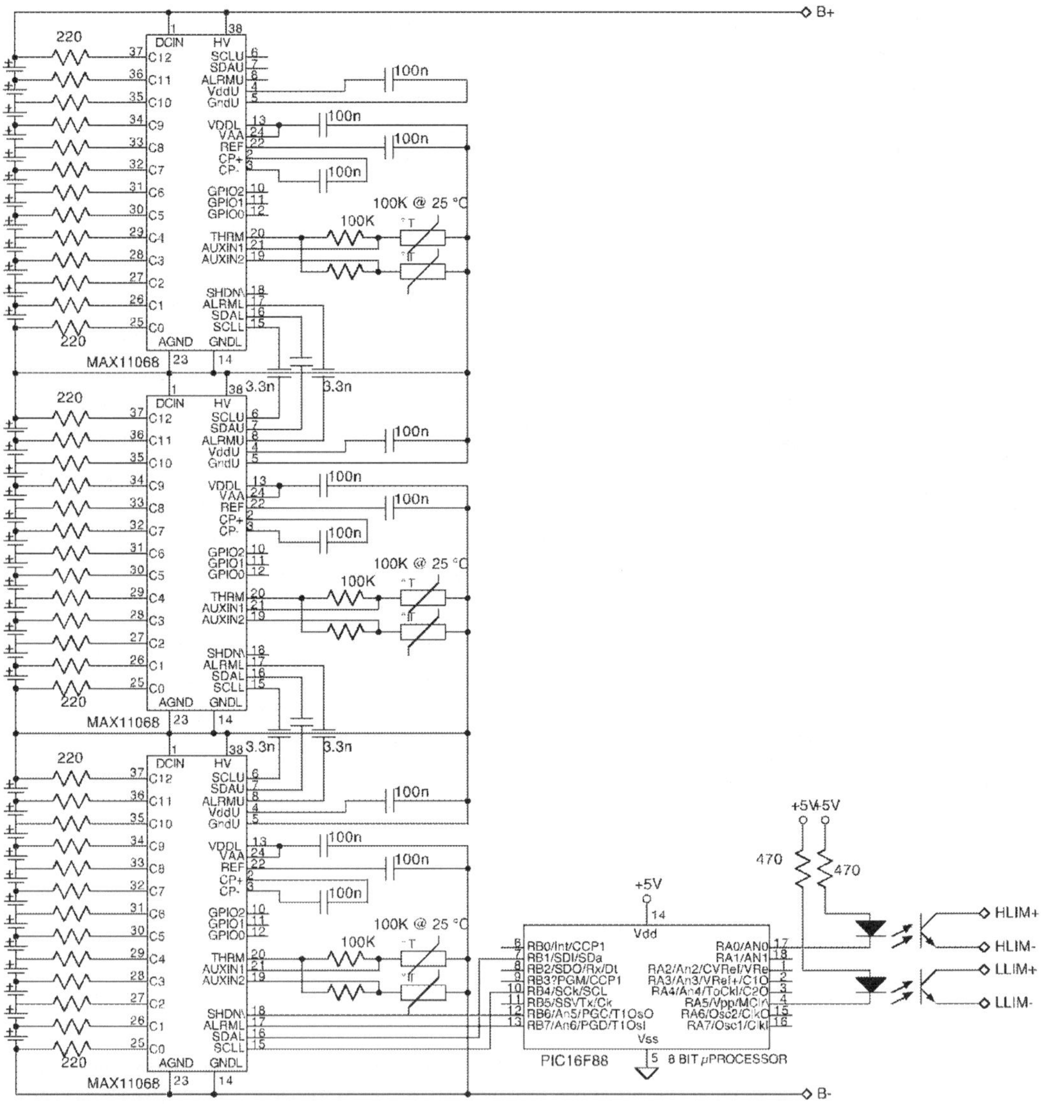

Figure 5.42 Maxim's MAX11068 in a 36-cell BMS.

outputs. This IC is not recommended for large pack designs due to its limited functionality and poor availability.

O2Micro's OZ890 O2Micro is rather secretive about its OZ890 BMS ASIC, which is used in some Chinese, inexpensive protectors for small batteries. Each IC can handle 13 cells in series (more than any other IC), and up to 32 ICs can be cascaded, for packs of up to 372 cells in series (why not 416 cells?). The link between adjacent ICs uses three lines, which require opto-isolators. The bus to the master is SMB. Balancing uses dedicated pins, for best performance.

While this IC's performance appears to be appropriate for large pack designs, it is not recommended due to its poor availability.

Texas Instruments' bq76pl536 Texas Instruments is a leader in ICs for small batteries (on the order of four cells), and, until recently, encouraged its clients to try to shoehorn those ICs into BMSs for large battery packs, with dismal results. Today I can say that TI finally "gets it," as it has announced two ICs for large packs, one of which, the bq76pl536, is indeed appropriate for large packs.

This IC handles between three and six cells, and is stackable for packs with up to 192 cells in series. Each IC has three interfaces, one for the next IC up (North), one for the next IC down (South) and one for the master controller (Host). (At most two of these ports are used in any given IC, always resulting in an unused port; the LTC6802-1 is better in this regard. It uses a single port that doubles as Host or South, depending on the setting of a Mode pin.) Each port has eight wires, which, in a distributed system, can be quite a lot wires going from cell board to cell board (by way of comparison, the LTC6802-1 uses only three lines). The links between adjacent ICs use direct connection (no isolators, diodes, or capacitors, unlike other ICs). The interface to the master controller, which includes an SPI interface, does need isolation. However, as each line is unidirectional, that isolation is easy to achieve with opto-isolators. Thanks to a 14-bit A/D, the accuracy of the measurement is excellent: +/-3 mV max. The IC includes an independent set of comparators to provide redundant protection. Though, whether it can be considered truly independent is debatable, given that it is contained in the same IC. This IC may be operated in standalone, to form the core of a simple BMS (Section 5.2.3.3).

A sample application of this IC is in an 18-cell BMS (Figure 5.43). External MOSFETs and power resistors balance the cells. Communication between adjacent ICs uses eight direct wires. The Host port of the lowest IC is connected to a microprocessor that handles the overall BMS functionality, and drives two opto-isolated limit outputs. This IC is recommended for large pack designs because of their performance, once it becomes available.

Texas Instruments' bq76pl537 The bq76pl537 is similar to the bq76pl536, except that it has provisions for active balancing, using TI's charge pump technology described earlier (Section 5.3.6). With the limited information available at this point, I have reason to believe that this IC suffers from the same limitations as the bq78PL114: its active balance topology cannot balance between cells that are not handled by the same IC. If so, this IC is not recommended for large pack designs because of its inability to balance the entire pack.

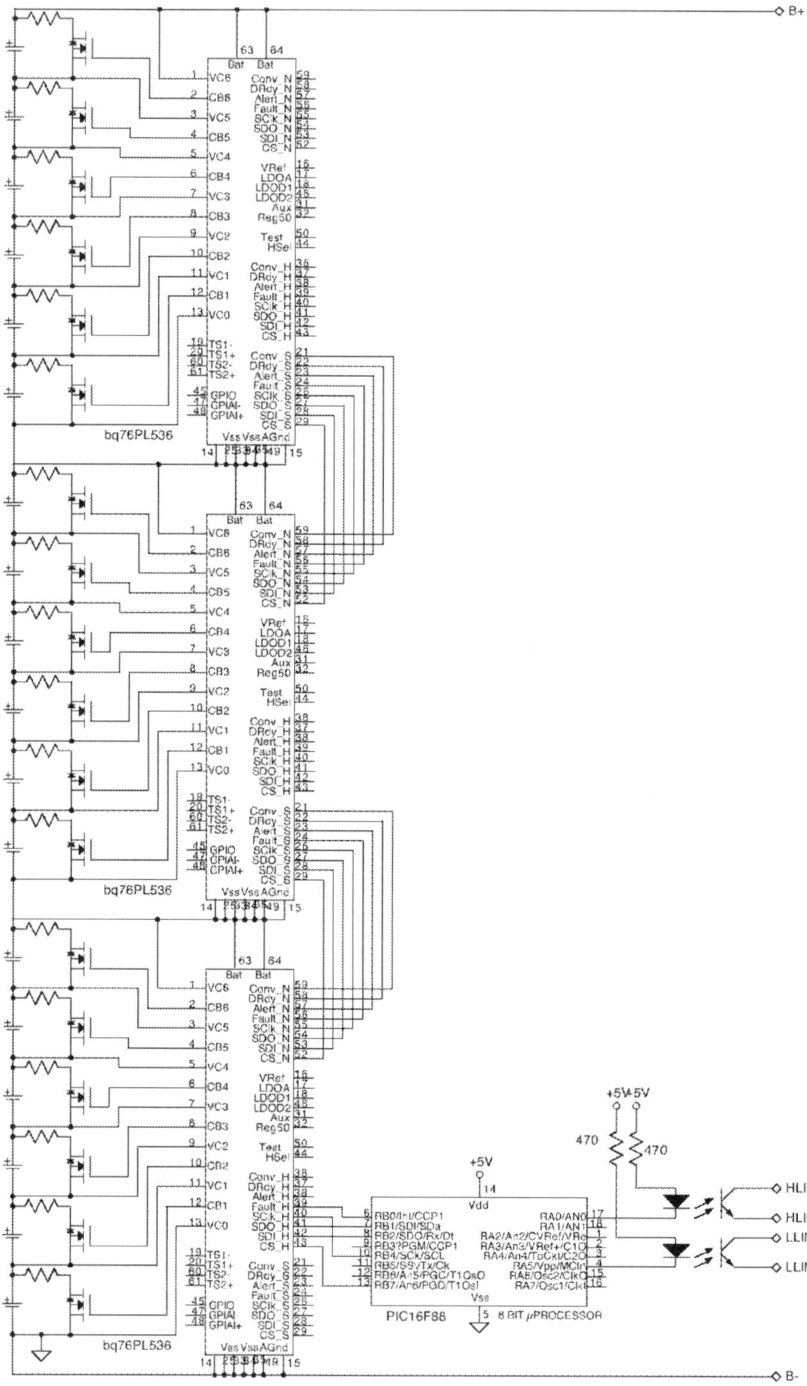

Figure 5.43 Example of an 18-cell BMS using Texas Instruments' bq76pl536.

Redundant Protection ASICs

Various manufacturers offer Fault Monitor ICs that can be used in conjunction with their BMS ICs to provide additional, redundant protection, should the primary BMS fail. We already saw two of these ICs in Section 5.2.2.2, in which we used them as the core of analog monitors: Linear Technology's LTC6801 and Maxim's MAX11080. Whether these ICs use analog comparators, or measure the cell voltages and compare them digitally to predefined limits, from the point of the user they behave the same. Should any cell voltage go beyond a specified range, they set an alarm, which can be used to shut down the battery pack. Other manufacturers include redundant protection directly inside a BMS ASIC (such as the bq76pl536). This may be viewed as advantage (simpler circuits, using fewer parts), or a disadvantage (a damaged IC or a broken trace will affect both functions in the ICs, and therefore defeat the intention of redundancy).

5.4.2 Current Measurement

We saw (Section 3.1.3) how the BMS benefits from knowing the battery current, and how it is often beneficial to measure the source current and the load current independently. We also saw that, unless other devices in the system are able to report the current to the BMS, the BMS must include one or more current sensors to measure the battery current. Current in AC circuits may be measured very accurately by current transformers, but BMSs are fundamentally interested in DC, so they cannot use current transformers to measure the current. That leaves the DC current sensor: current shunts and Hall effect sensors.

5.4.2.1 Current Shunt

Calculating current through a resistive element is done by measuring the voltage drop across it and, knowing its resistance, converting it to current (Section 3.1.3.1). For low currents (up to 20A) in PCB assemblies, various resistive elements can be used, including:

- PCB traces, with marginal accuracy due to large variations in the copper weight deposited on the PCB, and large errors due to self heating which affects the resistivity of copper;
- Low value, current sense resistors (typically used in switching power supplies);
- Solid wire loops (typically used in DVMs with a 10-A or 20-A current range).

For higher currents, larger resistive elements are used, including:

- Chassis-mounted, precision shunts (typically seen as part of an analog ammeter);
- Power wires (cables), with marginal accuracy, though better accuracy than for PCB traces.

To minimize losses (reduce the heat generated), current sensing uses a resistive element with the lowest practical resistance. A typical ammeter shunt for 500A has a resistance of 100 $\mu\Omega$, while a typical wire loop for a DVM has a resistance of 10 mΩ.

Because of the small voltage generated (on the order of 100 μV to 10 mV full scale), the signal from the resistive element must either be amplified locally, or be transferred to an amplifier using shielded, twisted pair wiring. The amplifier must be used in a differential topology, to produce an output voltage that is proportional to the difference in the voltage of its two inputs (Figure 5.44).

While the current shunt does not suffer from offset issues, its amplifier does. In order to minimize errors in the DOD calculations, an amplifier with minimal offset is required. Chopper stabilized op-amps are ideal in this application, because they feature offsets as low as 500 nV, and their relatively low frequency response is not a problem, given that the BMS samples the current no more often that once every 10 ms. Linear Technology is a major supplier of precision, chopper stabilized op-amps, though you may want to consider also Analog Devices, National Semiconductors, and Texas Instruments. Some part numbers to consider include: LTC2050, LTC2054, MCP6V01, LMP2021MFE, OPA333, and OPA734.

For a bidirectional current measurement, the amplifier needs to include an offset, such that, at 0 current, the output will be at the midpoint of the A/D converter's range (typically 2.5V). Of course, the offset could be skewed in one direction or the other if the current sensed is not symmetrical with respect to 0, so that a wider portion of the A/D input range is used.

To take advantage of the entire range of the A/D input, an amplifier with a rail-to-rail output should be used. If the amplifier is not able to generate 0V on its output, you may add a small offset to the amplifier, so that its output at 0 current is some known voltage above 0; then the offset can be subtracted out from the A/D reading in software. To maximize the use of the A/D converter, and achieve the best possible resolution, the gain of the amplifier must be such that at the peak current the output will reach a rail, without clipping.

5.4.2.2 Hall Effect Sensor

Hall effect sensors are easier to use as they are isolated and they already include an amplifier, so their output can be connected to an A/D converter, either directly, if their output is unipolar [Figure 5.45(a, b)] or through a resistive divider to bring its

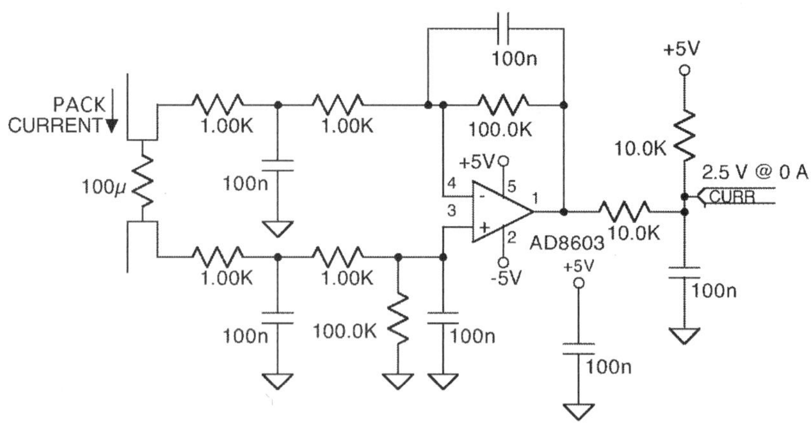

Figure 5.44 Shunt current sensor with chopper stabilized amplifier.

voltage to 2.5V when there is no pack current, if their output is bipolar [Figure 5.45(c)].

Hall effect sensors have an accuracy on the order of 1%. Error sources include initial offset, thermal drift, nonlinearity, and magnetic hysteresis (after a large current in one direction, the zero is affected).

Most Hall effect current sensors are either open loop or closed loop. In an open loop current sensor, the current produces a magnetic field, which is coupled to a linear Hall effect IC through a ferrite component. Inside the Hall effect IC, a Hall element produces a tiny voltage that is proportional to the magnetic field, and hence to the current. The Hall effect IC amplifies that voltage, attempting to compensate for errors, and filters out high frequency noise. The output of the IC is a voltage that is proportional to the sensed current.

A closed loop current sensor is similar, except that there is an additional coil (winding) with N turns wound on the ferrite component. The current sensor drives that coil with a low current such that the field from that coil is equal and opposite to the field from the sensed current, attempting to keep the output of the Hall effect sensor's element at 0, at which point the current in the coil will be $1/N$ of the current in the wire. The current sensor outputs a voltage that is proportional to the current in its coil, and hence to the sensed current. Closed loop sensors are more accurate, because the Hall sensor is operated within a narrow range, but are more expensive and draw more current (to power their coil).

To achieve a high accuracy over a wide range, two current sensors may be used measuring the same current—one sensor for low current and one for high current. Their outputs are sent to a processor with an A/D converter, to select the appropriate sensor based on the level of the current measured.

Various types of Hall effect current sensors and Hall effect IC are available.

- Chassis mounted, toroidal current sensor modules for power wires or bus bars.
- PCB mounted current sensor modules for power wire. They are mounted flush to the PCB, and the power wire goes through the sensor and through the PCB.

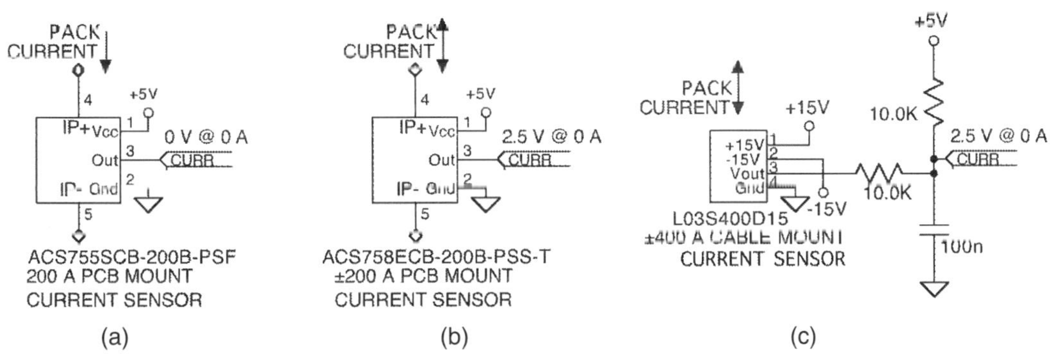

Figure 5.45 Hall effect current sensor circuits: (a) unipolar unidirectional, (b) unipolar bidirectional, (c) bipolar bidirectional.

- PCB mounted current sensor modules for bus bars. They are mounted vertically on the PCB, and a bus bar, parallel to the PCB, goes through the opening in the sensor.
- PCB mounted sensor modules with an integral bus bar that it is U-shaped, so that is can be soldered to the PCB.
- Integrated circuits that include within them a high current conductor. They are available from Allegro (ACS series), either in a small IC package (such as the surface-mounted SOIC-8 package) for currents up to 30A, or a larger package with two large tabs for the current, for currents up to 200A.
- Bulk Hall effect ICs with a linear output that sense magnetic fields, useful for custom-built current sensors. For example, they can be placed them in a gap in a ferrite toroid, through which a power wire is routed.

Manufacturers of Hall Effect current sensors include Allegro, LEM, Tamura, Honeywell, and CUI.

5.4.3 Evaluation

After measurement, the next BMS function is the evaluation of the state of the pack. Most digital BMSs implement this function, to a greater or lesser degree. This is purely a software function. Evaluation may include any of the following:

- Evaluation of cell operation within their SOA;
- SOC and DOD estimation;
- Resistance calculation;
- Capacity measurement;
- SOH estimation.

5.4.3.1 SOA Evaluation

This function evaluates whether each cell is operating within its safe operating area (SOA). If any cell is close to the edge of its SOA, the BMS reduces the charge current limit (CCL) or discharge current limit (DCL) below the maximum. If outside the SOA, the BMS reduces the CCL or DCL down to 0, and asserts the high limit (HLIM) or low limit (LLIM).

Not all BMSs implement a CCL and DCL, but most implement a HLIM and LLIM, though they may be called something else, such as high-voltage limit (HVL) or high-voltage cutoff (HVC), and low-voltage limit (LVL) or low-voltage cutoff (LVC). Protectors do not need to implement these limits.

CCL and HLIM

BMSs determine CCL and HLIM based on one or more of the following:

- Voltage of the most charged cell;
- Maximum and minimum cell temperature for charging;
- Value of the charging current;

• Total voltage of the pack.

Each of the above limits can have two thresholds: one at which limiting starts and one at which the limit is complete. For example, charging may be unlimited up to 50°C, proportionally limited between 50°C and 60°C, and completely disabled at 60°C. The CCL is reduced based on whichever of the above items is the limiting factor. If any of the above parameters is outside the SOA, the HLIM is asserted.

Maximum Cell Voltage The BMS reduces the CCL if the highest cell voltage exceeds a high threshold, reducing it gradually as that voltage, increases, until the voltage reaches a max voltage, at which point the CCL is 0 and the HLIM is asserted (Section 3.2.1.2). Hysteresis may be added to the HLIM, by deasserting it only when the highest cell voltage returns to below the high threshold (Figure 5.46). For the purposes of this evaluation, the voltage of the most charged cell may be used directly. Doing so may result in charging being limited prematurely because of a sudden peak in charging current (e.g., due to regenerative braking).

To avoid premature limiting, the cell voltage may be time averaged (Figure 5.47), to delay a reaction to a temporary peak and allow reaction only if the peak lasts for a while, or if it is of excessive amplitude. An averaging time constant

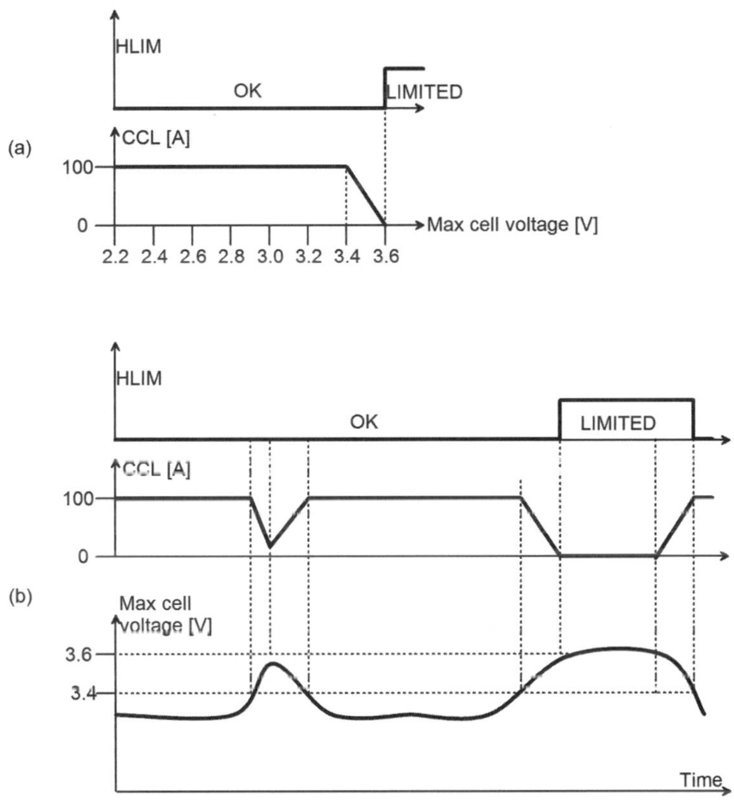

Figure 5.46 CCL and HLIM based on voltage of most charged $LiFePO_4$ cell: (a) transfer function and (b) plot.

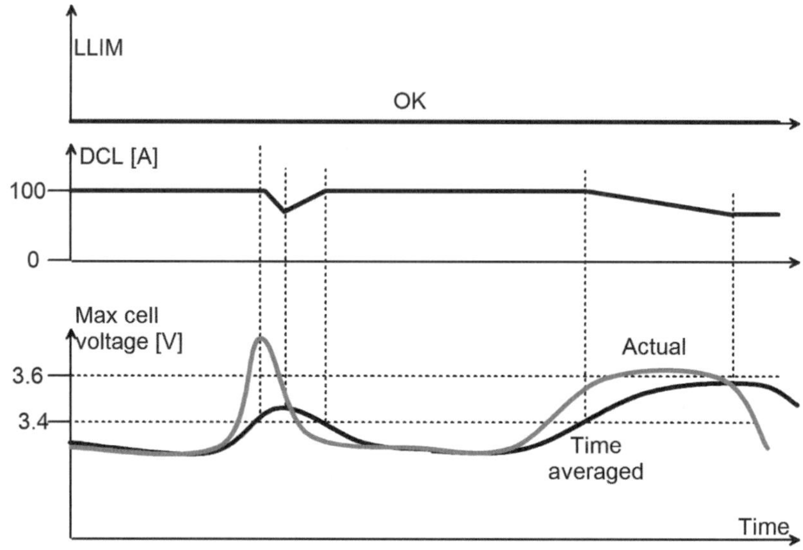

Figure 5.47 Time averaging of the maximum cell voltage can prevent premature limiting.

of 10 to 30 seconds is typical. Averaging is done with software, normally with a standard infinite impulse response (IIR) algorithm, which behaves the same as an RC low pass filter.

Another way to avoid premature limiting id for the BMS to estimate the cell's OCV by compensating the cell terminal voltage based on its model (Section 1.2.7). When using the cell's estimated OCV instead of the terminal voltage, charging is more rapid.

If the cell resistance and the pack current are known, the BMS can use the simplest DC model to get the estimated OCV by calculating the IR drop across the cell resistance (this voltage is negative, because the charging current is negative by definition), and adding it to the cell voltage [Figure 5.48(a)].

$$\text{OCV} = \text{Terminal Voltage} + I * R$$

Using a model that includes relaxation, the BMS can calculate the drop across the cell AC impedance, and subtract it from the cell voltage, to get an even more accurate estimate of OCV [Figure 5.48(b)]. The BMS will then limit the CCL and activate the HLIM based on the highest estimated cell OCV.

Cell Temperature Li-Ion cells may only be charged within a certain range of temperatures (this range is smaller than for discharging). On a first level, the BMS disables charging if any cell is too hot or too cold, to keep it within its SOA (Section 1.2.4). If the source of the charging current allows fine control of charging current, the BMS may request a gradual decrease of charging current as the cell temperature reaches either end of the cell's SOA range (Figure 5.49).

Battery Charge Current If the BMS is aware of the charging current, it may activate the HLIM should the current become too high (Section 3.2.1.1) (Figure 5.50).

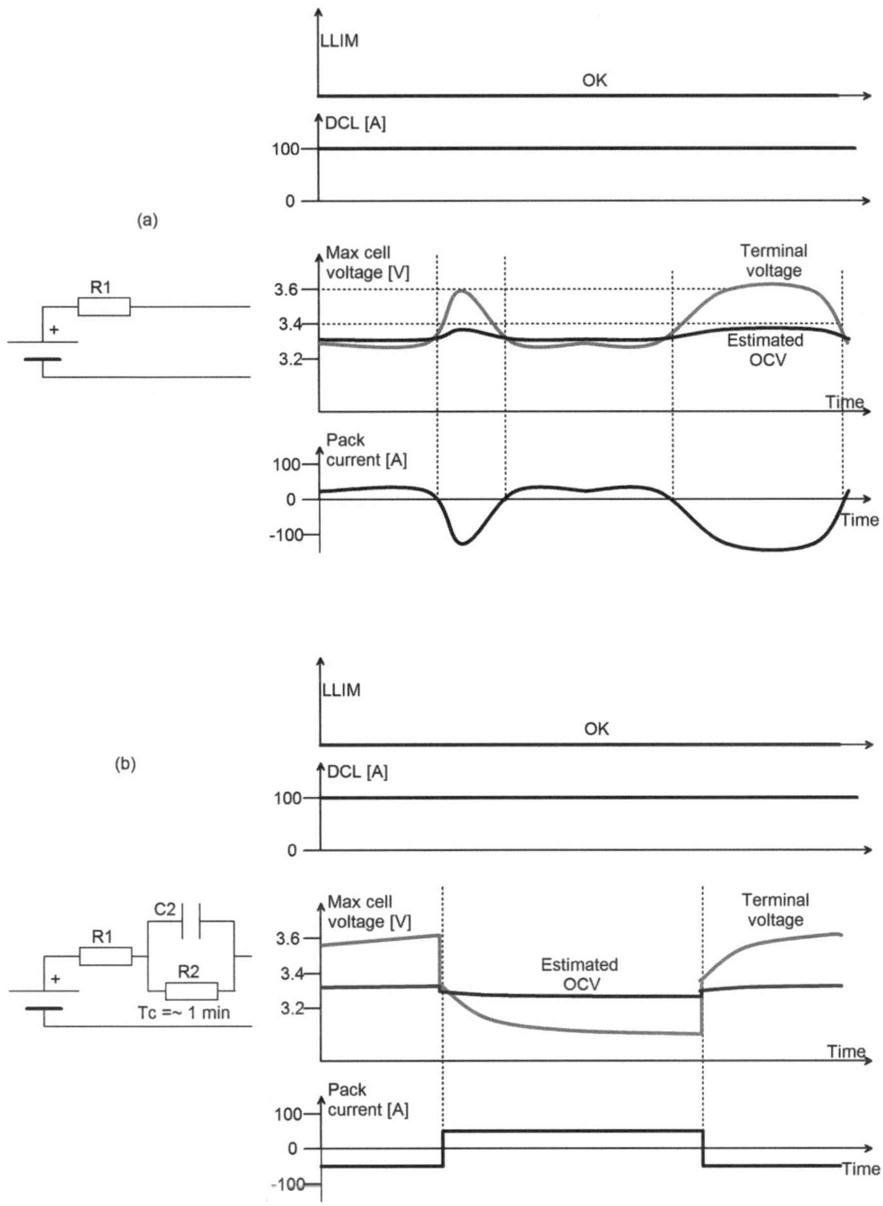

Figure 5.48 Using the cell model to estimate its OCV: (a) simple DC model and (b) model with relaxation.

A more sophisticated BMS may differentiate between a short pulse of current and continuous current, and reduce the CCL or even activate the HLIM should a current peak last too long. It can do so by integrating the excess current, and setting a limit if the integral exceeds a constant. That constant determines the response time to a pulse of current.

$$\text{Limit} = \int (\text{Actual Current} - \text{Continuous Limit}) / (\text{Peak Limit} - \text{Continuous Limit})\, dt > K$$

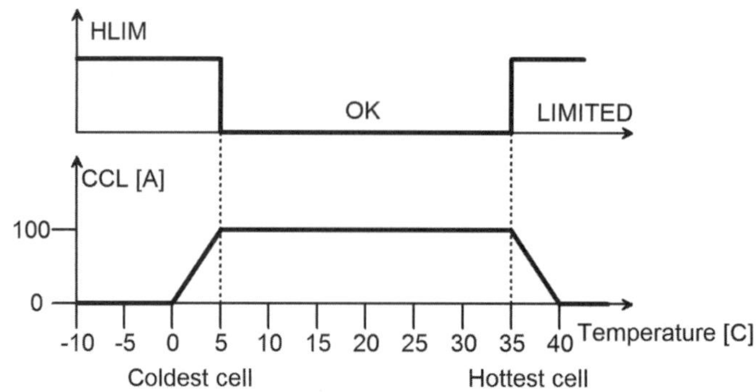

Figure 5.49 CCL and HLIM based on cell temperatures: coldest cell at the left, hottest cell at the right.

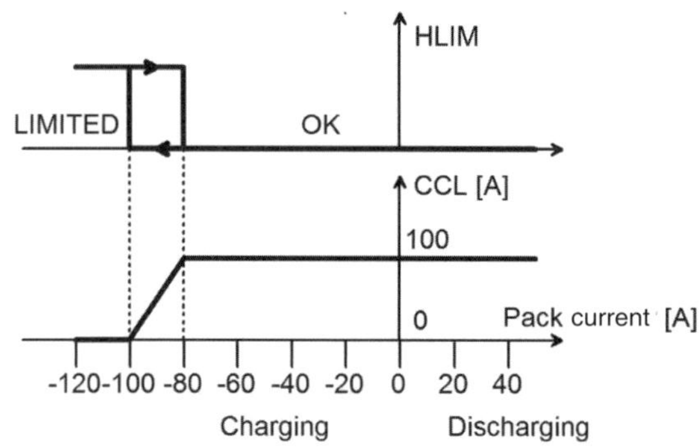

Figure 5.50 CCL and HLIM based on battery charge current.

Pack Voltage A BMS may monitor the pack voltage to prevent it from exceeding a certain range (Section 3.2.1.1). The BMS will lower the CCL as the limit is approached, and assert the HLIM when it is reached (Figure 5.51).

DCL and LLIM

BMSs determine DCL and LLIM based on one or more of the following:

- Voltage of the least-charged cell;
- Maximum and minimum cell temperature for discharging;
- Value of the discharging current;
- Total voltage of the pack.

The BMS determines DCL and LLIM is similar way as it determines CCL and HLIM, except that it uses the low cell voltage (Figure 5.52); the discharging temperature range, which is typically wider than for charging (Figure 5.53); the discharging current, which is typically higher than for charging (Figure 5.54); and the pack voltage (Figure 5.55).

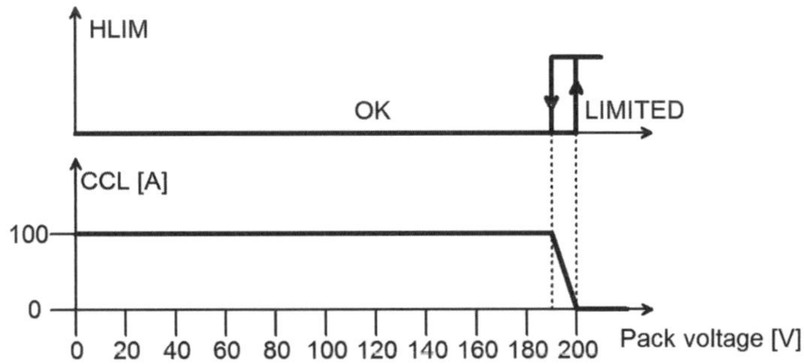

Figure 5.51 CCL and HLIM based on pack voltage.

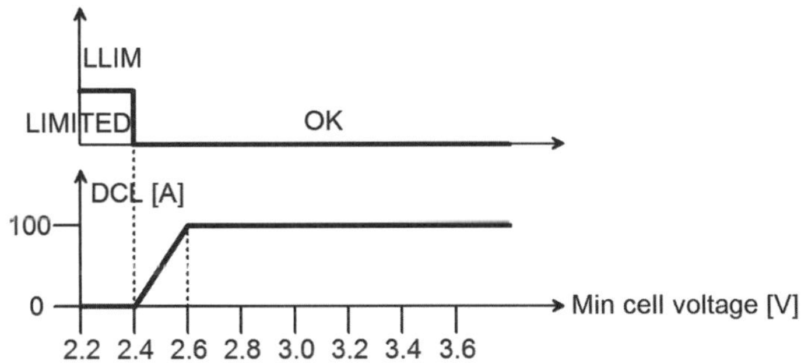

Figure 5.52 DCL and LLIM based on minimum cell voltage.

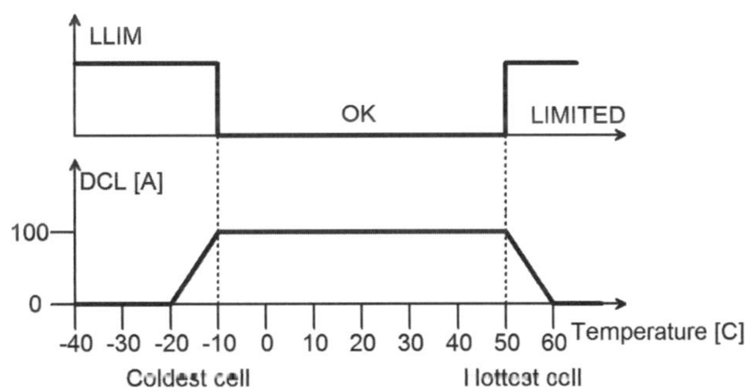

Figure 5.53 DCL and LLIM based on cell temperature.

5.4.3.2 SOC and DOD Estimation

Knowing the SOC and/or DOD may be useful to the user and to the external system, though it has no function within the BMS. Estimating DOD and SOC is probably the hardest task in the design of a BMS. No matter how good your algorithm is, it will never be sufficiently accurate when used in the real world. The fact that you are in good company (we all struggle with SOC estimation) may be of little consolation when you are dealing with an unhappy user of your BMS.

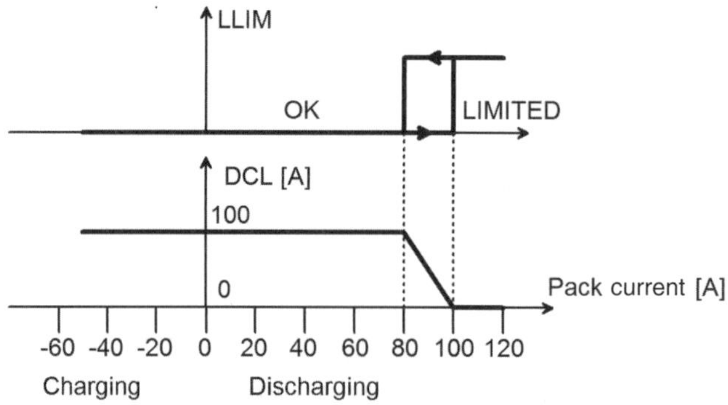

Figure 5.54 DCL and LLIM based on battery discharge current.

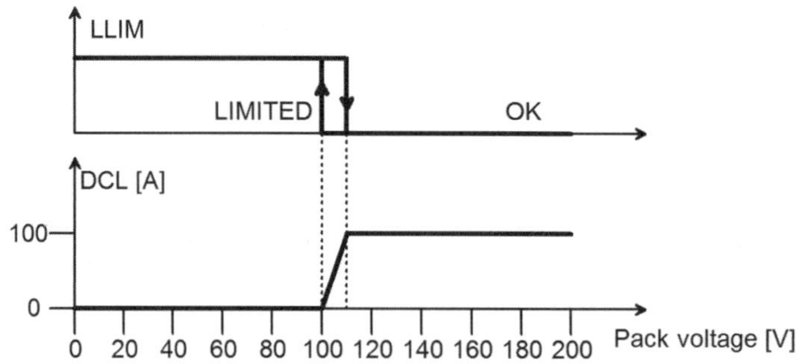

Figure 5.55 DCL and LLIM based on pack voltage.

If your main goal is to protect the battery pack, you will be relieved to know that SOC plays no part in managing the pack: the only function of SOC is to give guidance to the rest of the system and to the user. On the other side, a user is likely to be upset when the battery is suddenly empty, even though the BMS had until recently said that the battery still had charge left. Thus, in that regard, SOC estimation is important.

Earlier we saw that DOD and SOC are not quite the inverse of each other (Section 3.3.1.1). We also saw that combining voltage translation and Coulomb counting results in the best algorithm for SOC estimation (Section 3.3.1.3). At first sight, it would appear that we have enough information to design an effective SOC algorithm. Indeed, if applied to the laboratory environment, we do. Unfortunately, the limitations of the real world are such that the effectiveness of such a relatively simple algorithm will be severely limited, due to:

- Cells variations (cell-to-cell as well as within a cell);
- BMS hardware limitations;
- Limitations of the application.

Cell variations are the main limitation to an effective SOC estimation:

- *Resistance:* Under load, the terminal voltage is significantly different from the OCV. Resistance changes significantly with SOC, temperature, and age.
- *Relaxation:* After the current stops, the cell voltage takes a long time to reach the OCV.

The BMS hardware presents two significant limitations:

- Coarse resolution of cell voltage measurement limits the accuracy of voltage translation.
- Offset in the current sensor reading results in a drift in the Coulomb counting integral over time.

Every application presents a set of problems to SOC estimation:

- Some applications never fully charge the battery (e.g., HEVs), making calibration of the SOC at the top end challenging.
- Some applications never fully discharge the battery (e.g., HEVs) or rarely do (e.g., EVs), making calibration of the SOC at the bottom end challenging.
- In most applications the battery is not discharged from full to empty in a single swoop, making measurement of battery capacity unreliable.
- In some applications the battery is constantly being partially charged and discharged (e.g., HEVs), making Coulomb counting unreliable.
- In some applications the battery current is constant (e.g, backup power with constant load), making measurement of cell resistances challenging.
- In some applications, discharge cycles may seldom occur. The conditions of the battery may change significantly between cycles, so that previously measured capacity and resistance may be no longer correct by the time the next cycle occurs.

The ideal SOC estimation algorithm would work well with cells of previously unknown characteristics, and without the need for an initial learn cycle. It would be able to measure:

- Actual cell resistance continuously, even if the battery current were constant;
- Actual cell capacity on a regular basis;
- Cell SOC (and, from that, battery SOC) even if the cells are never completely charged and/or never completely discharged;
- Cell leakage and compensate for the subsequent loss of cell charge.

The few BMS gurus have come up with some clever approaches to SOC estimation by applying particular constraints to the problem. Most notably, the engineers at Texas Instruments have come up with a couple of great solutions that allow their ICs (again, under certain constraints) to estimate SOC within a few percentage points. Here are a few of those methods.

Impedance tracking is what the TI engineers call one of those technologies. It relies on having a very accurate model of the cells, and using it to know a priori what the cell resistance must be under certain conditions, from which the terminal voltage can be translated to OCV, and, from that, the cell SOC is estimated. This technology is essential for applications in which the BMS is implemented outside of the battery (system side battery management system), and the battery has no electronics. Once the battery is replaced, the BMS must quickly determine the SOC of the new battery just from the cell voltages. This approach has economical benefits for small consumer products, but, of course, is not appropriate for large battery packs, the focus of this book.

TI's impedance tracking works for BMS applications for which the cells used are known beforehand, and that may be the case in your custom BMS, but is certainly not the case with off-the-shelf BMSs. In any case, once a cells ages, all bets are off on cell resistance, so the technology's reliance on having an accurate model for the cell is less effective after some time.

Real-time resistance measurement is a slightly different method, one that has all the advantages of TI's impedance tracking, plus the advantage that there is no need for an accurate cell model in advance. However, resistance measurement is only possible if there is a variation in the battery current, and in many applications that is not the case. For example, when using a charger, the charging current is constant, and it may not be acceptable to turn the charger on and off just to measure the resistance. Even if doing so were acceptable, the cell voltage takes a long time to relax to the OCV (on the order of tens of minutes), making resistance measurement a time consuming process. Frequent measurement of cell resistance is most critical at the end of charge, because the resistance increases very rapidly at that point, reducing the usefulness of recent measurements.

The BMS in an HEV uses a different method of SOC estimation. Because the pack SOC hovers around 50% all the time, voltage conversion is useless, so SOC estimation must rely on Coulomb counting. When, invariably, calculated SOC drifts away from the actual SOC, a cell voltage will approach one end or the other end of the voltage versus the SOC curve. At that point, the BMS will notice that that cell's voltage is unexpectedly high or low, and will calibrate the calculated SOC based on it. As a matter of fact, once in a while the HEV may purposely allow the battery to become full, to get to the point that the cell voltage can be used for SOC calibration. Note that it is the HEV ECU that does so, not the BMS; the HEV knows enough about the entire vehicle to deduce when the vehicle is on a highway, has been on a highway for a long time, and probably will be on a highway for a long time to come, so the pack will not be needed for stop and go traffic in the near future. That gives the HEV time to play with the pack's SOC to calibrate it.

At this point I wish I could give you an ideal, universal algorithm for SOC estimation. Unfortunately, the best I can do is to suggest one that goes a ways towards that goal, but is less than perfect. First, the user must enter the following data in the BMS:

1. Battery nominal capacity.
2. Coordinates (SOC and voltage) of four points in the OCV versus SOC for the cells (Figure 5.56):
   - Full (F): V-full (e.g., 3.6 at 100% SOC);
   - Top (T): V-top and SOC-top (e.g., 3.4V at 95% SOC);
   - Bottom (B): V-bottom and SOC-bottom (e.g., 3.0V at 15% SOC);

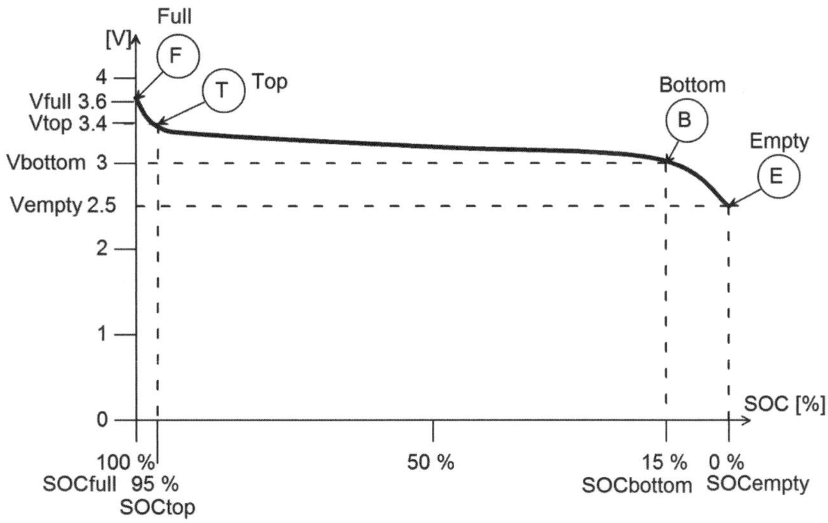

Figure 5.56 Points in an OCV versus SOC curve for the purpose of SOC estimation.

- Empty (E): V-empty (e.g., 2.5V at 0% SOC).

Then, the BMS uses this algorithm:

1. Whenever possible, attempt to measure resistance.
2. At all times do IR compensation of the cell voltage to get OCV.
3. If charging [Figure 5.57(a)]:
   - If all of the cell's OCVs are below V-top, integrate the battery current into the DOD [ah]. Given the nominal capacity, convert that DOD to SOC. If the SOC exceeds SOC-top, clamp it to SOC-top and convert to DOD.
   - Otherwise: convert the OCV to SOC, based on a straight line between SOC-top and 100% SOC, and between V-top and V-full. Given the nominal capacity, convert that SOC to DOD.
4. If discharging [Figure 5.57(b)]:
   - If all of the cell's OCVs are above V-bottom: integrate the battery current into the DOD [ah]. Given the nominal capacity, convert that DOD to SOC. If the SOC is below SOC-bottom, clamp it to SOC-bottom and convert to DOD.
   - Otherwise: convert the OCV to SOC, based on a straight line between SOC-bottom and 0% SOC, and between V-bottom and V-empty. Given the nominal capacity, convert that SOC to DOD.

5.4.3.3 Resistance Calculation

Knowing the resistance of each cell in the pack is useful for two things:

- To compensate for the IR drop on the cell voltage;
- As one of the parameters to evaluate the SOH.

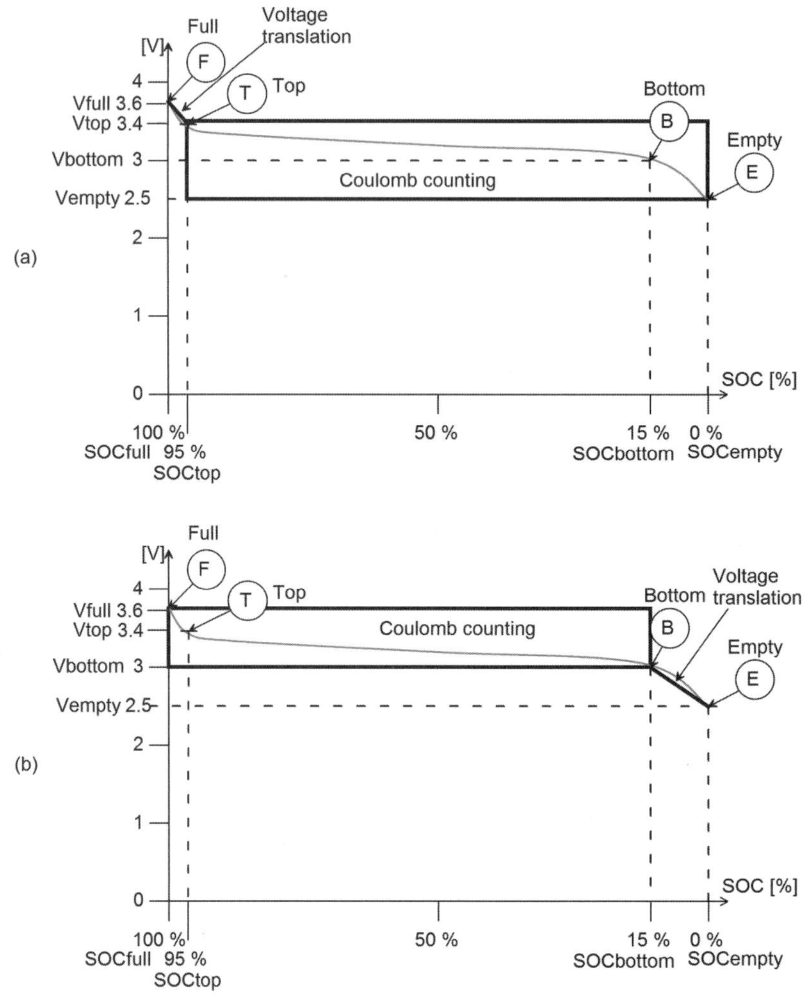

Figure 5.57 A possible SOC and DOD calculation algorithm: (a) charging and (b) discharging.

Resistance can be calculated as the slope of the line between two points in the cell voltage versus cell current curve (Figure 5.58). If the battery current changes significantly, then the BMS can measure the voltage at two levels of battery current, and calculate the resistance as:

$$R = (V1 - V2)/(I2 - I1)$$

Note the reversal of the order of the voltage values compared to the current values. This is because the voltage goes down as the discharging current goes up (discharging current is assumed to be positive). This method works just as well if one of the points is at 0 current, or if one of the points is while the other is discharging.

5.4.3.4 Capacity Measurement

Knowing the actual pack capacity is useful for two things:

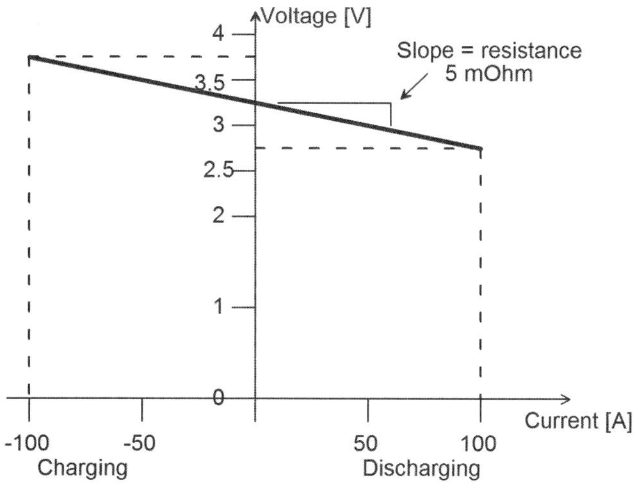

Figure 5.58 Resistance is the slope of the cell voltage versus current curve.

- To convert DOD to SOC;
- As one of the parameters to evaluate the SOH.

Actual capacity can be measured directly if the battery goes from fully charged (until the charging current ramps down to 0) to fully discharged (either at low current, or using IR compensation to know that the battery is really fully discharged), and that happens within a sufficiently short time that the error due to the offset in the current sensor integrated into the DOD is relatively small. In that case, the actual capacity is the DOD at the end of discharge (Figure 5.59). Of course, actual capacity could also be measured when charging from empty to full.

5.4.3.5 SOH Estimation

As we saw (Section 1.4.3), SOH evaluation is arbitrary because it does not measure a specific physical quality. A BMS may use a combination of one or more of the following parameters, with arbitrary weight factors, to evaluate the SOH: increase in cell resistance, decrease in actual capacity, number of charge/discharge cycles, self-discharge rate, and passage of time. Sometimes the "ability to accept charge" is listed as one of the parameters, but that can be translated to cell resistance or actual capacity, so it is not a separate parameter.

The easiest definition of SOH you may want to implement is based simply on number of cycles [Figure 5.60(a)]:

SOH = 100 * (1 − Cycle Number/Nominal Total Number of Cycles)

Of course, that begs the question: what constitutes a cycle? Consider that question from the point of view of a system that does many, shallow cycles, such as an HEV. In such an application, a cycle is pretty meaningless.

In a backup system that rarely gets discharged, the number of cycles is hardly a measure of SOH. In that case, you may want to evaluate SOH based on the passage of time [Figure 5.60(b)]:

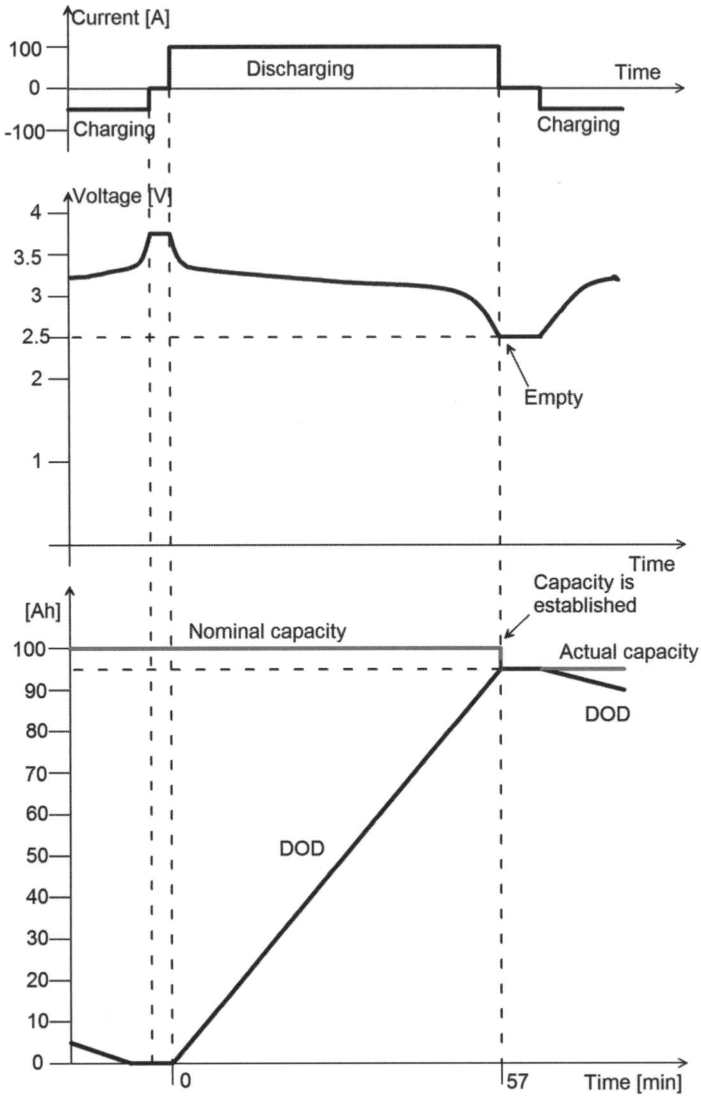

Figure 5.59 Measuring battery capacity.

$$SOH = 100 * (1 - \text{Age/Rated Calendar Life})$$

In systems that are able to measure the battery's capacity, you may want to evaluate SOH based on the relative actual capacity [Figure 5.60(c)]:

$$SOH = 100 * (1 - \text{Actual Capacity/Nominal Capacity})$$

Some systems (such as HEVs) use a pack as a power source (as opposed to an energy source), never fully using the entire energy in the pack, and therefore are not able to measure the pack's capacity. In such systems loss of pack capacity is inconsequential, so the fact that SOH cannot be based on capacity loss is moot.

In systems that are able to measure the battery's resistance, you may want to evaluate SOH based on the relative actual resistance [Figure 5.60(d)]:

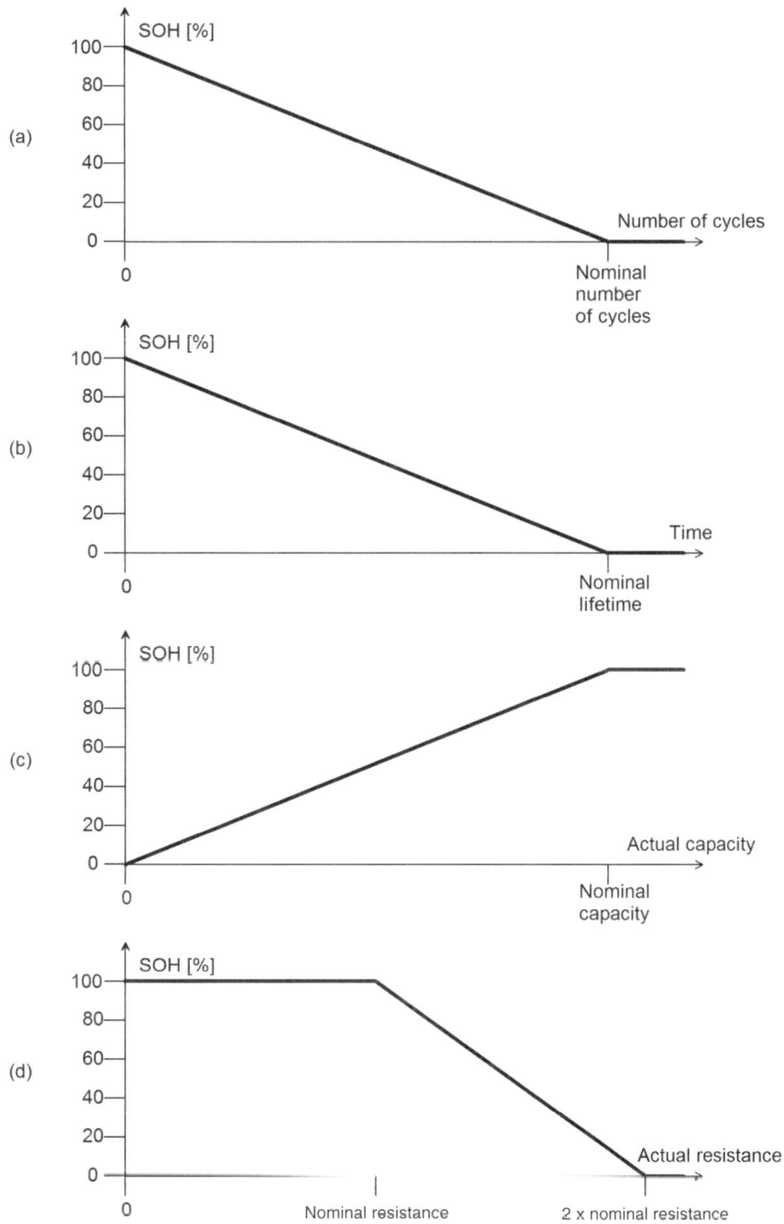

Figure 5.60 SOH: (a) versus number of cycles, (b) versus age of battery, (c) versus capacity, and (d) versus resistance.

$$SOH = 100 * (1 - \text{Nominal Resistance/Actual Resistance})$$

5.4.3.6 Diagram

A diagram of the evaluation process (Figure 5.61) shows its complexity. On the left are the three measurements: cell voltage, pack current, and cell temperature. On the right are 17 parameters that may be communicated to the system. In the middle are all the intermediate parameters, showing the various dependencies.

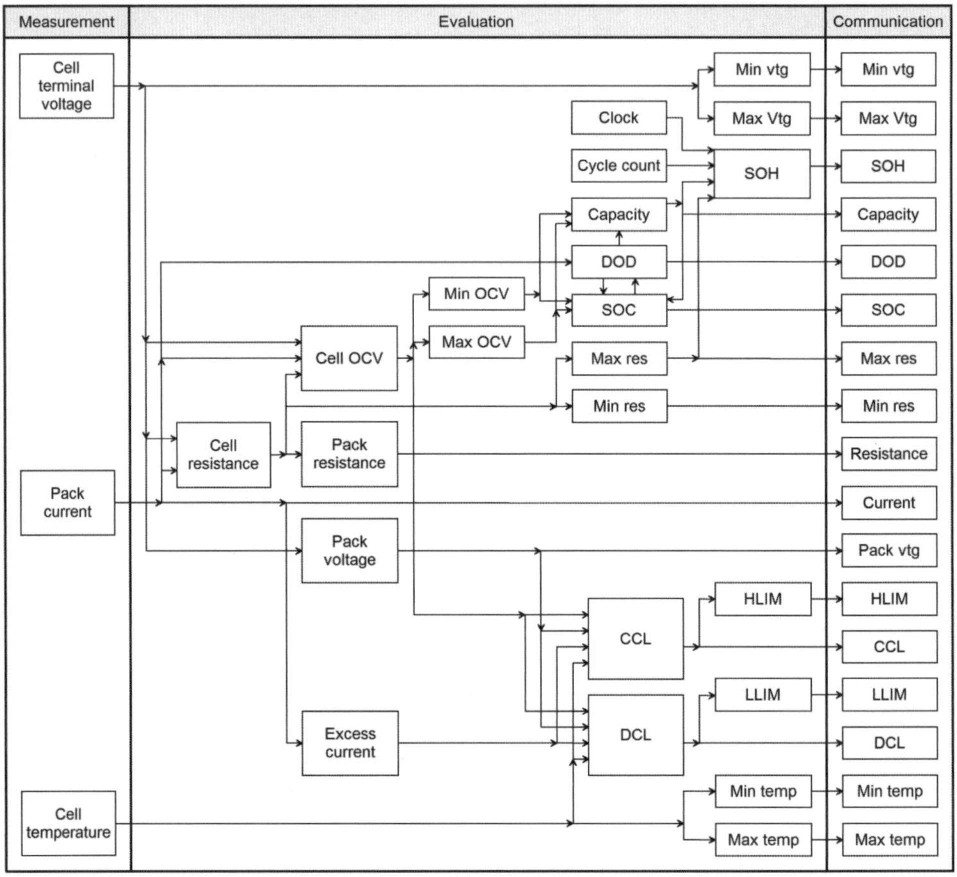

Figure 5.61 Evaluation process diagram.

5.4.4 Communications

After measurement and evaluation, the next BMS function is communication with the external system, at the very least for the purpose of protecting the battery by requesting reduction or interruption of battery current (and possibly to request heating or cooling). Meters do not implement this function, but all other types of digital BMS do.

Communication can be through wires with a dedicated function, or through data links. Wires dedicated to a specific function are easy to understand and to troubleshoot, which makes them desirable for prototypes, single run projects, or hobbyists' projects.

5.4.4.1 Dedicated Wire

However, a BMS with many functions will require large connectors with many pins, and wiring to it takes a lot of effort, which makes them impractical for production level devices. In those cases, relying on data links is more desirable.

Dedicated wire interfaces can be divided into two classes: digital (on/off) and analog.

Digital
By digital wire interface I don't mean just logic interfaces (the 0 and 1 kind), but any wire that can be in one of two states, such as open and grounded.

Outputs Digital outputs can be:

- Standard CMOS logic level [Figure 5.62(a)]: 0 volts or 5V.
  - This could be useful, but not terribly flexible.
- Open collector [Figure 5.62(b)]/open drain [Figure 5.62(c)]: either grounded or open.
  - Far more versatile, as it can be used to drive small loads (lamps relays) or drive logic input lines (with a pull-up resistor to a 5-V supply).
  - The fact that this output is referenced to ground can be a limitation: ground loops, no isolation.
  - Usually cannot handle too high a voltage (20 to 100-V maximum).
  - Only works with DC loads.
- Solid state relay (SSR)/opto-isolator [Figure 5.62(d)]:
  - Same as above, but is isolated.
  - Requires an extra wire for the return.
- Contacts: two wires that are either open or shorted together by a mechanical relay [Figure 5.62(e)].
  - This is the most flexible option, as it can be made to do what all other outputs can do.
  - Can work with AC.

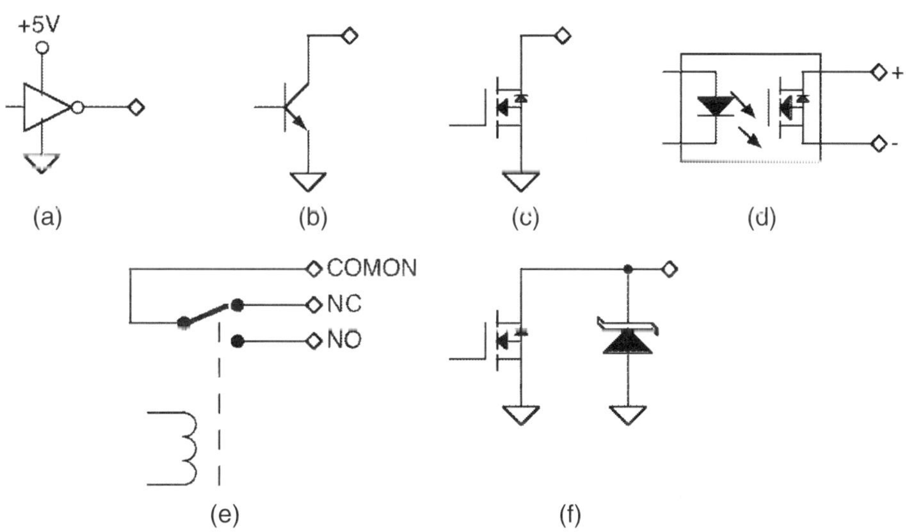

Figure 5.62 Output types: (a) logic, (b) open-collector, (c) open-drain, (d) SSR, (e) relay, and (f) TVS protection.

- With three wires, the relay's NC, C, and NO terminals can be made available, for additional versatility.
- Relays are more expensive than electronics, and have a long but limited life.
- Most relays contacts are designed to work either at low currents (dry contact) or high currents (power) but not both. There are ways of loading a power relay's output to work at low current (see Section 5.4.4.1)

If cost and number of wires is not an obstacle, using relay outputs is recommended. If cost or reliability is an issue, but number of wires isn't, an SSR is the next best choice. If the number of wires is an issue, open collector is the next best choice. In that case, the BMS outputs might as well be designed to handle a decent amount of current (such as 1A), which really does not cost much more than outputs that can handle only a small current (such as 10 mA).

Often an output is expected to drive a relay coil. When the output is turned off, the inductance in that coil will generate an inductive kickback on the output. As one cannot be sure that the relay is fitted with a diode across the coil to absorb that kickback, the output should include some protection mechanism against that kickback. This could be a capacitor, a diode to the supply rail, or a transient voltage suppressor (TVS). The capacitor is not a good choice, because it results in an inrush current when the output is turned on. The diode can be problematic because it routes the kickback into the BMS controller supply, possibly with bad results, and because it will not work if the relay is operated at a higher voltage than the supply voltage for the BMS controller. The TVS is the best option, as it has none of the disadvantages of the others, plus it has the advantage that it results in the most rapid drop in the relay coil current, improving its ability to turn off loads [Figure 5.62(f)].

Inputs Digital inputs can be:

- Standard CMOS/TTL logic level [Figure 5.63(a)]: 0 volts or 5V (actually, < 0.8V or > 2V).
  - This is a very versatile, as it can be used with logic outputs, open collector/drain outputs, relays, and SSRs.

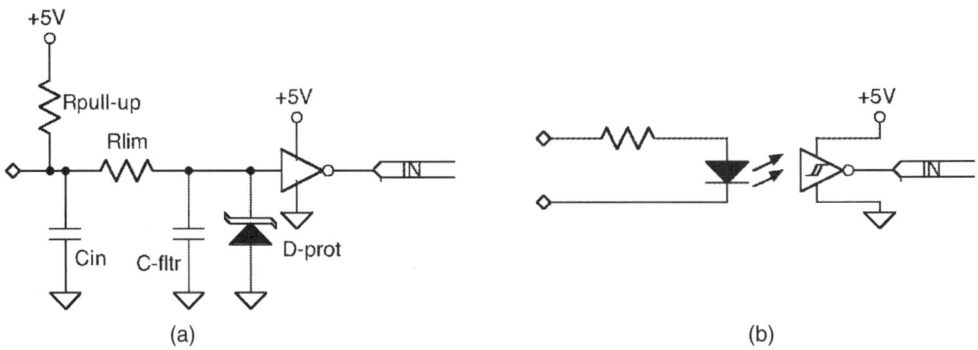

Figure 5.63 Input types: (a) logic and (b) opto-isolator.

- By adding protection on the input (noise filter, current limiting resistor: R-lim), they can be even more versatile, as they can be driven directly by 12V or 24 signals.
- By adding a pull-up to an internal supply (R-pull-up), they can be driven directly by switches to ground (open collector/drain outputs, relays, and SSRs).
- By adding a large capacitor (C-in), they can be driven directly by power rated switches to ground (when first closed, the switch discharges C-in and the resulting large current pulse cleans the switch's contacts).
- While they technically can only work with DC inputs, they can be made to work with AC. Internal filtering can be used (such as in software) to interpret a 50/60 Hz on/off rate to indicate that the input is being driven.
- They are ground referenced: they are not isolated, which may result in ground loops.

- Opto-isolator [Figure 5.63(b)]:
    - Same uses as above, but isolated.
    - Can be designed to work with AC (back-to-back LEDs in the opto-isolator, or a bridge rectifier driving the opto-isolator).

If cost and number of wires is not an obstacle, using opto-isolator inputs is recommended. Otherwise, a digital input is sufficient; just make sure it is protected from accidental application of excessive voltage (even line voltage or full pack voltage) by placing a resistor in series able to handle that voltage without damage.

Analog

Analog inputs used to read battery current were discussed earlier (Section 5.4.1.2). There may be other uses for analog inputs, but they are not common. Analog outputs, on the other side, can be quite useful.

- An SOC output can drive an analog fuel gauge (or a digital fuel gauge with an analog input).
- A CCL output can control an analog input in a charger to reduce current when required.
- A DCL output can be used to reduce the load current such as by reducing a throttle's range (Section 6.1.3.3).

Ideally, analog outputs can be generated by a D/A converter [Figure 5.64(a)]. Multichannel converters are readily available with 8-, 10-, or 12-bit resolution (usually 8 bits is sufficient for BMS applications). Typically, a microcontroller controls them through a standard I2C or SPI bus. Alternatively, an analog output can be generated form a PWM from the micro, filtered, then buffered by an op-amp in a voltage follower circuit [Figure 5.64(b)].

5.4.4.2 Data Link

Earlier I offered an overview of various data links and their uses (Section 3.4.3). Here we will look at the hardware, software, and protocols.

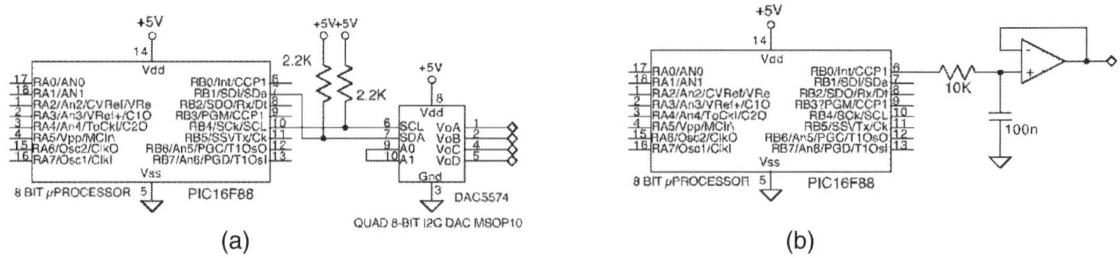

Figure 5.64 Analog outputs: (a) using a DAC and (b) using a PWM.

Serial

Implementing an RS232 port in a BMS is easy: most processors include a UART peripheral. The only other thing you need is an RS232 buffer. That buffer typically includes charge pumps to generate its own supplies to reach the higher voltages specified by the RS 232 standards, and protection against ESD. Today, the standard RS232 connector is a DE9[2] female, wired so that it can be connected directly to a Windows or Linux PC (Figure 5.65). Fewer and fewer PCs have a serial port, and most laptop computers do not have one, therefore a USB to RS232 adapter is often required between the BMS and the computer.

CAN Bus

The CAN bus, with its complex array of options, can be overwhelming at first. I am going to try to make it a bit easier by giving you a few tips.

Implementation First and most important: do not make your own CAN engine.[3] Instead, use a processor with a CAN engine already built into its hardware. For example, many microchip processors in the PIC18F series include a CAN engine, as does the AT89C51CC03C series from Atmel. A CAN engine is built by experts who are familiar with all the intricacies of CAN communications, and take care of them so you don't have to. With a CAN engine, the software's job is greatly simplified. First, you set the CAN parameters, then the CAN engine sends an interrupt when an interesting message is received. To transmit a message, you simply put it in a buffer, and the CAN engine will handle it from there.

Second, do not write your own CAN routines. Instead use library routines (chip manufacturers provide C routines for the purpose). They allow you to use a high-level call, instead of having to figure out which register bit does what.

Third, when baffled by the CAN bus settings, start from this example for a PIC processor running at 20 MHz, and then modify as appropriate.

2. The correct term is DE9, not DB9. The "E" stand for the shell size, and the "B" size shell is the one typically used in a DB25 connector.
3. The engineers of the ElCon chargers designed a CAN engine in software and used the RS232 lines to drive a CAN buffer. The result is a CAN implementation that works, but only if there are no messages on the CAN bus other than the ones it expects. The presence of more than one message per second on the CAN bus will freeze the charger's CAN engine.

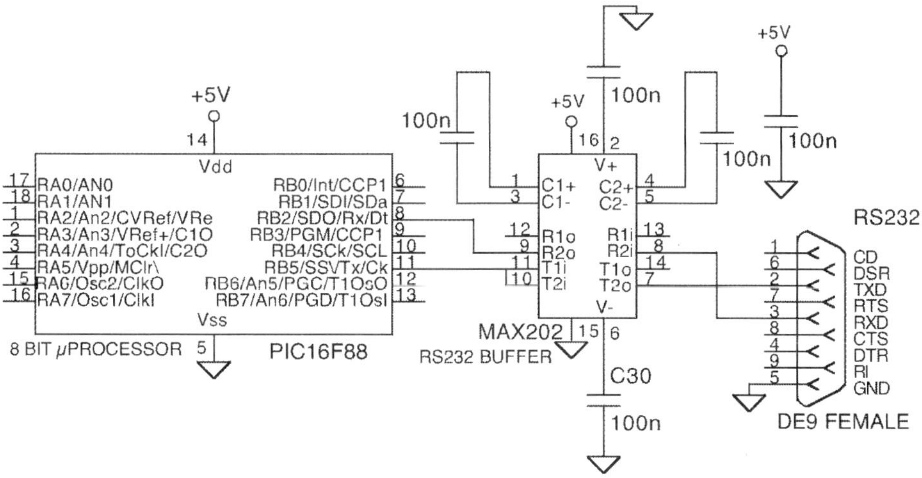

Figure 5.65 RS232 port.

```
; Baud rate registers
#DEFINE     BRP500_    0  ; Baud Rate Prescaler bit for 500 kHz
#DEFINE     BRP250_    1  ; Baud Rate Prescaler bit for 250 kHz
#DEFINE     BRP125_    3  ; Baud Rate Prescaler bit for 125 kHz

; Time Quanta
#DEFINE     PRSEG_     2  ; Propagation time 2 = 3 Tq
#DEFINE     SEG1PH_    7  ; Phase Segment 1 time 7 = 8 Tq
#DEFINE     SEG2PH_    7  ; Phase Segment 2 time 7 = 8 Tq
#DEFINE     SJW_       1  ; Synchronization jump width time 1 = 2 x Tq

; Flags
#DEFINE     SEG2PHTS_  1  ; Phase Segment 2 Time Freely programmable
#DEFINE     SAM_       1  ; Bus line is sampled three times prior to the
                            sample point
#DEFINE     WAKDIS_    1  ; Disable CAN bus activity wakeup feature
#DEFINE     WAKFIL_    0  ; CAN bus line filter is not used for wakeup
#DEFINE     ENDRHI_    1  ; CANTX pin will drive VDD when recessive
#DEFINE     CANCAP_    0  ; Disable CAN capture, RC2/CCP1 input to CCP1
                            module
```

Fourthly, to receive messages, do take advantage of the CAN engine filters. Do not do software-level filtering as you will run out of processing time. Also, I recommend that, instead of polling the CAN engine on a regular basis to see if a message was received, use interrupts to let the software handle a message as soon as it is received, lest you should miss messages as a new message overrides the receive buffer before the previous message was digested.

Finally, get the right tools. A CAN to USB dongle (try c-a-n.com, canusb.com, kvaser.com, rmcan.com) allows you to monitor messages on the CAN bus. Make sure the software that comes with the tool can do at least these two things:

- It has two display modes: a scrolling list by time or reception of received messages, with new messages added to the bottom, and a fixed list ordered by ID, with each new message replacing the contents of the line for that same ID.
- It can log messages to a file.

If working on a production vehicle, get an OBD-II adapter cable, as well as a scan tool (either standalone, such as from Actron, Equus, OTC, or as a dongle for a PC, such as Auto Enginuity).

Standard Traction Pack Messages I would like to propose a set of CAN messages tailored for large packs, which is already used by a few manufacturers, in the hope that it become a standard (Table 5.6). Note that:

- Period: 1s;
- Multibyte values are big-endian: the most significant byte (MSB) is in the lower numbered data byte (the left most byte in the table).
- The ID of the first message (ID0 in the table) is programmable (the default ID is 620h). The other messages use IDs following the first ID.

SAE J1939 Standard The SAE has defined a standard set of CAN messages (J1939) for use in heavy vehicles, for parameters such as coolant temperature and engine RPM. These message are quite complex, compared to the standard messages in the previous section. They have the advantage of being an established standard, but they don't seem to have been adopted in passenger vehicles. While the standard does include some parameters applicable to traction packs, messages that would report details of a Li-Ion pack are not included. Therefore, the designer using the J1939 standard is faced with creating proprietary messages just as well, yet with the added burden of the complexity of the J1939 format. To make it worse, the standard is not openly published, and it is priced in such way to be unaffordable to anyone but car manufacturers. Two devices on the CAN bus may both claim to be J1939 compliant, but unless we are talking about something simple like coolant temperature, they will not be really talking the same language, so one of them will have to be customized. Therefore, any J1939 support a BMS may offer, is no better, and is actually more troublesome, than defining a set of proprietary messages.

PIDs Parameter identifiers (PIDs) are codes that are used to request data in a vehicle through its onboard diagnostics (OBD) II connector. Since 1996, those codes must be sent through the vehicle's CAN bus. Typically, an automotive technician will use PIDs with a scan tool connected to the vehicle's OBD-II connector.

- The technician enters the PID.
- The scan tool sends the PID request to the vehicle's CAN bus through the OBD II connector.
- An electronic control unit (ECU) on the CAN bus recognizes the PID as one it is responsible for, and reports the value for that PID to the bus.

Table 5.6 Standard Traction Pack Messages

ID	*Bytes*	*Byte 0*	*Byte 1*	*Byte 2*	*Byte 3*	*Byte 4*	*Byte 5*	*Byte 6*	*Byte 7*
ID0+0	8	BMS data 1 (1)							
ID0+1	8	BMS data 2 (1)							
ID0+2	8	State (2)	Timer (3)	Flags (4)	Faults (5)	Warnings (6)			
ID0+3	6	Pack voltage (7)	Min voltage (8)	Min voltage ID (9)	Max voltage (8)	Max voltage ID (9)			
ID0+4	6	Current (10)	Charge limit (11)	Discharge limit (11)					
ID0+5	8	Pack energy in (12)	Pack energy out (12)						
ID0+6	7	SOC (13)	DOD (14)	Capacity (15)	00h	SOH (16)			
ID0+7	6	Temperature (17)	—	Min temperature (18)	Min temperature ID (8)	Max temperature (18)	Max temperature ID (8)		
ID0+8	6	Pack resistance (19)	Min resistance (20)	Min resistance (8)	Max resistance (20)	Max resistance (8)			

1. ASCII data, such as hardware or software rev level, serial number, model, date code. 2. State of system (defined by the BMS designer). 3. Power up time [s]. After 65,535 seconds, it overflows back to 0. Also useful as a clock to determine that the BMS is not hung. 4. Byte of flags, defined by the BMS designer, but may include: The LLIM is asserted; The HLIM is asserted; There is power from the load; There is power from the source. 5. Fault code (or flags), defined by the BMS designer, but may include: Driving off while plugged in; Internal communication fault; Charge overcurrent; Discharge overcurrent; Over temperature; Under voltage; Over voltage; High voltage leak to chassis; Contactor or precharge fault; Soft error. 6. Warning flags, defined by the BMS designer, but may include: Hot temperature; Cold temperature; High discharge current; High charge current; High cell voltage; Low cell voltage. 7. Total voltage of pack [V], unsigned, 0 to 65 kV. 8. Voltages [100 mV] of least charged cell and most charged cell, 0 to 25.5V. 9. ID of the cell that has the lowest/highest voltage/temperature/resistance. 10. Pack current [A], signed, positive out of pack, −32 kA to + 32 kA. 11. Maximum current acceptable (charge) or available (discharge), unsigned, 0 to + 65 kA [A]. 12. Total energy in or out of battery, since manufacture. Unsigned, overflows back to 0 [kWh]. 13. State of charge [%], unsigned, 0 to 100. 14. Depth of discharge [AH], unsigned, 0 to 65 kAh. 15. Actual capacity of pack [AH], unsigned, 0 to 65 kAh. 16. State of health [%], unsigned, 0 to 100 (100% = pack meets nominal specs). 17. Average pack temperature [°C], signed, −127°C to +127°C. 18. Temperatures [°C] of coldest and hottest sensors, signed, −127°C to +127°C. 19. Resistance [100 $\mu\Omega$] of entire pack, unsigned, 0 to 6,553.5 mΩ. 20. Resistances [100 $\mu\Omega$] of lowest and highest resistance cells, unsigned, 0 to 25.5 mΩ.

- The scan tool reads the response, and shows it to the technician.

Standard PIDs are defined by the SAE standard J1979, which, again, is priced to be affordable only by car manufacturers. The OBD-II_PIDs article in Wikipedia has quite a bit of information on PIDs obtained through reverse engineering (to the chagrin of some who would rather have you buy the standard).

Just like the J1939 standard, J1979 defines standard automotive items such as coolant temperature, but not many items that are applicable to an Li-Ion traction pack. Therefore, once more, the BMS designer is left with defining a proprietary set of PIDs.

PIDs may be used to request parameters such as:

- Cell:
  - Status, voltage, temperature, resistance SOC.
- Bank:
  - Status, any nonreporting or unconnected cells;
  - Minimum, average and maximum cell voltage, temperature, and resistance;
  - Total voltage and resistance.
- Battery:

- Status, any nonreporting or unconnected cells or banks;
- Minimum, average and maximum cell voltage, temperature, and resistance;
- Total voltage and resistance.
- Pack:
  - Status, any nonreporting or unconnected cells or banks;
  - Minimum, average and maximum cell voltage, temperature, and resistance;
  - Total voltage and resistance;
  - SOC, DOD, capacity;
  - SOH;
  - Power, energy in, energy out.
- System:
  - Status;
  - Power cycle number, time into the cycle;
  - Faults and warnings;
  - State of inputs and outputs;
  - Pack current;
  - Hardware and software revision.

PIDs may also be used to retrieve store fault codes, freeze data at the time of the fault, and to clear fault codes.

As a starting point, use this definition for PID messages:

- Send request messages at ID 0745h.
- Receive response messages at ID 074Dh (regardless, the received ID is 08h more than the transmitted ID).
- Use 8 data bytes, whether all bytes are used or not.

5.4.4.3 Display

With some exceptions, displaying the BMS status to a user is a function best left to the overall system. For example, in a vehicle, a dashboard display is expected to display information about the entire vehicle, not just the traction pack. Should you provide a display with your BMS, invariably you'll be asked to add to it the ability to display other parameters (speed, location, tire pressure) that have nothing to do with the BMS. When that happens, you will see why developing a display for the BMS is not ideal. Instead, the BMS should provide ways for the rest of the system to query its status (Section 5.4.4.2) and then display that data to a level of detail that is appropriate for the particular user (general user, design engineer, troubleshooting technician).

LEDs

LEDs may be built into a BMS and can be very useful for troubleshooting. For example, LEDs on distributed cell boards may indicate that they are operating properly or that they are balancing. LEDs on centralized BMSs may indicate balancing. LEDs on master controllers may indicate status: power in present, 5-V supply present, and the state of each input and output.

Unfortunately, LEDs are not considered acceptable in products intended for automotive applications. In that field, a voltmeter and a scan tool are the preferred troubleshooting tools, and any ECU that may benefit from the use of LEDs is sealed in a potted assembly or metal case, making it impossible to see

Fuel Gauge

One of the exceptions to the principle that displaying is not the BMS's job is a battery fuel gauge–type display visible to the user (such as on a dashboard). Such a display is a convenient option for use in vehicles retrofitted with a traction battery, especially in HEVs and PHEVs because they need both a true fuel gauge (for the gasoline tank) and a fuel gauge for the traction pack. A fuel gauge display can be a standard analog voltmeter driven by the BMS's SOC analog output. It could also use an LED bar graph, which the BMS could control in various ways: through a dedicated digital port; through an SOC analog output (the LM3914V LED bar graph driver can be used to convert a 0- to 5-V input to 10 LED drive outputs); or through the CAN bus (Figure 5.66).

Full Display

There are applications in which a full BMS display (alphanumeric or graphic) can be appropriate: prototype development and research labs. Items that could be displayed include:

- A general screen showing numeric data: pack voltage, current, SOC, temperature, cell voltage ranges, state (warnings and alarms). It may also show SOH, input/output state, SCL DCL, power, and energy (Figure 5.67).
- A graphic screen showing SOC as a bar graph, cell voltage distribution as a histogram (column graph). It may also show temperature, current, voltage and power as bar graphs.
- Specific screens for individual cells and batteries.

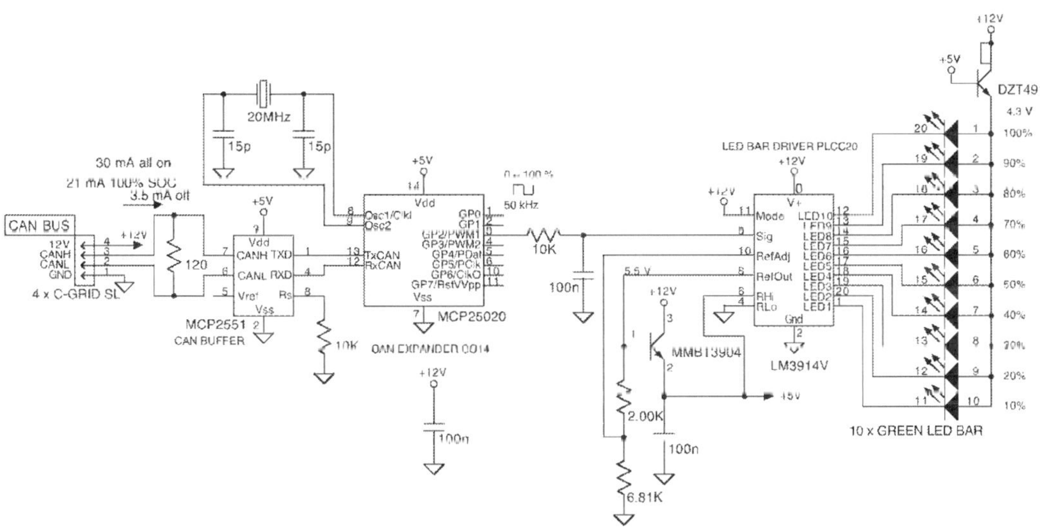

Figure 5.66 Example of a CAN bus driven SOC display.

Figure 5.67 Graphic screen for the Lithiumind BMS.

5.4.5 Optimization

After measurement, evaluation, and communication, the next BMS function is battery optimization, such as SOC balancing. Only balancers and protectors implement this function.

5.4.5.1 Balancing

In Section 3.2.3 we looked at balancing and compared passive and active balancing. Here we will look at the actual circuits that do balancing.

Passive

Passive balance is easy: all it takes is a simple resistor [Figure 5.68(a)]. Balancing could also use components other than resistors. Balancing can be done with a power transistor wired as a current source [Figure 5.68(b, c)]. It takes a few more components, but, for a given power handling capacity, a power transistor is cheaper than a power resistor.

At high power levels, using multiple resistors may be more expensive than a single higher power resistor or a single transistor. However, from a thermal management standpoint, it is better to have many sources of low heat than a single source of high heat (Figure 5.69). Radiated electromagnetic energy (light, infrared, RF) leaves the source more readily than convection can remove heat from a resistor, so less thermal management is required.

Lamps can be used as a balancing loads, either LEDs or incandescent bulbs [Figure 5.70(a)]. In reality, the advantage is not that great: light bulbs, especially small ones, are not that efficient. LEDs are more efficient that lightbulbs when considering just the visible spectrum, but in this application any energy that is radiated helps, so light bulbs are better in this regard. However, lightbulbs only last 5,000 to 20,000 hours, so, for reliability, LEDs are better. A problem with LEDs is that they operate at a lower voltage than Li-Ion cells, so they require a power resistor in series. Therefore, while radiating extra energy as light may sound attractive, today's light-emitting components are too expensive and not sufficiently efficient to justify using them as balancing loads.

Radio transmitters can also be used as loads [Figure 5.70(b)]. Their efficiency can be quite high, therefore offering the advantage we were looking for from lamps (reduction of local heat, as most energy is radiated). Their cost is relatively high, compared to a resistor, and there are issues of RF containment, as you don't want that RF to get out of the pack.

One side advantage of using an RF generator is that the RF could be modulated to allow a distributed cell board to communicate to the BMS controller wirelessly.

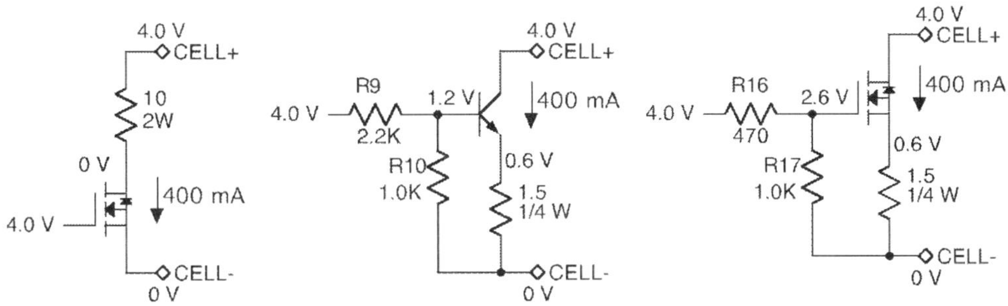

Figure 5.68 Balancing load: (a) resistor, (b) BJT, and (c) MOSFET.

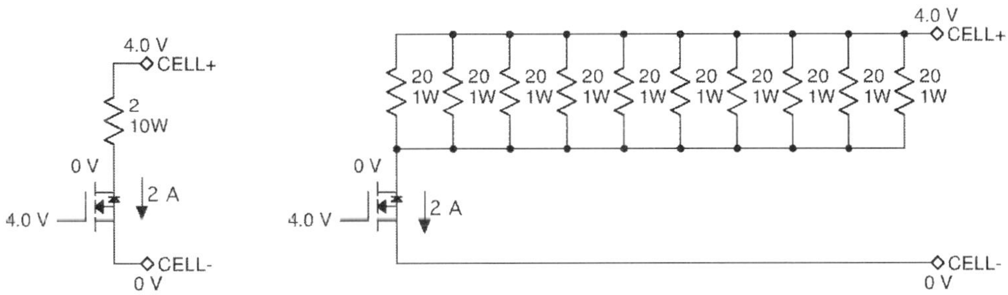

Figure 5.69 Power resistors as a balancing load: (a) single, large resistor and (b) multiple, smaller resistors.

Figure 5.70 Alternative balancing load: (a) incandescent lamp and (b) RF transmitter.

Spread spectrum transmission would allow all balance loads to be on simultaneously yet allow communication despite message collisions. Then, an RF receiver would be required on each cell board to allow the BMS controller to talk back to the cell boards; otherwise, the wireless advantage would be lost.

The cost and complexity of an RF generator is such that you might as well implement active balancing instead. In conclusion, power resistors or power transistors remain the only practical options as loads for passive balancing.

Active

Earlier we saw that active balancing is not as advantageous as one may think (Section 3.2.3.3).

As mentioned in Section 3.2.3.7, active balancing can be:

- *Cell to cell:* Energy is moved between adjacent cells.
- *Cell to battery:* Energy is removed from the most charged cells and sent to the entire battery.
- *Battery to cell:* Energy is removed from the entire battery and sent to the least-charged cells.
- *Bidirectional:* Either of the above two, as required.

Let us look at circuits to implement all four types of balancing.

Cell to Cell Cell-to-cell balancing requires a bidirectional DC-DC converted between each pair of adjacent cells in series. For N cells you need $N - 1$ converters. There are three types, based on the energy transfer component used:

- Capacitor;
- Inductor;
- Transformer.

Cell to Cell—Capacitor Capacitor-based active balancing is inherently very intuitive: it connects a large capacitor across one cell, then across the other cell. When connected to the cell with the higher voltage, energy will flow from the cell to the capacitor. When connected to the other cell, energy will flow from the capacitor to the cell. A ladder circuit with $N - 1$ capacitors can be built easily to handle N cells (Figure 5.71).

The switches between capacitors and cells can be relays or MOSFETs. Relays have practically zero resistance, which is good in terms of energy efficiency, but is bad in terms of inrush current when first connected. If using MOSFETs, two are needed in antiparallel to handle current in either direction. Both solutions suffer from the risk that two adjacent switches are turned on simultaneously, shorting across a cell, and causing serious damage.

What often gets overlooked by people considering this approach is that it has a fundamental physical limitation: connecting a cell (which is a voltage source) to a capacitor (also a voltage source) is the electrical equivalent of dividing by 0 (which is a serious mistake) and results in an energy efficiency never above 50%. The only reason it works is that there is a nonzero resistance in series (part in the cell, part in the switch, and part in the capacitor). When first connected, any voltage difference between the cell and the capacitor is dropped across this series resistance, determining the level of the initial inrush current pulse (may be on the order of hundreds of amps!), which then drops asymptotically as the two voltages equalize.

Your first thought may be to want to design the circuit for a higher series resistance, to reduce this current. Yes, it will do that, but it will also increase the time that the capacitor needs to be connected to the cell (reduce the sampling frequency), which will reduce the effective balancing current.

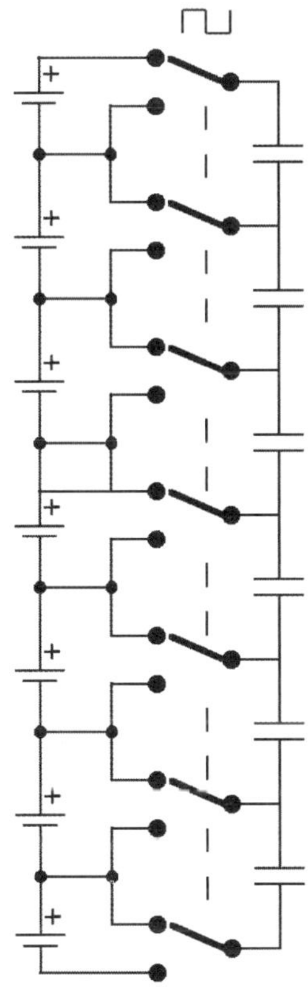

Figure 5.71 Capacitor ladder to balance cells.

Your second thought may then be to design the circuit for the lowest possible series resistance, and the ability to handle a higher inrush current. Yes, this will allow you to increase the sampling frequency, and therefore the effective balancing current. What may not be apparent, though, is that, regardless of the value of the series resistance, the amount of energy dissipated in heat by that resistance each time a connection is made is equal to the amount of energy transferred. If you double the resistance, the current will be halved, but it will last twice as long, so the energy dissipated is the same.

So, for example, if the circuit balances a 4-V cell using a 1-μF capacitor that starts discharged, when the switch is closed, 16 mJ will be transferred out of the cell, of which 8 mJ will be wasted in heat across the series resistance (regardless of its value) and 8 mJ will be transferred to the capacitor (Figure 5.72). When the capacitor is connected to the other cell, more energy will be wasted.

Capacitor based balancers are inherently 50% efficient, at best. This has to do with physics, and there is nothing that you can do about it. The only way to avoid this loss is to couple the voltage source of the cell to a current source (such as an inductor), as we'll see in the next section.

An inductor could be placed in series with the capacitor to form a resonant circuit (tank). Using 0-voltage or 0-current synchronous switching would result in a very efficient resonant cell-to-cell balancer.

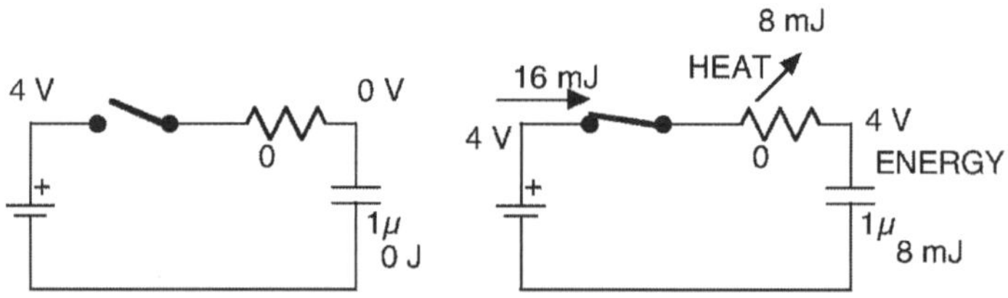

Figure 5.72 Losses in a capacitive cell-to-cell balancer.

Cell to Cell—Inductor An inductor-based cell-to-cell balancer is quite simple. It uses a bidirectional, inverting, DC-DC converter circuit. It works in two phases, first transferring energy from the most-charged cell to the inductor, and then from the inductor to the least-charged cell (Figure 5.73). The switching duty cycle is 50%, resulting in an inherent equalization of the voltages of the two cells.

This circuit doesn't suffer from the voltage-to-voltage disadvantage of a capacitor-based balancer because the inductor acts as a current source, connected to the cell as a voltage source; there is no inrush current.

The major disadvantage of an inductor-based balancer is that it is not isolated and will likely to be destroyed if the power connection between its cells is opened (by a safety disconnect, a fuse, or a loose connection) (Figure 5.74). Therefore, it is not appropriate for packs that have a midpack safety disconnect.

Cell to Cell—Transformer The circuit of a transformer-based cell-to-cell balancer can use any isolated DC-DC converter topology, so it can be either a forward converter or a flyback converter. A flyback topology is simpler and its output is a current source, which is great to transfer energy to the cell (Figure 5.75). A forward topology is less appropriate: it is more suited for really high power levels (but the power levels of a cell-to-cell balancer are quite low) and its output is inherently a voltage source (which we saw is a problem in balancers).

The transformer-based cell-to-cell balancer is more complex and less efficient than an inductor-based one, but can handle an open circuit between adjacent cells (Figure 5.76).

Cell to Battery We saw that, in general, cell-to-battery balancers perform the best (Section 3.2.3.7). They are most efficient because the rectifiers are at high voltage, they use low-voltage transistors, and the transistor drive is on the same side of the

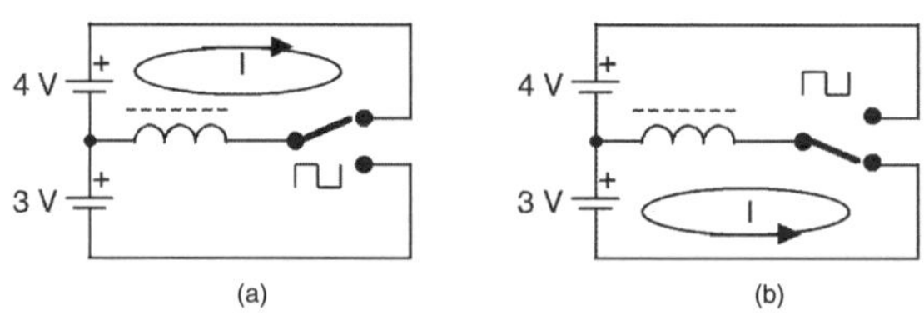

Figure 5.73 Inductor-based cell-to-cell balancer: (a) the inductor receives energy from the cell with the highest voltage; (b) the inductor dumps that energy into the cell with the lowest voltage.

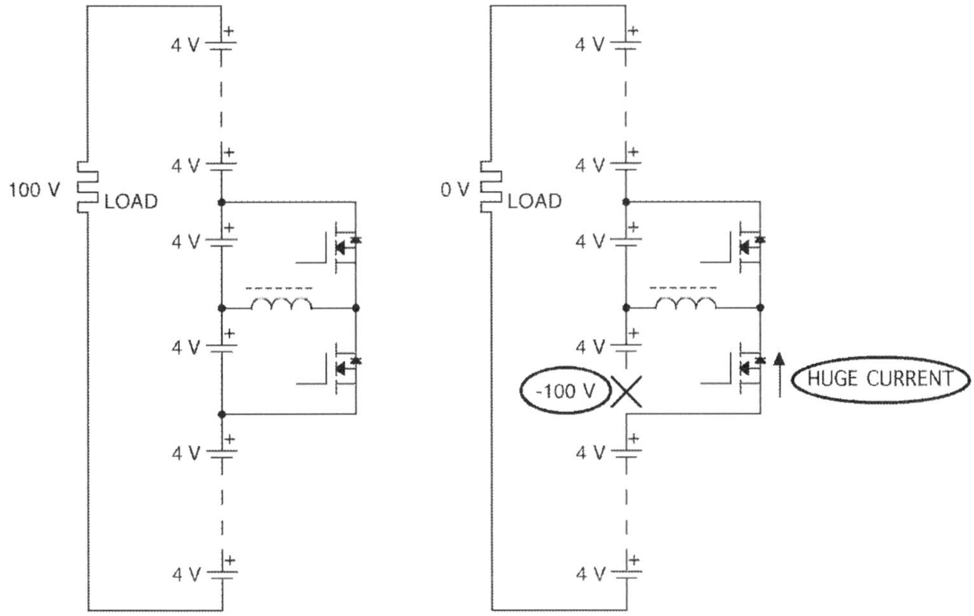

Figure 5.74 Open circuit between cells destroys an inductor-based cell-to-cell balancer.

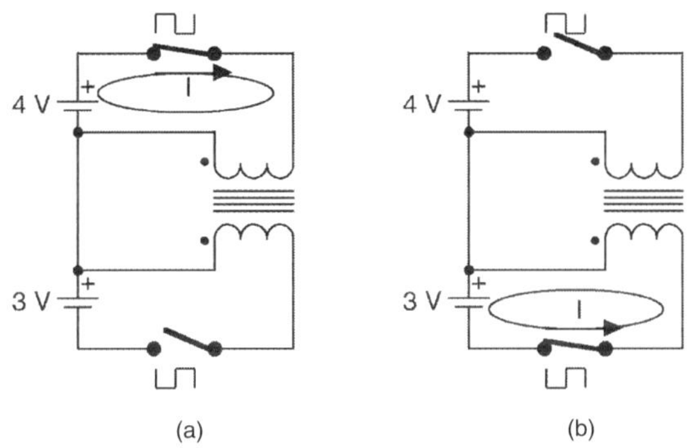

Figure 5.75 Transformer-based cell-to-cell balancer, using the flyback topology: (a) the transformer receives energy from the cell with the highest voltage; (b) the transformer dumps that energy into the cell with the lowest voltage.

isolation barrier as the circuit that decides whether balancing should be on. Their circuit can be the very simple and reliable, classic multivibrator circuit, widely used in low-cost DC-DC converters (Figure 5.77). The outputs of all the DC-DC converters are connected together (the rectifiers in each DC-DC converter isolate the converters that are turned off from the ones that are turned on) and are fed to the two overall battery terminals. The DC-DC converter output must be current limited (a current source) so that it can feed the battery regardless of the battery voltage.

To balance the battery, the BMS turns on the DC-DC converters of the cells that should be discharged a little. This solution especially suitable for distributed BMSs: a DC-DC converter can be added to each cell board, and when the cell circuit

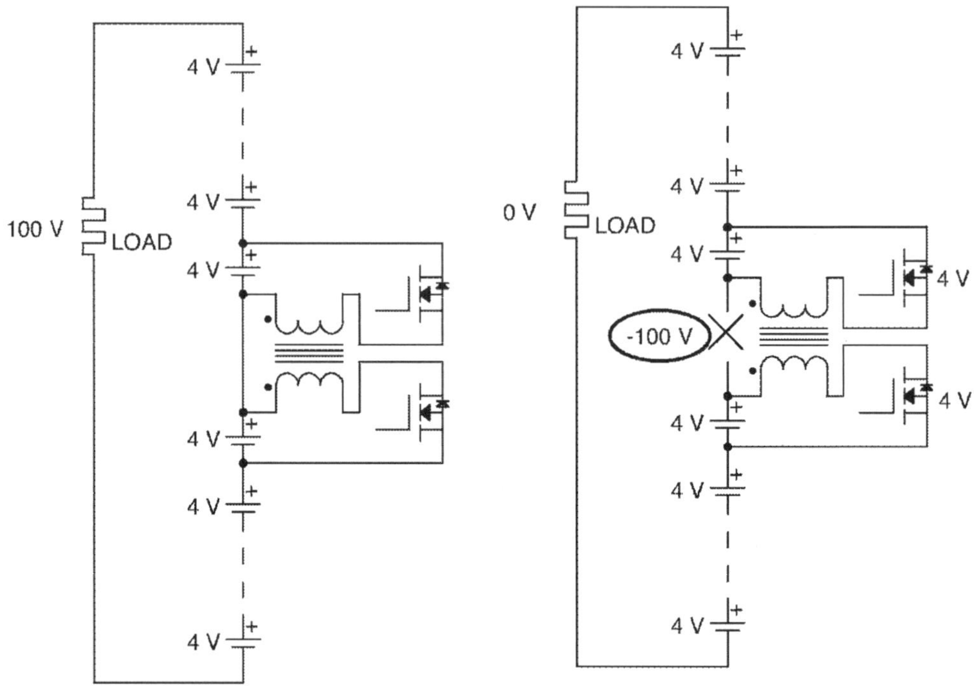

Figure 5.76 A transformer based cell-to-cell balancer can handle an open circuit between cells.

decides that its cell needs to release some charge, it turns on its DC-DC converter. The DC-DC converters will survive an open circuit in the battery, because they are isolated.

Battery to Cell The circuit of a battery-to-cell balancer is the mirror image of the cell-to-battery balancer: the transistors are on the high-voltage side, and the rectifier diodes on the low-voltage side. A typical circuit uses the flyback DC-DC topology (Figure 5.78).

A disadvantage of this balancer is that high-voltage transistors are required (which tend to be less efficient) and the rectifiers are inefficient, because their voltage drop is a significant portion of the cell voltage. A solution to the rectifiers waste problem is to use synchronous rectifiers: MOSFETs that are actively turned on when they should be conducting (Figure 5.79). A further refinement is to replace the rectifiers with a switch that is not only synchronous, but can also be turned off when a cell is full (instead of turning off the DC-DC converter input) (Figure 5.80). That way, a cell board can control its own charging, without having to cross the isolation barrier.

Finally, multiple DC-DC converters may be replaced with a single, high-power, multiple-output DC-DC converter, using a single, large transformer with many secondaries, in which each output is controlled independently by each cell circuit (Figure 5.81). A more subtle disadvantage is that this balancer has difficulties charging those cells in the pack that have a higher resistance. This is unlike bulk charging, in which all the cells receive exactly the same current (regardless of

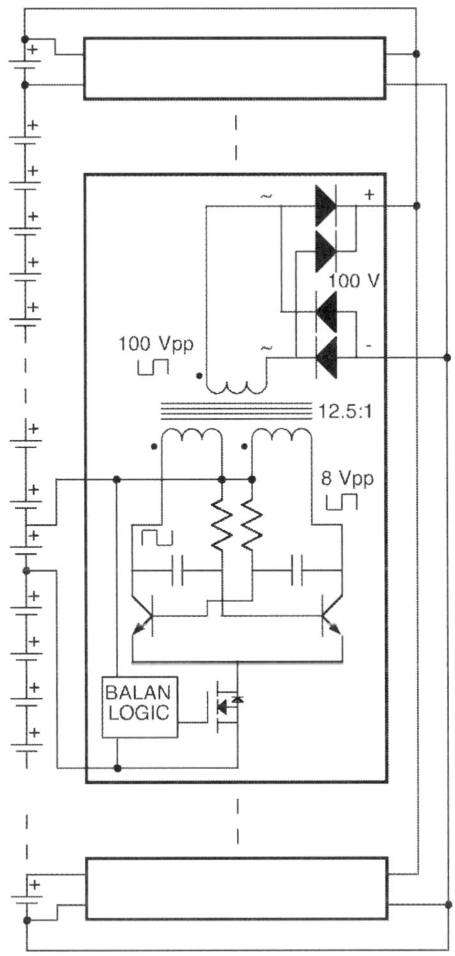

Figure 5.77 A cell-to-battery balancer using a multivibrator DC-DC circuit.

their resistance), with parallel charging all the cells receive the same voltage, and cells with high resistance draw less current from that voltage, taking longer to charge.

Bidirectional A bidirectional balancer has the advantages of both cell-to-battery and battery-to-cell balancers; it can be operated in any combination, simultaneously discharging some cells while charging others. It can be implemented with multiple, independent bidirectional DC-DC chargers [Figure 5.82(a)], or with a single, large DC-DC converter, using a multitap transformer and synchronously switching power electronics [Figure 5.82(b)].

The common bus of the DC-DC converters may or may not be connected to the overall battery terminals. Each approach has disadvantages:

- If connected to the pack terminals [Figure 5.83(a)], the DC-DC converters must be designed for a particular pack voltage which complicates the design and reduces flexibility.

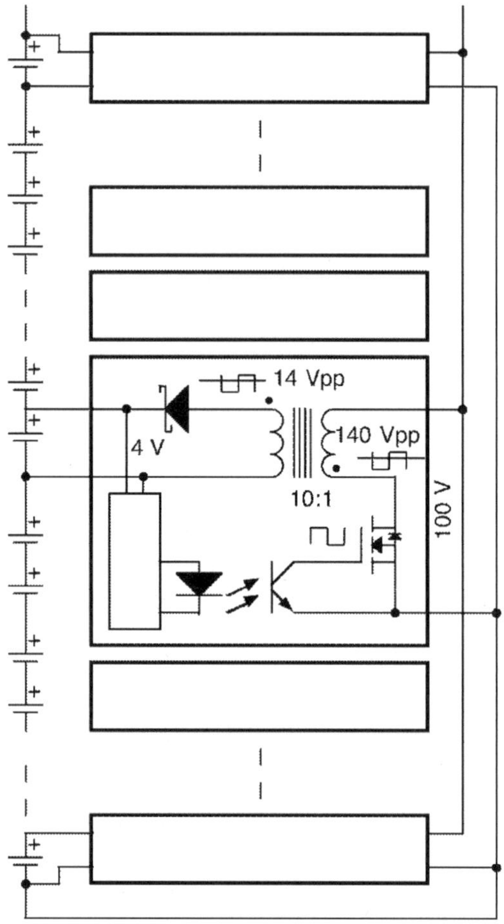

Figure 5.78 A battery-to-cell balancer circuit using the flyback DC-DC topology.

- Otherwise [Figure 5.83(b)], energy has to flow through two DC-DC converters from a fuller cell to an emptier one, reducing the efficiency of the energy transfer.

Just like in the battery-to-cell converter, this topology has difficulties charging those cells in the pack that have a higher resistance.

5.4.5.2 Redistribution

Redistribution (Section 3.2.4) uses the same circuits as parallel active balancing, though usually at higher power levels. The real difference between balancing and redistribution is in the software, which must know which DC-DC converter to turn on and when, so that all the cells are always at the same SOC. The algorithm must be optimized for the application. The most likely application of redistribution is land-based backup power, such as grid connected, peak shaving applications, and backup power for telecommunication plants or server farms.

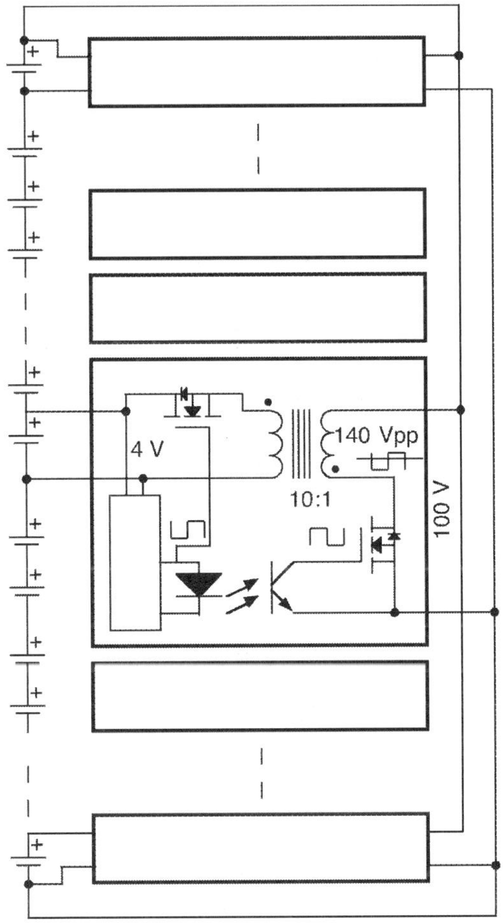

Figure 5.79 A battery-to-cell balancer circuit with synchronous rectifiers.

In order to do redistribution, the BMS needs to know each cell's SOC and capacity. The SOC can be set using the balancing techniques described above. But, to enable the BMS to measure cell capacity, the cells must go through a full charge or discharge cycle, which is not always desirable.

After the BMS knows each cell's capacity, it can start regular operation in redistribution mode:

- When charging, turn each DC-DC converter on and off with a duty cycle such that the power into each cell is proportional to its capacity. This results in all the cells reaching 100% SOC simultaneously.
- When discharging, turn each DC-DC converter on and off with a duty cycle such that the power out of each cell is proportional to its capacity. This results in all the cells reaching 0% SOC simultaneously.

Here is possible algorithm for a grid connected, peak shaving application, using cell-to-battery or bidirectional DC-DC converters.

Measure the capacity of each cell:

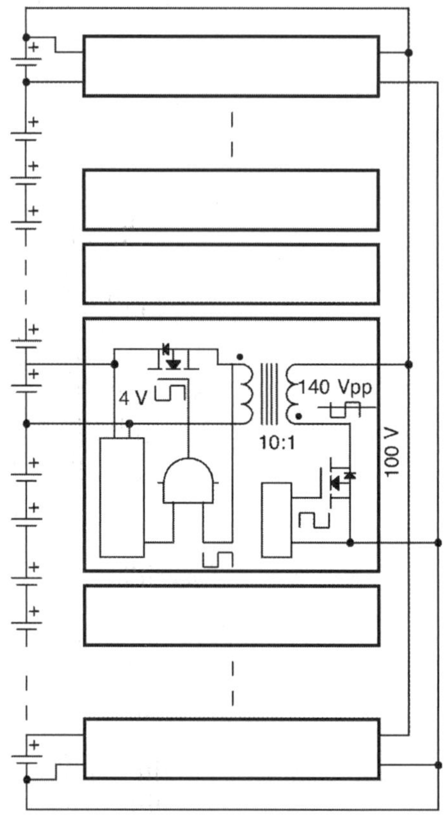

Figure 5.80 A battery-to-cell balancer circuit in with the cell board controls its own charging.

1. Balance the pack at the top. Remove energy from the most charged cells, then get a little more energy from the grid with the bulk power electronics to complete charging all the cells.
2. Wait for a time when peak shaving in not likely to be required (late evening).
3. Clear the battery DOD (set it at 0 Ah).
4. Discharge the pack into the grid using the bulk power electronics (without transferring any charge out of individual cells), integrating the current to keep track of the battery DOD.
5. When any cell reaches its minimum voltage, set each cell's DOD at the battery DOD value.
6. Turn off the bulk power electronics.
7. Balance at the bottom. Discharge each cell into the grid using its own DC-DC converter, integrating its current to increment its DOD, until empty.
8. Note the capacity of each cell as its DOD.
9. Calculate the capacity of the pack as the average of all the cell capacities.
10. Start redistribution mode (described above).
11. Recharge the pack.

For systems using battery-to-cell DC-DC converters, a possible algorithm is the following.

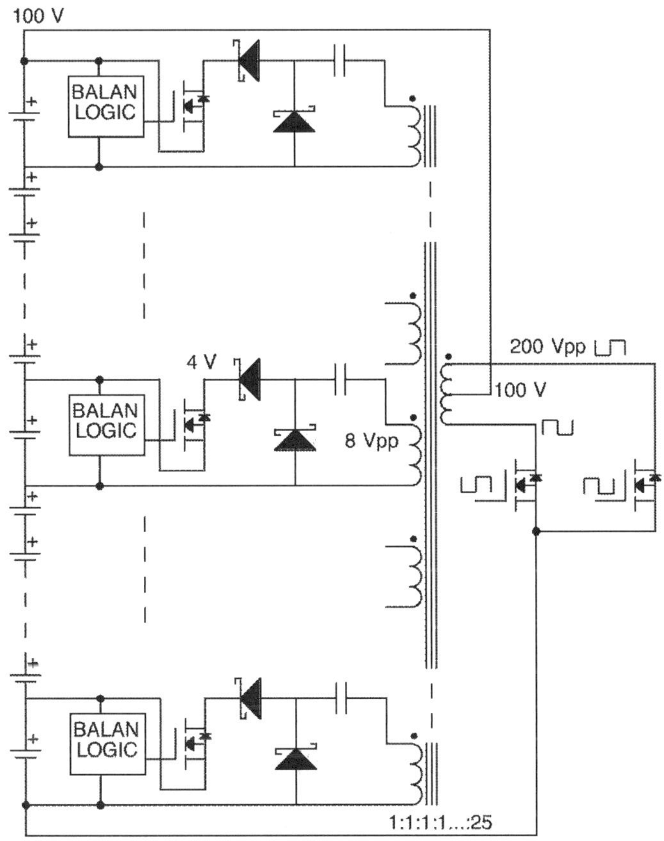

Figure 5.81 A battery-to-cell balancer circuit with a single DC-DC converter.

Measure the capacity of each cell:

1. Wait for a time when peak shaving in not likely to be required (late evening).
2. Discharge the pack into the grid using the bulk power electronics, until any cell reaches the minimum voltage.
3. Turn off the bulk power electronics.
4. Balance the pack at the bottom. Add energy to the least charged cells, then send a little more energy to the grid with he bulk power electronics to complete discharging all the cells.
5. Clear the battery DOD (set it at 0 Ah).
6. Charge the pack from the grid using the bulk power electronics (without transferring any charge into individual cells), integrating the negative of the current to keep track of the battery DOD.
7. When any cell reaches its maximum voltage, set each cell's DOD at the battery DOD value.
8. Balance at the top. Turn off the bulk power electronics, then charge each cell from the grid using its own DC-DC converter, integrating its current to increment its DOD, until full.
9. Note the capacity of each cell as its DOD.

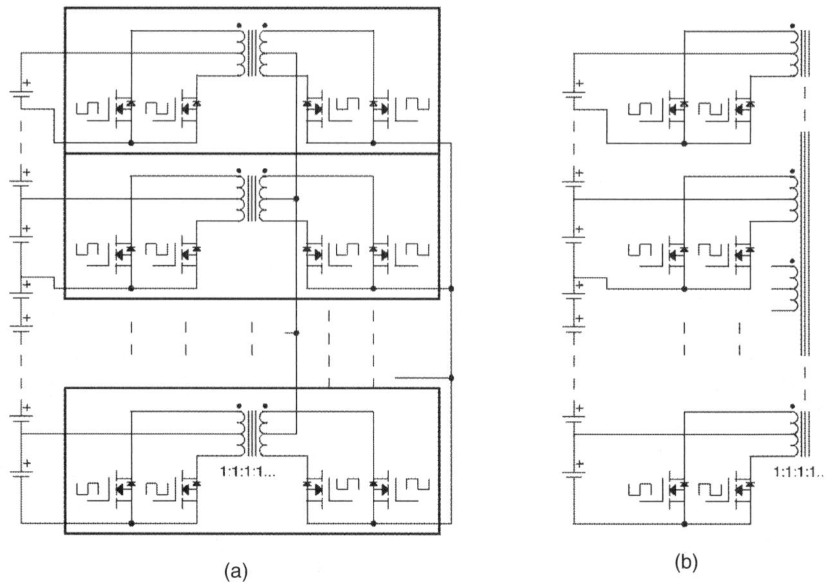

Figure 5.82 A bidirectional balancer circuit: (a) with multiple DC-DC converters, and (b) with a single DC-DC converter.

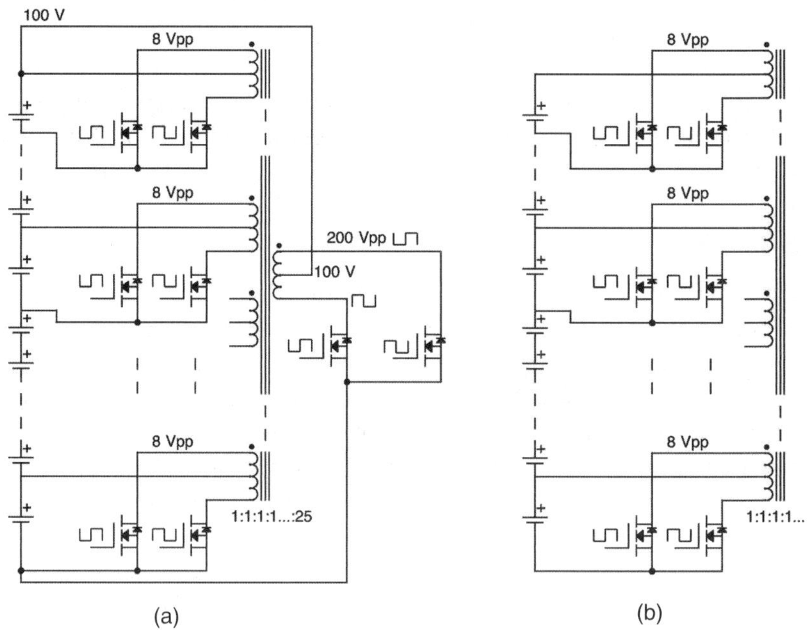

Figure 5.83 A bidirectional balancer circuit: (a) with battery connection and (b) without battery connection.

10. Calculate the capacity of the pack as the average of all the cell capacities.
11. Start redistribution mode (described above).

Here is a possible algorithm for a backup application for a server farm, using cell-to-battery DC-DC converters.

Part 1 is to do top balancing while waiting for a blackout.

1. Remove energy from the most-charged cells.
2. Get a little more energy from the grid with the bulk power electronics to complete charging all the cells.
3. Clear the battery DOD (at 0 Ah).

Part 2 is to power the server farm during the blackout, trying to measure the capacity of each cell:

1. Discharge the pack into the server farm using the bulk power electronics (without transferring any charge out of individual cells), integrating the current to keep track of the battery DOD.
2. If and when a cell reaches its minimum voltage, set each cell's DOD at the battery DOD value.
3. Tell the servers to do an orderly shutdown.
4. Power the little remaining load through the cells DC-DC converters, integrating each converter's current to increment that cell's DOD, until empty.
5. Note the capacity of each cell as its DOD.
6. Calculate the capacity of the pack as the average of all the cell capacities.
7. Start redistribution mode (described above).

Part 3 is to wait for power to be restored, and recharging the pack. Part 4 is to operate normally, handling the next blackout, but this time using all the energy in all the cells.

5.4.6 Switching

After measurement, evaluation, communication, and balancing, the next BMS function is switching the battery current to directly prevent operation outside its SOA. Only protectors implement this function.

The switch in a protector must be able to handle the peak load current for a short time and the average load current continuously. It also must be able to withstand the total battery voltage when it is open. Also, in large packs, in which the battery must be isolated from the control circuit, the switch must not bypass that isolation.

Switches can be implemented with contactors, transistors (typically MOSFETs, but some times IGBTs), or solid state relays (SSR). The latter is simply an isolated MOSFET, enclosed in a single module—an SSR saves design effort, but it has the same advantages and disadvantages as the MOSFET solution.

Table 5.7 compares the relative merits of each technology.

Transistors can withstand a high voltage or they can carry a high current, but they are not very efficient in application where they have to do both. Therefore, transistors are best for smaller batteries, while contactors are better for large packs (Figure 5.84).

Table 5.7 Comparison of Contactors and Transistors as a Protector Switch

	Contactor	*MOSFET*	*IGBT*	*SSR*
Voltage	200V to 1 kV	50 to 500V	600 or 1,200V	50 to 100V
Current	100A to 1,000A	30 to 100A[a] (up to 1,000A possible)	30 to 100A[a] (up to 1,000A possible)	30 to 100A
Turn Off Speed	< 100 ms	< 10 us	< 1 ms	~ 1 ms
Power Loss	~ 1W in the coil, ~ 0.1% of load power in the contacts	~0.3 % of load power at up to 200V—more at higher voltages	~ 1% of load power	~50W at nominal current
Isolation from Control Circuit	Inherent	Requires isolator	Requires isolator	Inherent
Isolation from Load When Open	Yes	No	No	No
Reliability (Normal Conditions)	Good	Excellent	Excellent	Excellent
Sturdiness (Abnormal Conditions)	Excellent	Good	Good	Good
Weight	High	Medium[b]	Medium[b]	Medium[b]
Volume	High	High[b]	High[b]	Medium
Cost	~$100 to $500	~$10 to $100[b]	~$10 to $100[b]	~$100

[a]Two transistors are required to handle both charging and discharging current.
[b]Including heat sink.

5.4.6.1 Contactors

A contactor works well because it can switch high voltage (such as 200 Vdc) and carry high current (such as 200 Adc), in either direction, efficiently (the heat wasted is on the order of 1W to 10W). Contactors are inherently isolated from the control (the coil). When two contactors, one on each pack terminal, are open, they totally isolate the pack [Figure 5.84(a)]. But contactors are large and expensive.

While contactors carry current just as well in either direction, they are better able to interrupt the current in just one direction. That is because they may include a mechanism to interrupt the arc that forms when opening an inductive circuit, which physically works only in one direction of current.[4] That is why many contactors specify a "+" contact terminal and a "−" one. Assuming that the load is inductive and the charger is not, and that the load current is significantly higher than the charging current, then the contactor must be oriented so that the load current flows into the "+" terminal of the contactor.

4. The contactor includes a magnet to quench the arc that forms when the contacts are opening a DC circuit. The field and the current in the arc are orthogonal to each other, resulting in a force in the direction that is orthogonal to both the current and the magnetic field, pushing the arc off to the side, against an insulating blade, and interrupting the current. The relay is physically designed for the arc to go in that particular direction. If the current is in the opposite direction, the arc will also move in the opposite direction, where the contactor does not have provisions to slice the arc.

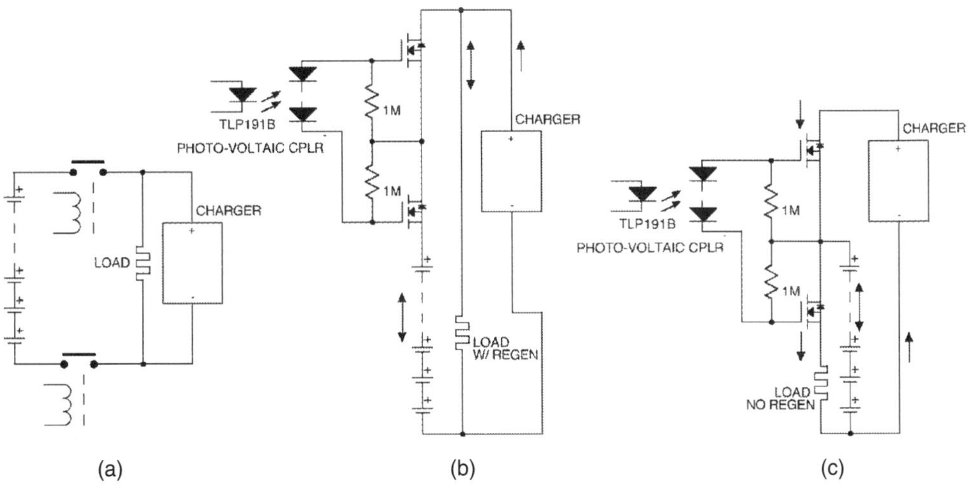

Figure 5.84 Switch in a protector: (a) contactors, (b) two MOSFETs in antiseries, and (c) two separate MOSFETs.

When selecting a contactor, note that there are three values for the load current (peak, average, and interrupted) and two ratings for the contactor current (carrying and breaking). The average load current must be within the contactor's carrying current rating. This rating is determined by the contactor's contact resistance when on, and its ability to dissipate heat. In most cases the peak load current can be ignored when selecting a contactor, because it has practically no effect. One example is a contactor with a carrying current rating of 100A, and a handle a load current that is, on the average, 80A, even though it may have peaks of 500A.

The contactor's breaking current rating (which may be lower than its carrying current rating) is for applications in which the contactor is the only means by which the full load current is interrupted, such as a direct connection to a DC motor. Such applications are less common today, as high-power loads are no longer driven directly, but through an electronic controller that can reduce the load current to 0 before the contactor is opened. If so, the "breaking current" rating of the contactor is generally moot. This rating only enters in effect if the contactor is ever in the position of having to interrupt the load current in an emergency, under full load current. Asking the contactor to do so decrements the contactor's number of cycles, which is rated to be on the order of 10,000 or higher. In a well designed system, such emergency interruptions should be rare, and the emergency is probably such that the loss of one contactor cycle out of 10,000 is the least of one's concerns.

5.4.6.2 Transistors

Transistors are small and relatively inexpensive, but they have several disadvantages compared to contactors:

- They are not isolated from the control.
- They do not isolate the load from the battery when open.
- They are unidirectional.

Transistors are unidirectional—they are intended to switch current in only one direction. In the other direction, MOSFETs and IGBTs include an intrinsic diode, which will allow current to flow in the opposite direction unimpeded. So, for example, a protector that uses a single MOSFET to interrupt the load current will not be able to prevent current from flowing from the load to the battery (for example, during regeneration). To interrupt current in both directions, a protectors needs either two transistors connected in antiseries (back to back) [Figure 5.84(b)], or two separate transistors, one for current to the load and one for current from the charger [Figure 5.84(c)].

Transistors are not isolated from their drive. If used in a BMS for a large Li-Ion pack, isolation must be provided between the control circuit and the transistor. In the examples above, a photovoltaic coupler is used to provide that isolation: when the coupler is powered, the light from its LED shines onto a few photodiodes in series, which generate very little current, though sufficient voltage to control the transistors' gates.

Often, many MOSFETs are placed in parallel, to increase the switch's current handling and efficiency. (Only rarely are they placed in series, to increase the switch's voltage handling; higher voltage transistors are selected instead.) A protector may have 10 MOSFETs for the charging current, and 10 more for the discharging current, all mounted on a heat sink.

MOSFETs are the transistor of choice in protectors, because they can be paralleled easily, and, in so doing, the switch's efficiency increases. Thanks to the positive coefficient of their resistance versus temperature characteristic, MOSFETs are particularly appropriate for paralleling. Should any of them carry more than its share of current, it would get hotter, increasing its resistance and resulting in more current going through its mates, balancing the load among all MOSFETs. As MOSFETs are paralleled, the switch resistance goes down in inverse proportion to the number of transistors, as does the power loss, resulting in greater efficiency.

For high voltages and very high power levels, IGBTs may be more appropriate than MOSFETs. IGBTs do not act as nicely when paralleled, as their characteristics do not ensure that they will share current equally. Also, on a first order, the power loss does not decrease as more IGBTs are paralleled, because the voltage drop across them is a constant, regardless of current. For paralleling, IGBTs must be binned at the factory and then used in matched sets. The main reasons to parallel IGBTs are to handle the total current and to spread out the heat generation. While switching losses are a concern in switch-mode power circuits, they are not an issue here, as these transistors are switched only once per power-up cycle.

Transistors are not as forgiving of short circuit current as contactors, so a protector may include circuits sensing the transistor current, to shut them off immediately if the current is excessive.

5.4.7 Logging

Logging is a job best left to loggers. A logger placed on a CAN bus can record not only BMS data (pack voltage, current) but also data from other devices (vehicle speed and GPS location) in a way that they can be correlated. Compare that to the limited information that a logger built into a BMS could record. Clearly a system-wide logger is more useful than a BMS based logger.

Still, it may be useful for a BMS to record error codes, and possibly a snapshot of the conditions at the time the error occurred (freeze data). That data should be available upon request, such through a PID request on the CAN bus (Section 5.4.4.2).

5.5 Cell Interface

A lot of thought must go into how a BMS is connected to the cells, to ensure reliability and ease of manufacture.

5.5.1 Nondistributed

Nondistributed BMS use $N + 1$ wires to connect to N cells (the so-called spaghetti) (Figure 5.85). We already saw the disadvantages of this approach (Section 2.3.4). Here I would like to talk about the advantages and implementation.

Wiring directly to a cell is usually straightforward. Whichever method is used to connect cells in series can be extended to connect a BMS tap wire to that same point. For example, for prismatic cells (whose terminals use bolts), the wire can be crimped to a ring terminal with a stud ID that is appropriate for the bolt size, and a barrel size that is appropriate for the wire gauge. To minimize the resistance of the cell-to-cell connection, the ring terminal should be placed between the bolt head and the power connection between cells (not between the cell and the bus bar).

Figure 5.85 Spaghetti in a centralized traction pack for a Prius PHEV conversion. Note the wire harness (with 80 or so white wires) routed from the BMS (bottom) to the various batteries.

To reduce spaghetti wiring's risk, a fuse or a resistor should be placed on each wires, as close as possible to the connection to the cells. Unfortunately, the resistance of these devices can compromise slightly the accuracy of the voltage measurement. Additionally, a resistor will not work if the BMS includes balancing, forcing you to use a fuse instead.

If using fuses, they must be rated for DC and for the full pack voltage, which can make them very large and expensive. The cheapest solution is fuses for DVMs, as they are rated 1 kV, yet they are small and readily available. If using resistors, they must be sized to dissipate the heat generated if the full pack voltage were applied across them. For example, for a 300-V pack, you could use 100 kΩ, 1W, leaded resistors (RC32 size). The hassle and cost of using a fuse or a resistor is such that very few companies go to the trouble (except for prototypes), preferring instead to rely on good design and good manufacturing practices to ensure that no shorts occur.

To meet certain safety standards, all wires connected to a traction battery pack, including BMS tap wires) must be orange. BMS tap wires do not need to be big: 24 AWG will do fine. However, their insulation needs to be tough (to avoid cuts) and rated for the pack voltage. Wire rated for 600V is readily available. For higher voltages, Teflon or silicone insulation can be used. Probe wire for DVMs, which is rated at 1 kV, is readily available, but it is too soft and it is not available in orange.

For safety, the battery pack must be designed so that the spaghetti wire is bundled and retained in such way that it doesn't come in contact with any high voltage conductor as it is routed to the BMS.

5.5.2 Distributed

In some regards, the requirements for distributed systems are simpler, as there are at most a few wires between the cell boards and the BMS master. These wires (or, more often, cables) need to follow the same requirements as the spaghetti wiring described in the previous section, except that they should not be orange, because they are low voltage, referenced to ground.

On the other side, a way of mounting the cell boards to the cells must be devised, for which there are no standard solutions. A lot of design work has to go initially into designing cell boards, but the end result can be very effective, reliable, low cost, and easy to assemble. Therefore, designing a distributed BMS from scratch is only recommended for BMSs that will be mass produced, not for one-off projects.

One approach is to mount the cell board in the vicinity of the cells (not directly to the cells) and use short pigtail wires to connect to the cells, with the techniques described in the nondistributed section above.

A better approach is to mount the cell boards directly to the cells. In that case, the mounting method depends on the format of the cells. The next sections will show actual examples of installation to each cell format.

5.5.2.1 Small Cylindrical Cells

Typically, small cylindrical cells are welded together with nickel foil (rarely, with copper foil). The foil can be shaped to have additional tabs for connection to the cell boards, to provide both mechanical mounting and electrical connection to the cell voltage (Figure 5.86). This way the cell boards will be held in intimate contact with the cells allowing them to measure the cell temperature with an onboard thermistor.

The cell boards do not take any additional space, as they are thin, and the electronic components fit in the space between cells (Figure 5.87). If the cells are contained in a frame, that frame can also include provisions to hold the cell boards. A single cell board for multiple small cylindrical cells can ease assembly considerably (Figure 5.88). The cells are welded together in a block, the cell board is placed against the side of the block of *N* cells, and *N* + 1 nickel tabs are folded down and soldered to the cell board. Installing the cell board can take just as couple of minutes.

Figure 5.86 Cell boards mounted directly small cylindrical cells.

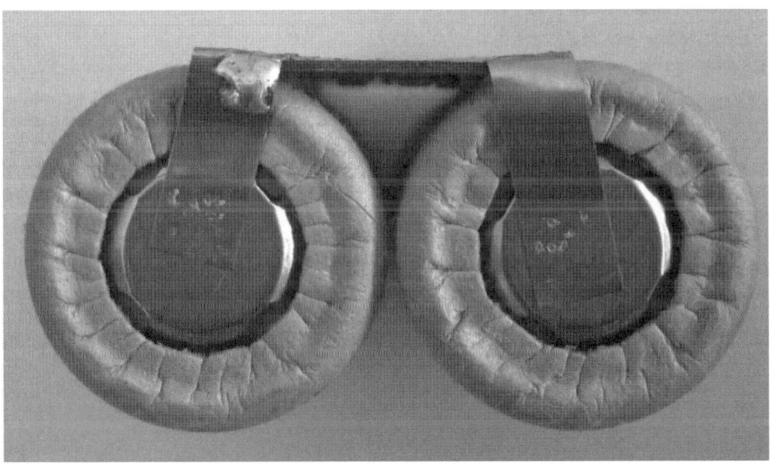

Figure 5.87 Cell board components nestled in the space between adjacent small cylindrical cells.

Figure 5.88 Single cell board for multiple small cylindrical cells, placed next to the cells (15 cells, 16 solder pads).

5.5.2.2 Prismatic Cells

For prismatic cells there are two basic approaches: direct PCB connection or ring terminals. Mounting a PCB directly to a prismatic cell's terminals is very attractive: a single PCB assembly functions as both a cell board and as two terminals. The board is designed for a particular cell size, with two holes of the right size and with the correct spacing to match the cell (Figure 5.89). The cell board is dropped onto the cell, on top of the power bus bars, and bolted down. There are some concerns about this approach:

- The bolt squeezes the PCB, whose thermal expansion coefficient does not match the coefficient for steel, which can result in a loose connection over time. Also, the PCB material (FR4) may be compressed over time, again resulting in a loose connection.
- The quality control with some prismatic cell manufacturers is not sufficient to guarantee a consistent spacing between terminals (I have seen a 5-mm cell-to-cell variation in that spacing).
- Mechanical vibrations between cells will be transferred through bus bars to the cell board first, and the cell second, stressing the cell board.

Figure 5.89 Example of a cell board with a PCB directly mounted to a cell: EV Power's distributed balancer. (*Source*: Rod Dilkes, EV Power, 2010. Reprinted with permission.)

Using ring terminals (which are perfect for mounting to cell terminals with bolts) eliminates all the concerns listed above. Two ring terminals may be mounted directly to the PCB (Figure 5.90). In this example you can see that the PCB is designed for two different cell sizes. By breaking the PCB along perforations, its length can be reduced, allowing a shorter center-to-center distance between the ring terminals. However, the PCB is still going to be subject to stresses, as described in the last point above.

A further refinement is to have only one ring terminal mounted to the cell board, to provide electrical connection and mechanical support, while the ring terminal for the other connection is at the end of a wire (Figure 5.91). This way there is no stress on the cell board due to vibration between cells. Another advantage of this approach is that the same cell board can be adapted to a variety of prismatic cells simply by changing the size of the ring terminal and the length of the wire. However, that introduces another stress, as now the cell board is cantilevered. To minimize that concern, the cell board must be small and with few, light components so that its moment of inertia is minimized.

The connection between the ring terminal and the PCB is an area of concern: if the ring terminals are soldered to the PCB, the adhesive that holds the copper to the FR4 material will be stressed. It is better if the ring terminal is press-fit to the PCB, so that it is holding onto the FR4, not onto the copper. The solder joint between the terminal and the PCB is there only to provide an electrical connection, not mechanical support.

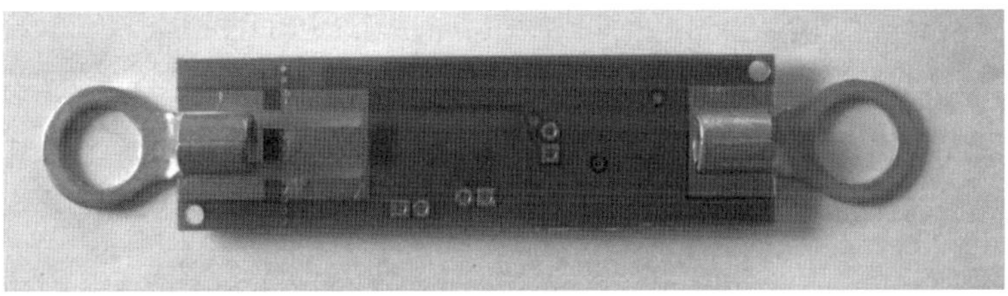

Figure 5.90 Cell board for prismatic cell with two ring terminals mounted to the PCB.

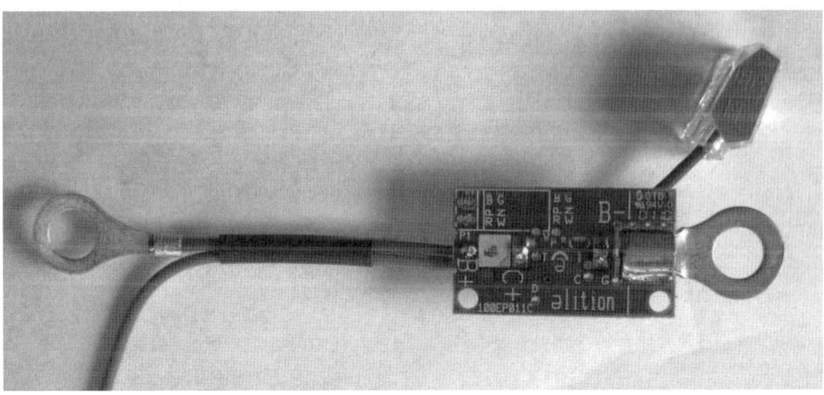

Figure 5.91 Prismatic cell board with a ring terminal on the PCB and one on a wire.

A single cell board for multiple prismatic cells can ease assembly considerably. The stresses involved are such that mounting the cell board directly to the cells is not advisable. Instead, the cell board can have $N + 1$ ring terminals for N cells (Figure 5.92).

5.5.2.3 Large Cylindrical Cells

Just like prismatic cells, large cylindrical cells also use bolts for connections. However, those bolts are at opposite ends of the cell, so a prismatic cell board will not work. Instead, a version of the single ring terminal, prismatic cell board can be used, except that the ring terminal on the wire must be on a longer wire, long enough to reach the other terminal at the other end of the cell. In this example, a cell board for two adjacent cells has two ring terminals for one end of those two cells, and a long wire with a ring terminal for connection to the bus bar that connects those cells at their far end (Figure 5.93).

5.5.2.4 Pouch Cells

Pouch cells are trickier to use. The best cell board design for them is a single cell board for multiple cells. The cell board has slots for the cells' tabs and contains mul-

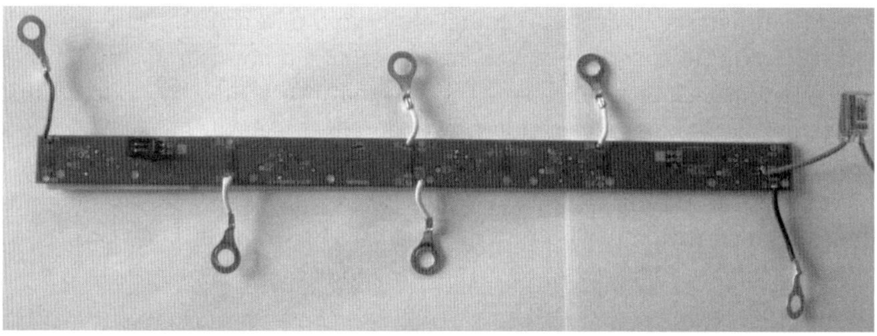

Figure 5.92 Single cell board for multiple prismatic cells (five cells, six terminals).

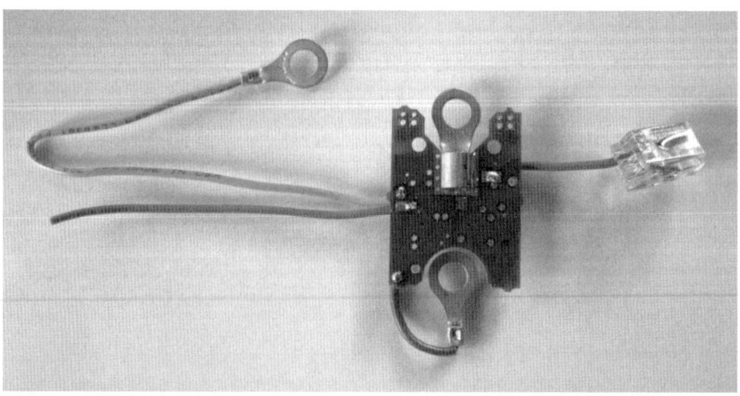

Figure 5.93 Cell board for two large cylindrical cells.

tiple sets of cell board electronics. The tabs of the cells are slipped into the slots, and may be soldered in place (Figure 5.94).

The PCB itself may be relied on to interconnect the cells. If the currents are too high for that, then the tabs may also be joined directly. In most cases, the resistance of the PCB connections is orders of magnitude lower than the resistance of the cells. Here is an example from an actual cell board:

- Cells: pouch cells with an 80-mΩ resistance;
- PCB: 2-oz copper, double sided;
- Trace between adjacent cells: 8 mm × 10 mm;
- Trace resistance between adjacent cells: 0.2 mΩ;
- PCB's I^2R loss at 200-A peak current: 0.1W per junction between cells;
- Cell I^2R loss at 200-A peak current: 65W per cell.

Therefore, in this example, the PCB (whose resistance is 1/400th of the cell resistance) is far from being the limiting factor to the current handling of the battery. In other cases, with very high currents and very low cell resistances, the PCB cannot be relied on to carry the current, and the cell tabs must be connected directly together for that purpose.

There are some concerns that soldering the cell's tabs will damage the cells. Cell manufacturers are divided on this regard. Some worry that the heat from soldering the cell's tabs will melt internal layers in the cell; others have no such concerns, and will sell battery packs with pouch cells soldered directly to PCBs.

The main disadvantage of a single cell board for multiple pouch cells is the effort required to replace a cell. From that point of view, individual cell boards are preferable. To handle the narrow distance between adjacent pouch cells in a stack, a tiny cell board can be placed vertically, in the same plane as the cell, between the tabs. In this example, eight cell boards are stacked, spaced so as to match the pitch between cells (Figure 5.95). They are fitted with nine flags, one for each node in an eight-cell battery. The stack is then dropped onto a stack of eight cells, and the flags are squeezed between the tabs when the battery connections are made.

Figure 5.94 Cell board for multiple pouch cells (12S7P arrangement).

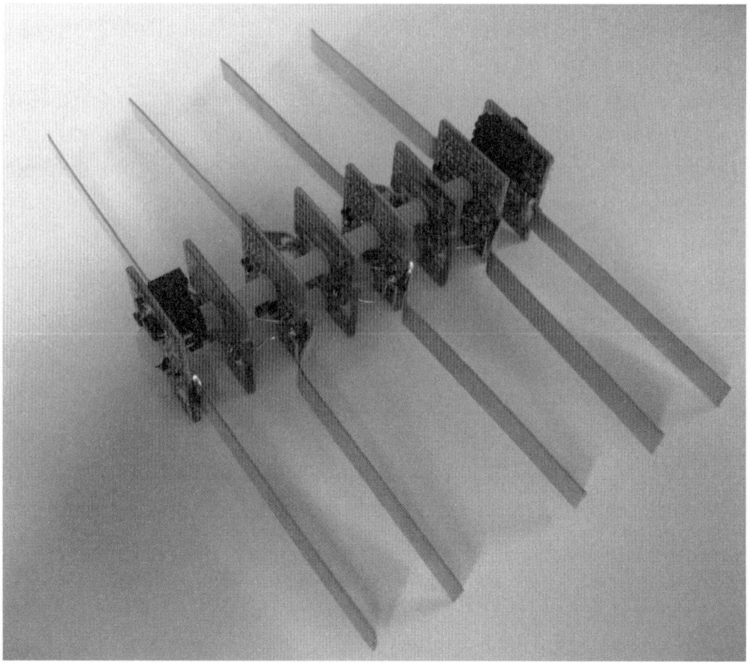

Figure 5.95 A stack of eight individual pouch cell boards.

5.6 Distributed Charging

The characteristics of distributed charging were described earlier in Section 3.2.5. To reiterate an important point, parallel charging inherently balances all the cells, with no risk of overcharging, though it has difficulties charging high-resistance cells, because they receive less current than the other cells, so charging takes longer than when using bulk charging (charging in series).

The beauty of this approach is offset by the low efficiency of many low-voltage chargers, because small chargers tend to be less efficient than large ones, and low-voltage chargers are less efficient that high-voltage ones because the voltage drop of rectifier diodes (~1V) is a larger portion of the output voltage in low-voltage chargers (~20%) than in high-voltage chargers (~1%).

The simplest way of doing distributed charging is to use many individual, small chargers [Figure 5.96(a)]. This solution requires little design effort, but the cost of many small chargers quickly adds up. A less expensive alternative is to use a single large charger with a single transformer and multiple low voltage secondaries [Figure 5.96(b)].

Feedback to set the output voltage must be taken from the primary, or from a secondary dedicated just to voltage feedback, because sampling the voltage of one of the output connected to a cell, which happens to have a low voltage will lead to other outputs being too high and overcharging their cells.

Initially, this circuit will charge all the cells equally. As cells get charged, this circuit will charge fewer and fewer cells, ending with just the lowest cell being charged by itself at the very end. More and more power is available to charge the remaining

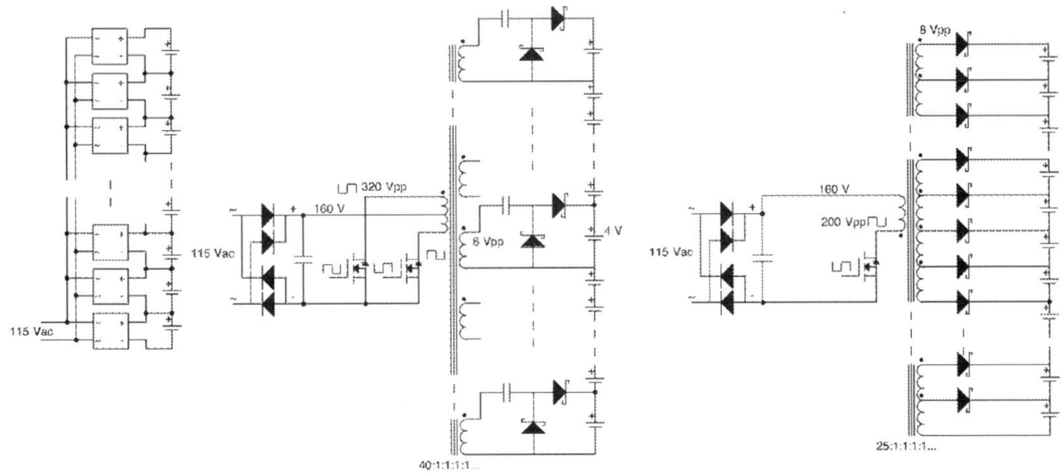

(a) (b) (c)

Figure 5.96 Distributed-charger circuits: (a) individual chargers and (b) single charger with individual secondaries. (c) Single charger with one secondary with multiple taps.

cells, so this circuit can charge a pack more rapidly than a bulk charger of the same power.

Instead of a transformer with multiple secondaries, it is possible to use a transformer with a single secondary with multiple taps [Figure 5.96(c)]. The flyback topology must be used. With this circuit, charging is rather uneven. Initially only one diode will operate, and just the more negative half of the battery will be charged, bringing the battery further away from balance. Given enough time, eventually all the voltages do reach equilibrium and all the cells will be charged. Finally, as some cells reach full charge before others, balancing starts occurring naturally. Therefore, this circuit is not recommended, due to its poor performance.

CHAPTER 6

Deploying a BMS

Whether an off-the-shelf product or a custom design, a BMS must be properly deployed to do its job well. It must be installed properly, configured for your application, and fully tested.

6.1 Installing

BMS installation depends to a great degree on the particular pack, the BMS itself, and the external system. Therefore, all I am able to do is to present a few tips, some general concepts, and a few examples.

6.1.1 Battery Pack Design

Careful attention early on to the design of a battery pack will save significant effort during its production and avoid problems during its use.

6.1.1.1 Pack Layout

Battery packs should be designed so that its batteries are directly accessible. It should not be necessary to remove one battery to reach another one. It would be unfortunate if in the process of removing a good battery to reach one that needs to be repaired, the good battery would be damaged.

An appropriate location for a BMS electronic assembly (such as a BMS master) is outside the batteries but inside the pack. This makes the BMS accessible for testing and service, while reducing its exposure to the electrical noise typically experienced inside batteries.

Cells in Parallel Versus Batteries in Parallel

You may be faced with the choice between designing a pack that uses multiple batteries in parallel, versus one that uses more cells connected directly in parallel. The fact is that using parallel batteries has few real advantages and significant disadvantages in performance and cost.

Often, packs are resigned with multiple batteries in parallel, for the sake of using a modular design and to enable expansion of the pack (such as to give the user the option of trading range for weight). While this may be a valid reason to design a pack that places batteries in parallel, one must also consider the many reasons why doing so will increase the cost and reduce the performance of the pack.

You may design a pack with multiple batteries in parallel because you believe that the reliability will be improved: should one battery become defective, it could be disconnected, and the product will be able to operate without that particular battery. That may be true in theory. However, in practice, I have never seen it work as planned. A vehicle with a defective battery is brought back to the shop to remove that battery, defeating the original goal of keeping the vehicle operational. The real difficulties will be when the battery is reinstalled. The difference in the SOC levels between the old battery and the new one may cause some challenges, including a phenomenal inrush current, and an excessively long time to balance the pack.

Not only does a pack with multiple batteries in parallel have limited advantages, but it also suffers from significant penalties in the form of reduced pack performance, due to higher sensitivity to cell-to-cell variations, and higher cost. Instead, cells should be connected directly in parallel (lattice network), as doing so simplifies the design (reducing the pack's cost) and provides major improvements in performance in the presence of weak cells.

Using *N* batteries in parallel requires *N* times as many BMS tap points (or *N* times as many cell boards). For example, a pack arranged as a 4S4P requires a BMS that can support 4 tap points (or 4 cell boards) [Figure 6.1(a)]. The same cells used in a pack that is divided into 4 batteries that are then connected in parallel (a $4 \times$ 4S1P arrangement) requires a BMS that can support 16 tap points (or 16 cell boards) [Figure 6.1(b)].

A pack with a few low-capacity cells, distributed randomly, has more capacity if it uses blocks of cells wired directly in parallel than if it is split into multiple batteries

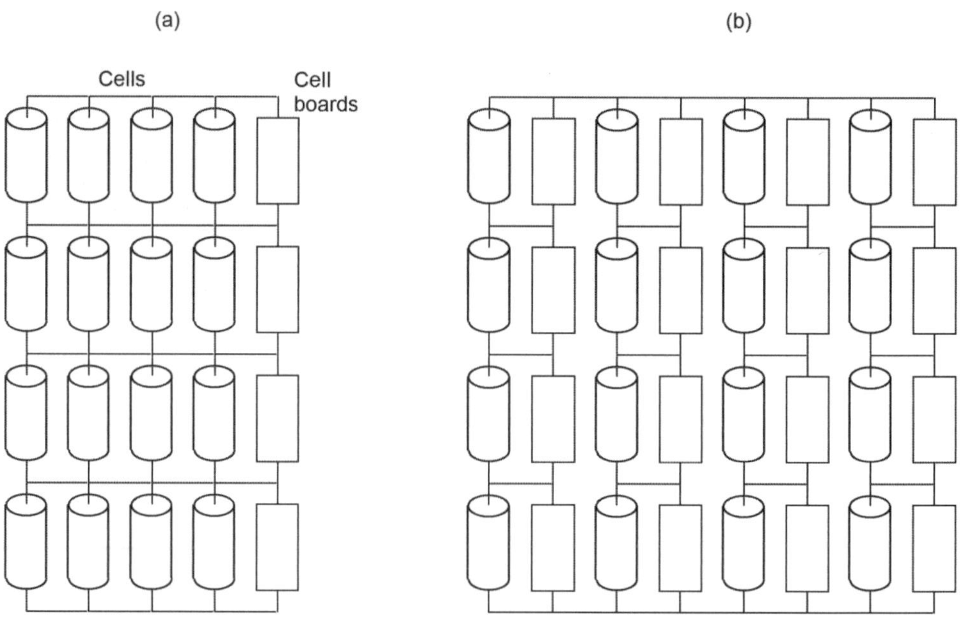

Figure 6.1 Cells directly in parallel (a) require fewer cell boards than (b) batteries in parallel.

in parallel. (We assume that there are only a few low-capacity cells, so that it is statistically improbable that a particular cell block would contain more than one low-capacity cell.) That is because the low-capacity cells are not likely to be all in the same cell block, so their effect is not additive. Each cell block with a low-capacity cell does limit the capacity, but each of these blocks limits the capacity to the same value. For example, let us look at a pack with 16 each, 2 Ah cells, 4 cells in series.

With cells connected directly in parallel [Figure 6.2(a)], each block has a capacity of 8 Ah. If a block includes a cell whose capacity is reduced to 1 Ah, that block's capacity is 7 Ah. The capacity of any other block with a low-capacity cell is also 7 Ah. So the pack's capacity is 7 Ah, no matter how many blocks have a low-capacity cell. If the pack is split into 4 batteries in parallel [Figure 6.2(b)], each battery that includes a low-capacity cell will have a 1 Ah capacity, so the total pack capacity is only 5 Ah.

A pack with just 1 low-capacity cell has more capacity if it uses blocks of cells wired directly in parallel than if it is split into multiple batteries in parallel. That is because cells in parallel support each other, so that a low-capacity cell is not a major limiting factor. In the example above, with cells connected directly in parallel [Figure 6.3(a)], if a cell's capacity is reduced to 1 Ah, its set of parallel cells is the limiting factor to the pack's capacity, which is reduced from 8 Ah to 7 Ah.

On the other side, if the pack is split into four batteries in parallel [Figure 6.3(b)], they will be depleted at the same rate, until a 4-Ah charge is extracted from the pack. At that point, the 1-Ah cell will be empty, its voltage will drop, and the BMS will shut down the pack. Therefore, the effective capacity of the pack is 4 Ah.

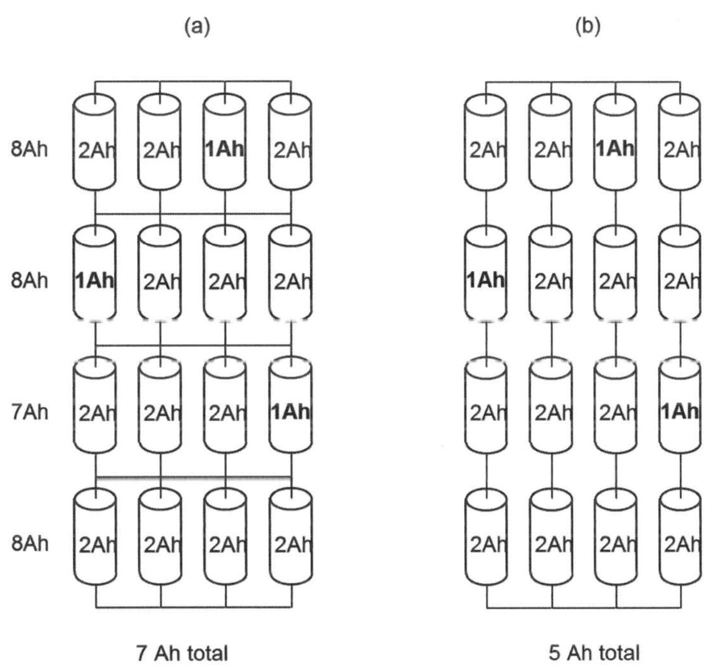

Figure 6.2 Multiple low-capacity cells. (a) Their effect is small when placing cells directly in parallel, and (b) is significant with batteries in parallel.

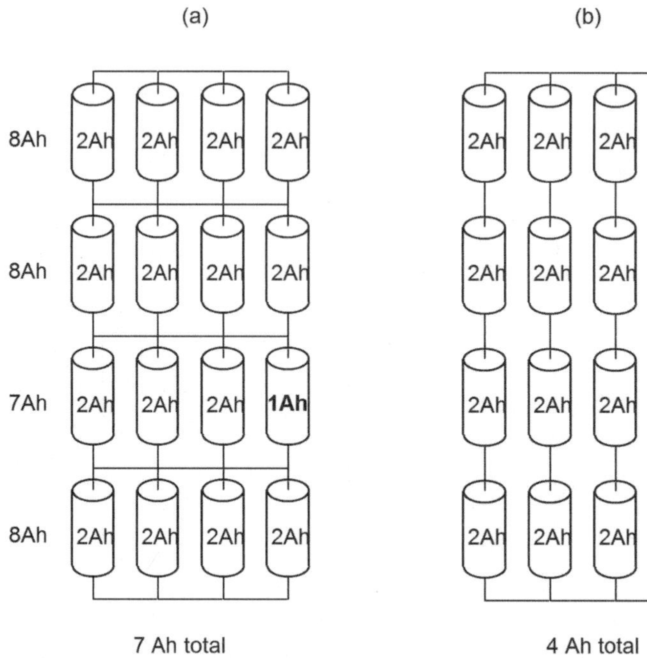

Figure 6.3 A single low-capacity cell. (a) Its effect is reduced by placing cells directly in parallel. (b) The cell compared to using batteries in parallel.

(Actually, it is really not quite as drastic, because when the low-capacity cell is nearly depleted, its resistance increases, reducing the current drawn from its battery, allowing the other batteries to be discharged a bit further.)

A pack with a bad cell (such as a high-resistance cell) is able to continue operating if the cells are wired directly in parallel, but will be shut down immediately if it is split into multiple batteries in parallel.

That is because, with cells connected directly in parallel [Figure 6.4(a)], the other cells will support the bad cell. Of course, the pack's capacity will be reduced somewhat, but at least the pack will remain in operation.

On the other side [Figure 6.4(b)], a pack with a bad cell, in a battery by itself, is quickly shut down. As the pack is discharged, little current flows in the battery with the bad cell, due to its high resistance. Therefore the voltages of the other cells in that same battery remain constant. Consequently, as the pack voltage is reduced during discharge, the delta in the pack voltage is applied directly to the bad cell. Its voltage is reduced drastically, and may even reverse. When the BMS notices that the bad cell's voltage has dropped below the low voltage cutoff, it shuts down the pack. Therefore, placing cells directly in parallel reduces the parts count and the effects of cell-to-cell variations, resulting is a less expensive, more reliable, and better performing battery.

6.1.1.2 Battery Layout

When designing a battery, consider carefully how cells are placed and oriented within it. Do spend a great deal of time upfront planning each battery and the pack,

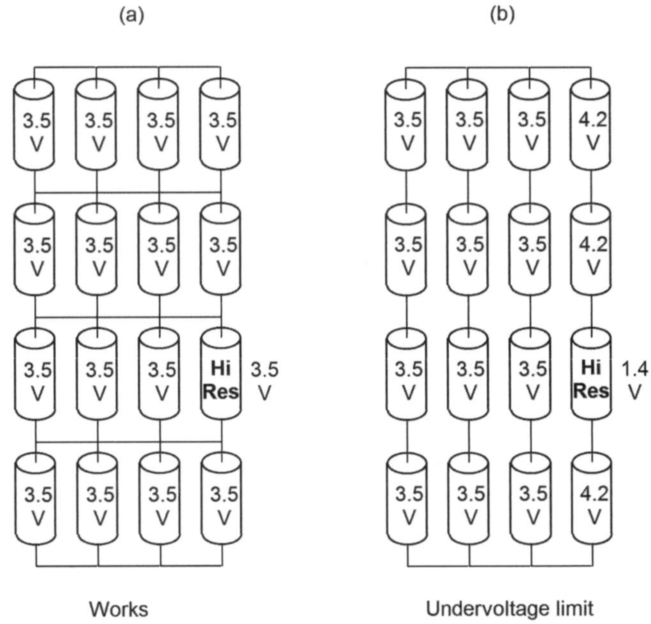

Figure 6.4 A pack with a bad cell. (a) The pack will keep on working if the cells are placed directly in parallel. (b) The BMS will shut down the pack if split into batteries in parallel.

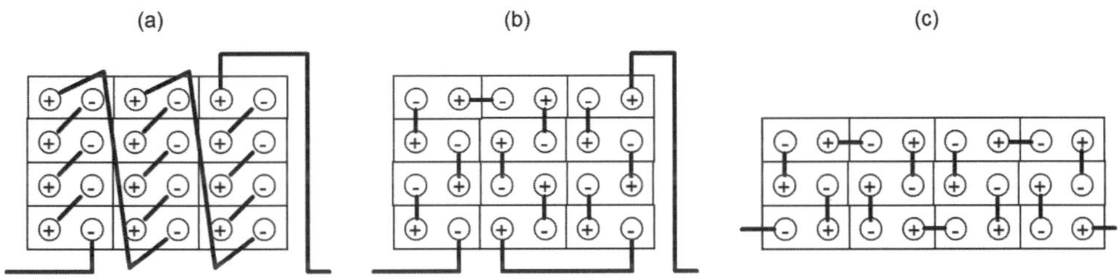

Figure 6.5 Battery layout: (a) original, (b) optimized by flipping cells, and (c) optimized by changing from an even number of cells per row to an odd number.

to minimize the meandering of power cables, avoid areas with narrow clearance between exposed conductors, and make the power wires exit a battery in a most direct route. Just flipping cells 180°, or going from an even number of cells to an odd one, can improve the power routing drastically (Figure 6.5).

Use a 3-D CAD program to experiment with cell placement, or, more simply, use mock-ups to simulate the battery in the application. You can make mock-ups of prismatic cells out of building insulation panels, and place them in an actual vehicle (Figure 6.6).

A minute spent planning a battery meticulously can save an hour building the first one, and, in the long run, can improve manufacturing considerably. If you are not adept at visualizing and rotating objects in 3-D in your mind, find another person who can do so. As a matter of fact, do ask others to look over your design regardless. They will see things that your familiarity with your design will keep you from seeing.

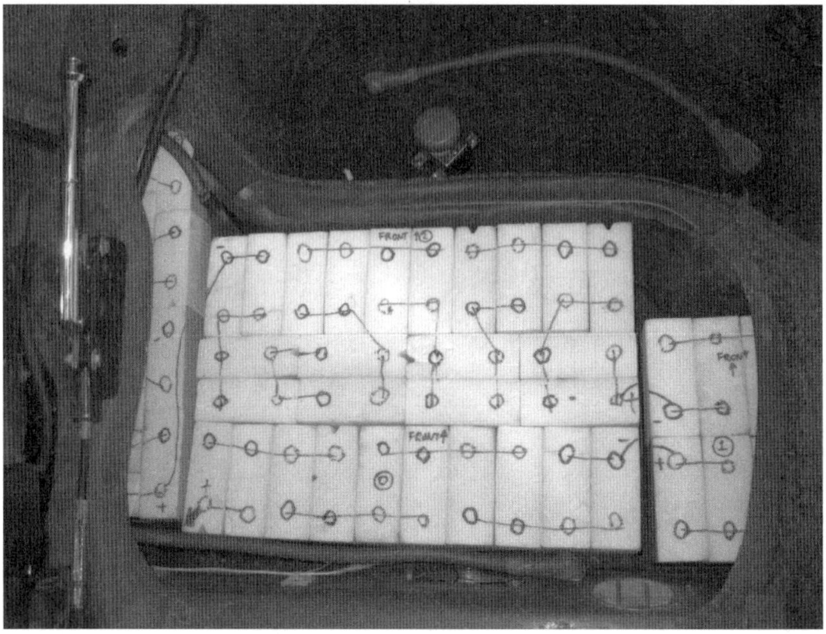

Figure 6.6 Mock-up prismatic batteries in a Sparrow EV.

6.1.1.3 Battery Assembly

Here are a few tips on battery assembly.

Balancing Before Assembly

It is far better to build a battery that is balanced from the start, than to balance it by hand after it is built, or ask a BMS to do so (which can take literally weeks, depending on its level of balance current). One way to build a balanced battery is to use cells that are known to have been delivered from the manufacturer all at the same SOC, and known to all have the same history.

Another way to build a balanced battery is to charge each cell individually to its top cutoff voltage (such as 3.6V or 4.2V, as appropriate for the cell chemistry), before you assemble the batteries. Better yet, you could connect all the cells in parallel and then charge them. Not only is this faster (assuming you are using a high-current charger), but if you let the cells sit for a day or so, swapping charge, until there is no current flowing among them (and therefore no errors due to variations in internal resistance), all the cells will be exactly at the same OCV, and therefore essentially at the same SOC (Figure 6.7).

Cell Containment

Unlike cylindrical cells, pouch and prismatic cells expand whenever their SOC is high (Section 1.2.1). That expansion must be contained in order to avoid permanent deformation and delamination of the internal layers. The easiest way of accomplishing this is to use cells that have not yet been fully charged, stacking them flat against each other, clamping the stack with thick metal plates at either end, and strapping the stack so that it may not expand (Figure 6.8). The metal banding technology used in packaging pallets works well for this.

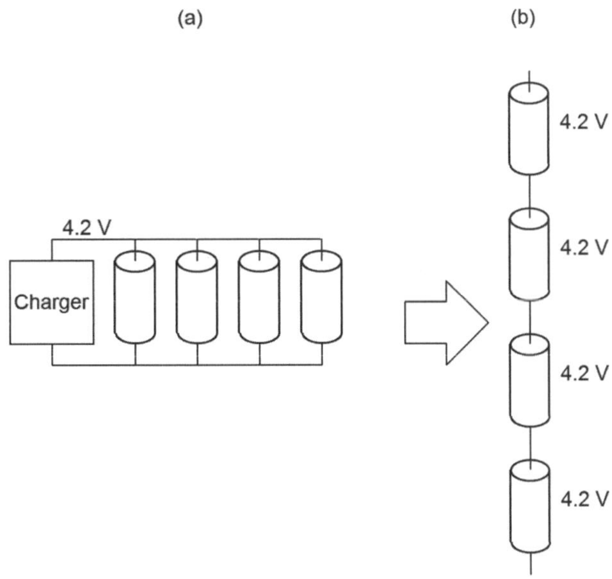

Figure 6.7 (a) Balance the cell in parallel, and (b) then connect them in series in the battery.

Figure 6.8 Prismatic cell stack.

Cell Isolation

Usually it is obvious whether the outer case of a cell is conductive (small cylindrical), or isolated (prismatic). What may not be obvious is that, while the plastic bag of a pouch cell appears to be isolated, actually it may not be: the fused seams along the edges of Kokam pouch cells are conductive, and there is a voltage between them and the terminals. Victor Tikhonov of Metric Mind found that the hard way when he used aluminum heat sinks around a pouch cell stack. Cells were discharged to 0 volts. Therefore, do insulate pouch cells' edges if used in a conductive container.

6.1.1.4 BMS Placement

When designing a battery, consider the ease of installation of BMS tap wires (nondistributed BMSs) or cell boards (distributed BMSs); consider access to them after the pack is installed, for service and troubleshooting; and consider the visibility of LEDs on cell boards (if any).

If your BMS uses cell boards, plan to leave space for them, especially if the cell boards use high-current balancing and therefore radiate a significant amount of heat.

If a BMS can be connected to multiple points in a cell block, choose the best one from an access standpoint. For example, if cell boards could be placed on one side of a battery or the opposite side, choose the side that, when the battery is first opened, lets you immediately have access to the BMS connections.

6.1.1.5 Power Circuits

Proper design of the power connection and routing in a battery is essential for safety.

Power Connection Fastening

There are two schools of though on battery terminal fastening: one says that you should use just a flat washer, the other one that you should also use a split washer (SEMS). The flat washer argument comes from the automotive industry, which states that, in the presence of vibration, proper torquing is what keeps a bolt reliably fastened for years. A flat washer is the ideal element to receive the force from the bolt's torque, which deforms the washer slightly, to achieve complete contact area.

The split washer argument (Figure 6.9) comes from battery people, who have used that technique successfully for a long time. Their argument is that a split washer continues imposing a force keeping the various elements in the terminal under constant pressure, against different expansion rates of the various metals (e.g., copper ring terminals, brass split washers, and steel bolts). The flat washer solution relies on having a complete contact among the various surfaces and the split washer solution relies on having a constant pressure maintaining contact under varying conditions. (I go along with the split washer school.)

Regardless, what both schools agree on is that a flat washer is required, and that serrated lock washers (internal or external star) are not acceptable. Lock-nuts (such as Nylon insert nuts) do remain in place when subjected to vibration and thermal expansion, but may only be used once. The risk is too high that they may end up being reused by an inexperienced technician, so do not use them.

Your arm is not calibrated. Always use a torque wrench to tighten terminal bolts to the torque specified by the engineer. While a single loose bolt holding a battery down may not cause damage, a single loose bolt in a power connection can be disastrous. When a power connection starts getting loose arcing results, which may start a fire. A bolt coming undone may allow a power cable to flop around, possibly

Figure 6.9 Power connection fastening.

shorting across another battery terminal and starting a fire, or may make a connection to the chassis, creating a shock hazard. Two design features can reduce that risk: create an antirotation feature to prevent a power cable from rotating and thereby loosening its bolt (such as by securing the cable along the side of the battery), and add a feature that prevents a cable that should become disconnected from moving away from the terminal.

In power connections, use all the parts that the engineer specified, and only those parts. Not doing so may cause arcing, and may start a fire. The PHEV Prius fire described in the introduction (Section 1.2.3) was initially caused by missing flat washers, which resulted in a power connection being clamped onto a plastic sheet instead of the washers. In the summer heat, the plastic softened, removing pressure on the contacts. Arcing ensued, which ignited nearby combustible material. This overheated Li-Ion cells, which exploded. The cells' innards made contact with the pack's metal case, creating a short circuit between two different batteries with about 100-V difference in potential. This created plasma, which cut through the metal case, allowing a fountain of sparks to escape that ignited the carpeting over the pack. The fire then moved on to the rest of the upholstery, burning down the car [1]. No one got hurt. For want of a washer, a Prius was lost. The manufacturer of that PHEV conversion, Hybrids Plus, ceased to exist soon after that.

You have to be careful with aluminum cell terminals (typically found in prismatic cells). You must either use aluminum bus bars, or, to use copper connections, remove the top oxidation layer on the terminal with sand paper and immediately treat that area with a deoxidant such as Noalox.

When using prismatic cells, check the terminals to make sure that the large nut retaining the inner stud does not extend past the stud. If it does, the electrical connection will be through the nut, not directly to the stud. This results in additional resistance as the current has to flow through the thread lock between the stud and the nut (Figure 6.10).

Current Splitter

An often overlooked trick of the trade when dealing with large currents and small current sensors (both in the sense of physically small or in the sense of low current handling capability), is that it is easy to split the current into multiple paths, some going through the current sensors and some not. For example, using two identical wires connected at both ends, each wire will carry half of the current. If only one of these wires goes through the opening in a cable-mounted current sensor (Figure 6.11), that sensor will only need to handle half the current (and will report half of the current). Extending the concept, for example, it is possible to use a 100-A current sensor to measure the current in a 1,000-A circuit.

Source and Load Current Sensing

In many applications the battery current has two separate paths:

- From a source, such as a charger (or to it);
- To a load, such as a motor controller or an inverter (or from it).

Note that in some applications, especially grid connected systems (Section 6.1.3.4), current can flow to the source as well as from the source: vehicle to grid

Figure 6.10 Prismatic cell terminal. A center stud is secured with a large nut. The nut must not extend past the stud. The aluminum stud must be treated with deoxidant if it is to be used with copper connections.

Figure 6.11 Running only 1 of the wires in a bundle splits the current seen by the current sensor.

(V2G), vehicle to home (V2H), distributed resources (DR), and peak shavers (to supplement the grid from a battery in case of peak demand).

Note also that in many applications, especially for electric vehicles, current can flow from the load due to regenerative braking.

In applications with two current paths, it is often beneficial to measure each current independently, especially if their magnitudes are significantly different. One current sensor is placed on a power wire between the source and the battery, while the other one is placed on a power wire between the battery and the load. For example, in a typical EV that is charged overnight and is driven for 1 or 2 hours, the charge current is an order of magnitude lower than the peak load current. In this example, a small, unidirectional current sensor can be placed in series with 1 of the power wires from the charger to sense the charging current, while a large, bidirectional sensor can be mounted on the power wire to the motor controller, to measure the load current to the motor controller (and back from it in case of regenerative braking).

Current Shunt Wiring

Current shunts produce a very small voltage, proportional to the current. Care must be taken to avoid overwhelming that small voltage with electrical noise, or losing accuracy from improper connections to the shunt. If the current sensor amplifier is not mounted directly on the shunt, use a twisted pair, shielded cable between the two (Figure 6.12). Ground the shield only at one end (typically at the amplifier end).

Use kelvin connections for the sense wires. The sense connection points must be different from the power connection points. Additionally, the sense connection points must be "inside" the power connection points, so that the uncontrolled resistance at the power connections has no effect on the resistance between the two sense points.

Figure 6.12 Connections to a shunt current sensor.

Contactors and Precharge

When initially connecting a pack to a capacitive load (such as a motor controller), an inrush current occurs as the load capacitance is charged up to the pack voltage. With large, low-resistance batteries, and loads with large capacitors across the input, the inrush current can easily peak at 1,000A. A precharge circuit limits that inrush current, without limiting the operating current. A precharge circuit between a pack and its load is required if any of the following are true:

- The load has input capacitors that will be damaged by the inrush current.
- A fuse will blow if subjected to the inrush current.
- Contactors (high-power relays) will be damaged by the inrush current.
- The pack cells are not rated for the inrush current.

Function A precharge circuit (Figure 6.13) consists, at the minimum, of:

- A precharge resistor, to limit the inrush current (R1).
- A contactor (K2) across the precharge resistor to bypass it during normal operation.

Additionally, the precharge circuit may have:

- A precharge relay (K1), to keep the load from being powered through the precharge resistor when the system is off.
- A contactor (K3) in series with the other pack terminal, to isolate the load when the system is off.

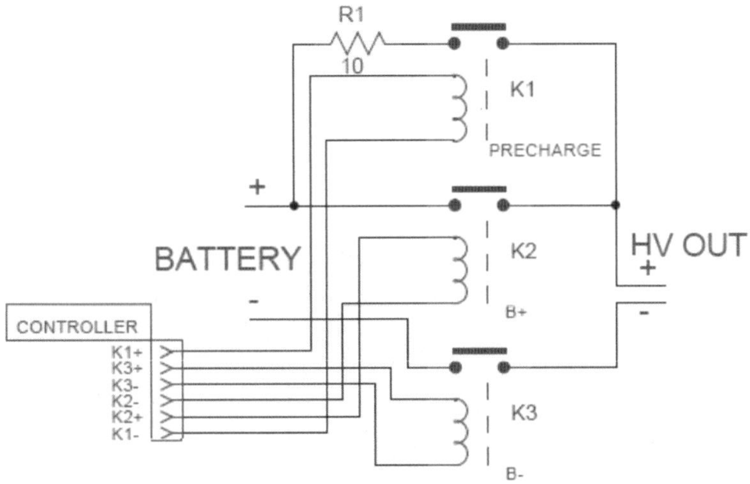

Figure 6.13 Typical precharge circuit.

Typically, the precharge resistor is on the positive terminal of the pack, though it could just as easily be on the negative terminal. While you are free to use any designators you wish, the ones in this schematic (R1, K2, K2, and K3) appear to be industry standard, so you are encouraged to use them as well.

Operation In the most basic form, the precharge circuit operates as follows (Figure 6.14):

- *Off:* When the system is off, all relays/contactors are off.
- *Precharge:* When the system is first turned on, K1 and K3 are turned on to precharge the load through R1, until the inrush current has subsided.
- *On:* After precharge, contactor K2 is turned on, and the load is powered directly from the battery. Relay K1 may be turned off to save coil power.

Additionally, the precharge circuit may be used in combination with sensors to detect problems with ground isolation, short circuits across the load, and problems with components in the precharge circuit itself. Such methods are specialized and are usually proprietary to the BMS manufacturer.

Contactors Selection In a well designed system, under normal operation, the contactors are not required to be able to interrupt the full load current, because the system will reduce the load current to 0 before the contactors are opened. Therefore, the rating of the contactor's contacts needs to be sufficient just for the average load current. For example, a contactor with a 50 A breaking current rating and a 100-A carrying current rating will work with a load that draws 100-A (that is average, not peak—it will work with peaks much higher than 100A, as long as the average is no more than 100A). The contactors must be rated for the maximum pack voltage, because that is the voltage that appears across the contacts when they are opened.

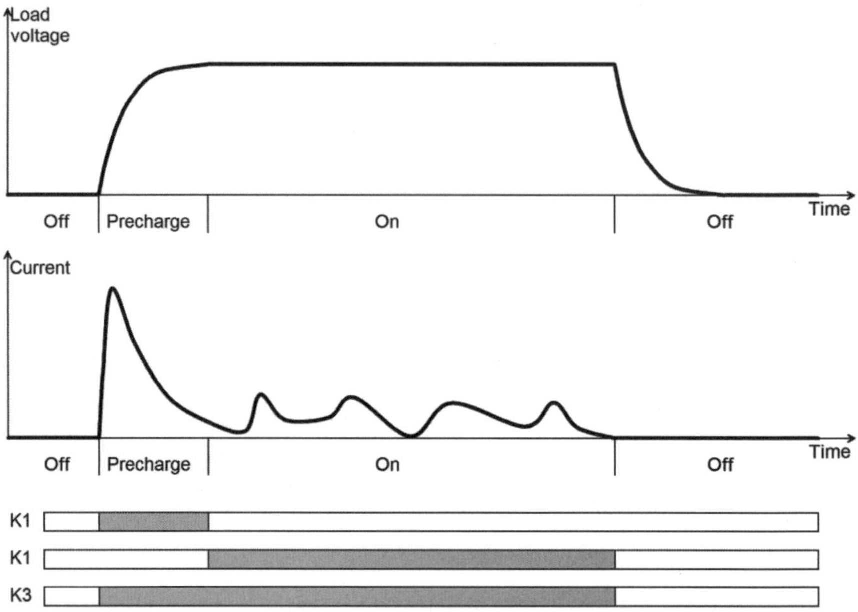

Figure 6.14 Typical precharge waveforms.

Contactors must be rated for DC operation. To interrupt any arcing across the contacts as they are opening, an AC rated contactor relies on the fact that the current waveform goes through 0A at every 0 crossing of the line frequency waveform. That is not the case with batteries, which operate at DC. DC-rated contactors incorporate ways of quenching the arc that forms when turning off an inductive load (Section 5.4.6.1). DC contactor manufacturers include Gigavac, Kilovac (Tyco), Omron, Curtis/Albright, and Panasonic.

Precharge Relay Selection The precharge relay needs to be rated for the full pack voltage, because, when the system is off, the full pack voltage appears across the relay contacts. An AC relay may be used because, by the time it is turned off, the contactor across it (K2) is turned on, so the current through the precharge relay is 0. The relay needs to be able to handle the peak of the inrush current, but, since the average current is low, and the breaking current is nearly 0, the current rating of the relay is not very critical, just as long as its contacts are rated to switch a power load (not a "dry circuit").

In real-world applications, this relay will be operated, at most, 100,000 times. Given that relays tend to be rated for a higher number of cycles (and at full load), the precharge relay will easily operate well within its ratings.

Resistor Selection The resistance of the precharge resistor is chosen based on the capacity of the load and the desired precharge time. The precharge surge current reaches $1/e$ of its initial value after a time of $T = R * C$. The rule of thumb is that the current is reduced to a manageable value after a time of approximately $5 * T$. So, for example, if the desired precharge time is 500 ms, and the load capacity is 10,000 μF, then:

$$R = T/C/5 = 500 \text{ ms}/10{,}000\ \mu\text{F}/5 = 10\Omega$$

The precharge resistor needs to dissipate as much energy as the energy stored in the load's input capacitors. So, for example, with a 100-V pack voltage and a 10,000-μF capacitance, the energy in the charged capacitors (and therefore the energy dissipated by the precharge resistor during turn on) is:

$$E = (C * V^2)/2 = (10{,}000\ \mu\text{F} * 100\ V^2)/2 = 50 \text{ joules}$$

The power dissipated by the precharge resistor is equal to that energy divided by the precharge time. For example, with a precharge time of 500 ms:

$$P = E/T = 50\text{J}/500 \text{ ms} = 100\text{W}$$

At the very beginning of the precharge, the instantaneous power will be quite high:

$$P = V^2/R = 100^2/10 = 1{,}000\text{W}$$

During precharge, the resistor will not need to dissipate any significant power (it will not get hot). But, it will be stressed by that high, sudden power. That is why the resistor needs to be very sturdy and high power, yet it doesn't need a heat sink.

Some resistor manufacturers specify the peak power dissipation of their products. For example: "Overload: 5 times rated wattage for 5 seconds." In that case, a 50-W resistor will handle 500W (well above the 100W of the example above). Ultimately, you should ask the resistor manufacturer if a particular resistor will work in your application, and try the resistor in the application. Wire-wound resistors are recommended, typically encased in ceramic, cement, or extruded aluminum. Inductive resistors are acceptable. For example:

- *Tubular wire wound:* Ohmite 270 series, Vishay/Dale NL series;
- *Cement wire wound:* Xicon PW-RC series;
- *Aluminum extrusion wire wound:* Ohmite 89 series, Stackpole KAL series.

6.1.2 BMS Connections to Pack

The way a BMS is connected to the cells and measures the battery current depends greatly on the particular BMS, though, certain cautions apply in general.

6.1.2.1 Wires and Cables

Route BMS wires and cables out of each individual battery and out of a pack so that they are securely fastened, not exposed to excessive wear and tear, and out of the way of power cables and electrically noisy lines. The criteria for installing low-voltage communication cables, high-voltage communication wires, and high-voltage tap wires are different, and are described next.

High-Voltage Communication Wires

Place high-voltage communication wires close to the power connections and away from the chassis. This is not simply for safety, but specifically to reduce electrical noise pick up, maximize the capacitance between these wires and the power connections, and minimize the capacity between these wires and the chassis (Figure 6.15). This is because the switching electronics in the load are likely to electrically bounce up and down the power wiring (and these BMS wires along them) at a fundamental frequency of 5 to 50 kHz and a magnitude of a few hundred volts peak to peak. As long as the BMS wires bounce up and down exactly the same as the power connections they are monitoring, they will not be exposed to differential noise. Consider the capacitance between the BMS wires and the power wires as 1 capacitor, and the capacitance between the BMS wires and the chassis as a second capacitor. These two capacitors form a capacitive voltage divider with the BMS wire at the midpoint. As long as the capacity to the power wire is significantly higher, the BMS wire will not experience much differential noise. If the capacity to chassis becomes significant, the BMS wire will experience a differential noise of many volts peak to peak, which will prevent the BMS from operating properly.

The ideal placement of daisy-chain links between cell boards is along the power connection (but not wound around it—that would add inductance). This minimizes the area of the loop antenna (Figure 6.16) that is formed by the power connection and the daisy-chain wire, and therefore minimizes the noise pickup.

Low-Voltage Communication Cables

Low-voltage communication cables are referenced to ground, and, in a large pack, must remain isolated from the chassis. Routing these cables is subject to two con-

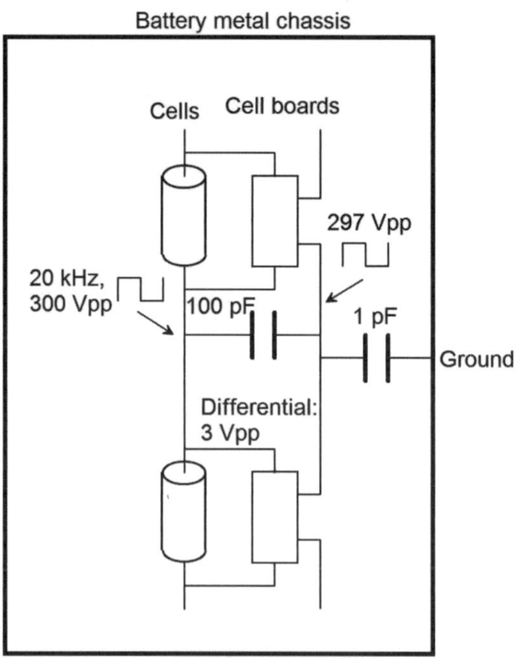

Figure 6.15 Capacitances between a BMS wire and a power wire, and between it and chassis.

Figure 6.16 The loop antenna area: wide loop area (left—bad) and small loop area (right—good).

cerns: isolation and noise immunity. Low-voltage communication cables must withstand a hi-pot voltage (such as 2.5 kV) for safety reasons. That requires attention to routing, cables insulation, and any additional, secondary insulation.

To improve noise immunity, these cables may use twisted pairs and may need to be shielded. The shield must be connected only at one end. They are typically grounded to the BMS master and isolated from the chassis, or grounded to the chassis at a single midpoint, and otherwise isolated at both ends. These cables should be routed far from power cables (to minimize the capacitance between them), and close to the chassis (to maximize the capacitance between them).

High-Voltage Tap Wires

Isolation is much more of an issue for high-voltage tap wires, as they are routed along points of significantly different voltage. High-voltage tap wires must also withstand a hi-pot voltage for safety reasons, if they are able to come in close proximity to the chassis (ground).

If cell blocks (multiple cells in parallel) use a single power connection between adjacent ones, connect the BMS to the same cell to which the power connection is made [Figure 6.17(b)]. In the event that the connection between parallel cells is broken, if the BMS is connected to one of the outlying cells, it will not be able to protect the cells that are still connected to the power because it is looking at a cell that is not being used and therefore remains within its SOA [Figure 6.17(a)].

Electrical noise is not as much of an issue for tap wires (in a nondistributed BMS), because they operate at DC and the BMS includes input filters.

6.1.2.2 Placement Across Adjacent Cells

Certain cell boards, and many cell-to-cell balancers, may not be placed across adjacent cells if there is a possibility that the power connection between those cells may be opened (such as by a fuse, contactor, or safety disconnect) (Figure 6.18). Should that connection open while the pack is under load, the entire pack voltage (and of the opposite polarity) will appear across that gap, and be applied to the electronic

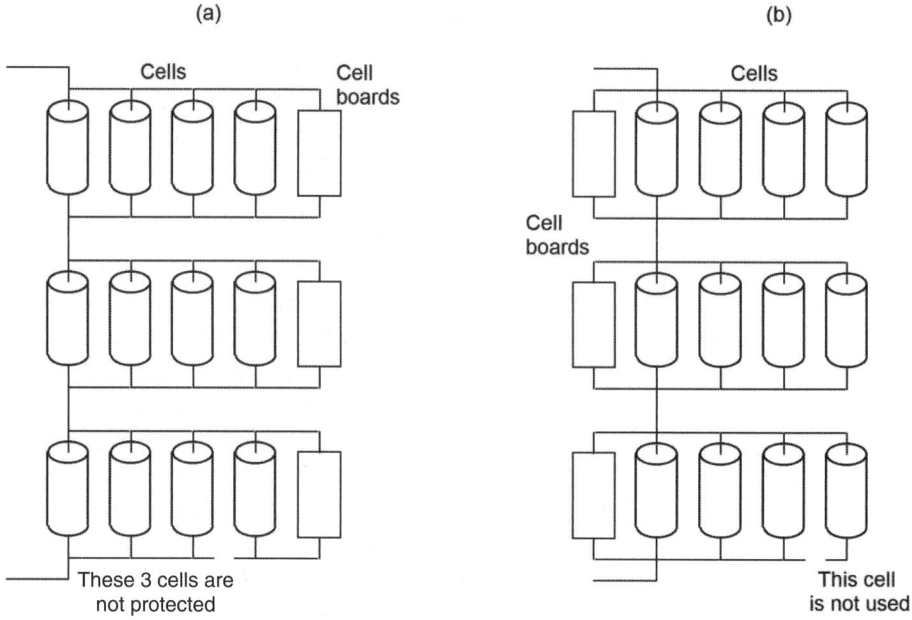

Figure 6.17 Cell monitored by the BMS: (a) unsafe and (b) safe.

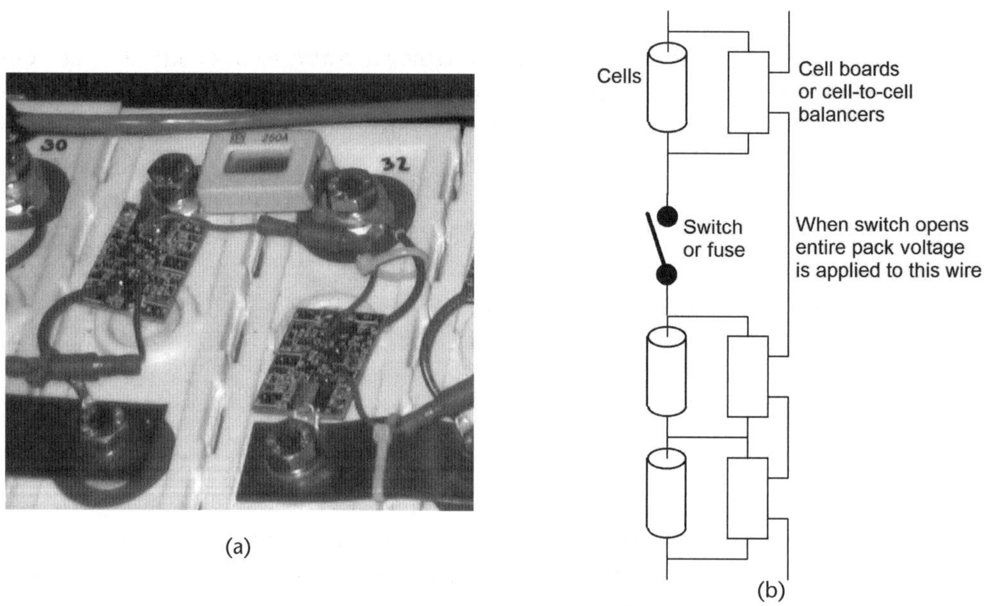

Figure 6.18 (a) If the fuse opens, the cell board across it will be destroyed. (b) An electrical circuit showing the open switch.

assembly across it, destroying it. Therefore, do not place such electronic assemblies across power connections that may open unexpectedly under load.

6.1.3 BMS Connections to System

Too many people have blown up a pack on the first day they were able to use it, because they were too eager to take their new toy for a spin, before they finished connecting the BMS. (I am not talking about just college students and EV hobbyists; professional automotive manufacturers have done it too.) Do resist the temptation and don't make that same mistake. Make sure that the BMS is properly set up to control the load and the charger, and test that it does before you try using the pack.

While generally the BMS acts directly on the devices that charge and discharge the pack, in professional systems a central controller assumes the command. It communicates to all the devices (charger, BMS, load), gathers information from each, and sends commands to each. In particular, in production vehicles, a vehicle control unit (VCU) performs that function. The VCU may receive a CCL message from the BMS, saying that, for example, the load current should not exceed 100A, and yet the current is now 105A. The VCU tells the motor controller that it should operate at a maximum torque of X N-m, and start decreasing that limit until it sees that the pack current is below 100A. Not only does the VCU talk the two different languages that the BMS and the motor controller speak, but it also translates between the two different ways of thinking of the BMS and the motor controller. The BMS thinks in terms of pack current, while the motor controller thinks in terms of motor torque or power (regardless of input current).

6.1.3.1 General Requirements

Except for packs that use protectors, which are able to interrupt the pack current by themselves, it is of the utmost importance that the BMS be able to control the system to interrupt the pack current. There are various ways that can be done, but typically they involve one of the following:

- Sending a control signal or message directly to the external device to shut it down (or to a central controller to do the same).
- Controlling a power switch directly to disconnect the pack from the device, or to remove AC power to the device

We'll see examples of this in the next sections.

6.1.3.2 Chargers

The way that a BMS and a charger are connected together depends a lot on the level of sophistication of both devices. The most basic way for the BMS to control a charger is through a power relay (or contactor) to disconnect the AC power from the charger's input (Figure 6.19). It is a reliable solution, and can be applied universally. An alternative is to disconnect its DC output, but that would require a DC rated relay, and those are harder to find and more expensive. Thus, disconnecting the AC input is best. If a contactor coil requires a lot of power, the BMS may not be able to drive it directly, so a helper relay may be required. Also, a helper relay may be required to flip the polarity of the BMS's output, in case it is normally off and comes on when charging should be stopped. Alternatives to relays are solid state

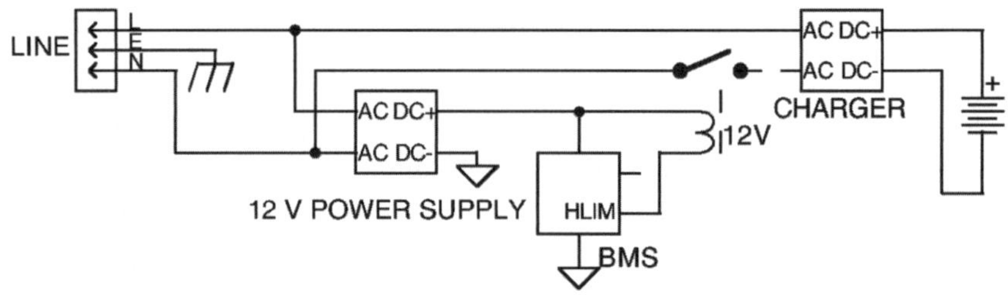

Figure 6.19 Controlling a charger's AC input.

relays (SSRs), which are more reliable, and often cheaper than a contactor of the same rating.

At the next level of sophistication, some chargers can be shut down with a control signal (often, the particular control input used was meant for some other function, but is pressed into service as an on/off control input).

Specifically, the Zivan[1] charger has an input for a temperature sensor; you can stop the charge by emulating a hot temperature sensor connected to that input, which lowers the output voltage. That is not the same as shutting down the charger, but it can have the same effect. Manzanita Micro's PFC charger has a RegBus input that includes two pins that disable the charger when shorted together. In both chargers these inputs are electrically hot, so isolation is required. That can be done with a relay, an opto-isolator, or an SSR (Figure 6.20).

At an even higher level of sophistication are chargers with an analog input for a fine control of the output. When a CCCV charger finishes charging a balanced pack, the current will be gradually reduced. The BMS may wish to control the charger current at other times through its CCL analog output. In many cases, the charger's control input is not ground referenced, and may expect a variable resistance (not a voltage). A simple electronic circuit (Figure 6.21) may be used to provide isolation and output a variable resistance.

At the top end are chargers and BMSs that are perfectly matched to communicate seamlessly to each other through a data link. A Brusa charger and an Elithion BMS can be mated through a CAN bus (as can the ElCon Charger and a Guantuo BMS, but only as long as there are no other messages on the CAN bus). The Brusa charger has to be configured to operate in a CAN mode, which disables its charging profile and makes the charger a slave of the BMS. Then, not only will the Elithion BMS be able shut down the Brusa charger, but also it will be able to control its output current and voltage. Additionally, the Brusa charger will measure the pack current and report it on the CAN bus, meaning that there is no need for an external current sensor to measure the charging current (Figure 6.22).

I am puzzled by how often designers place a contactor between the battery and the charger, since I have yet to see an instance in which that was needed. It is far better to control the charger through a low-voltage signal, or by disconnecting its AC input. Thus, controlling the charger is not a good reason to use a contactor between the battery and the charger. Chargers are designed not to allow any significant reverse current from the battery back into the charger when unpowered, so that is not a good reason either. Chargers have a large capacitance across their output, so

1. Zivan is Italian and is pronounced "ZEE-van."

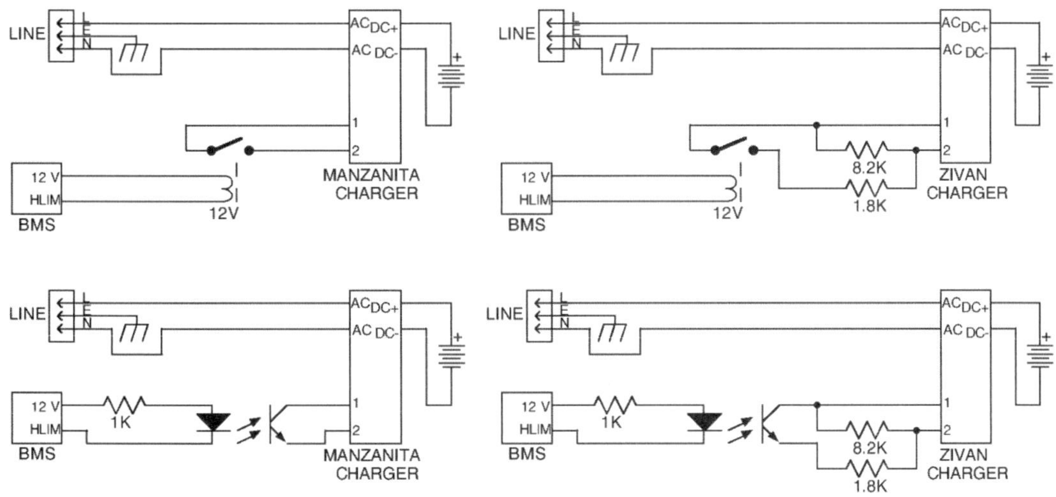

Figure 6.20 Controlling the Manzanita Micro and the Zivan chargers directly.

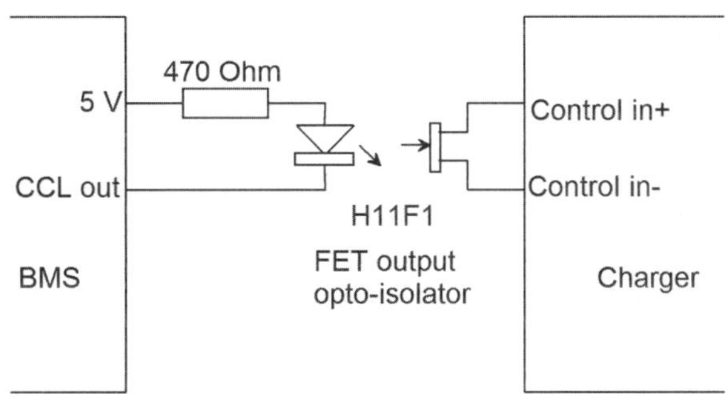

Figure 6.21 Using the BMS's CCL output to reduce the charger current by emulating an isolated, variable resistor.

connecting and disconnecting them each time the battery is to be charged results in an inrush current each time, and there is no need for that. Some chargers may be damaged if they are suddenly disconnected from the battery under load, so there is a good reason *not* to use a contactor in that location, in case the contactor may suddenly open. If the contactors are controlled by a VCU, now the VCU has to remain powered while charging, adding drain to the low voltage (12V) system.

My recommendation is to use contactors only between the battery and the load, and only if the load requires it. However, if an off board charger is used without an inherently safe DC charging port, a contactor is required between the battery and the charging port, to isolate the terminals in the charging port until a charging cable is mated to it.

6.1.3.3 Motor Controllers

At the most basic level, a BMS can disable a motor controller by disconnecting the pack through contactors. At the next level of sophistication, with a motor controller

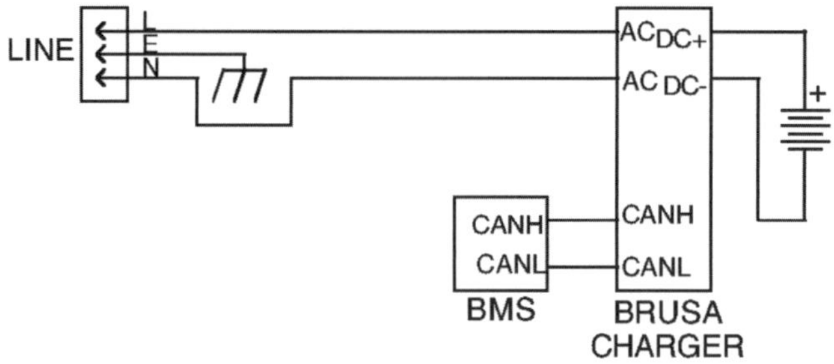

Figure 6.22 Controlling the Brusa charger.

that uses a pot as speed control (throttle), the BMS can reduce (and eventually remove) the available torque to the throttle. In a vehicle, a gradual reduction in the available torque is safer than forcing a sudden loss of power. (This method will not work with Hall effect speed controls, which have to be powered at 5V at all times.)

Typically, the throttle in electric vehicles is a two-wire or a three-wire pot. The top lead of a three-wire pot (the one to which the wiper is connected at full throttle) is connected to a fixed voltage from the motor controller (typically 5V), either directly or through a resistor. Instead of feeding that lead from the motor controller, feeding it from the DCL output of the BMS (Figure 6.23) will result in a reduction in available throttle range as the traction pack nears empty.

At an even higher level of sophistication are motor drivers with a disable input. Motor controllers that allow regenerative braking (as all AC motor inverters do) should have two separate control inputs, one to disable forward power, and one to disable regeneration, because there are times when the pack can be discharged but cannot take in any more charge (Azure's DMOC does that, UQM's PowerPhase does not).

Motor controllers and BMSs that are perfectly matched to communicate seamlessly together through a data link would be ideal. No pair of commercially available BMS and motor controller today can do so; a VCU is required to translate between the two (Figure 6.24). The CalMotors' PCM is such a VCU, allowing their GP series AC motor inverters to work with the Elithion BMS.

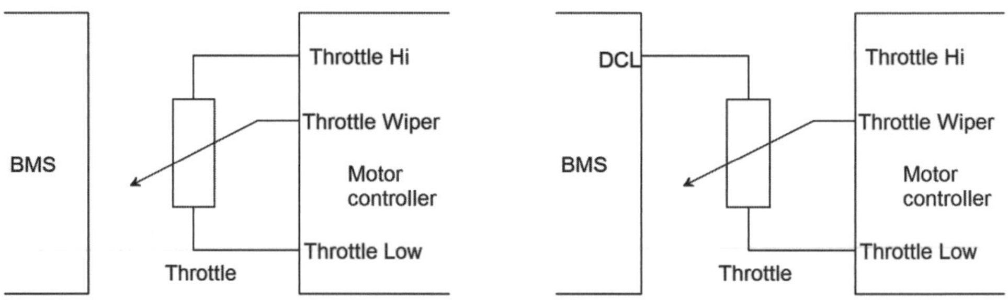

Figure 6.23 BMS analog interface with motor controller.

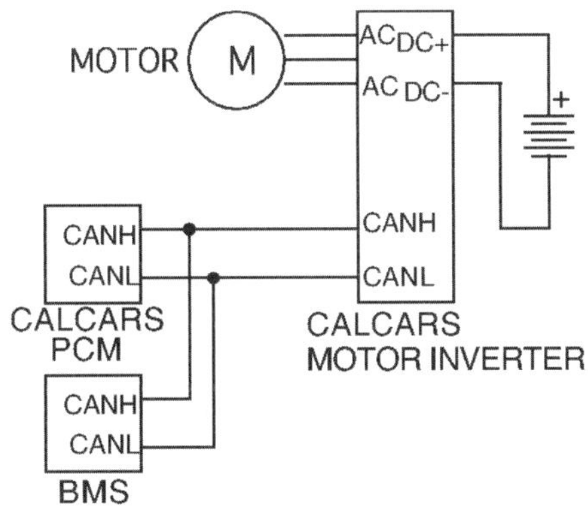

Figure 6.24 Elithion BMS interface with CalMotors motor controller.

6.1.3.4 Grid-Connected Inverters/Chargers

Grid-connected bidirectional inverters/chargers are either the combination of a charger (typically lower power) and a high-power inverter capable of feeding the grid or a local AC power line with energy from a pack. They are used for backup power, or for peak shaving. A BMS in a pack used in conjunction with an inverter/charger differs from a standard BMS in just one small detail: it does not assume that current can only flow from the charger to the pack. Its sensor for the charging current must be bidirectional and its software must be able to handle signed values for the charge current and power.

In a land-based system, the inverter is likely to be the only load, so that any BMS will work. The inverter is like any other load (no different from an AC motor inverter), so the BMS is able to handle current flowing in or out of the pack to the inverter.

In a mobile application [vehicle to grid (V2G) or vehicle to home (V2H)], the pack could be powering either the motor controller or the inverter. In that case, the BMS sees the motor controller as the load and the charger/inverter as the charger, which means that the BMS does have to be able to handle a bidirectional current to the charger/inverter.

BMSs that include a data link to report status are preferable, as grid based applications typically require knowledge of the pack conditions before they request a discharge into the grid. Other than these small details, any BMS will work in a grid-connected, bidirectional application, just as well as any other application.

6.1.3.5 Fuel Gauges

In an EV conversion, it may be desirable to retain the original, analog fuel gauge, though driven by the pack's SOC from the BMS. A simple electronic circuit (Figure 6.25) can convert a 0- to 5-V SOC output to the signal that the fuel gauge expects.

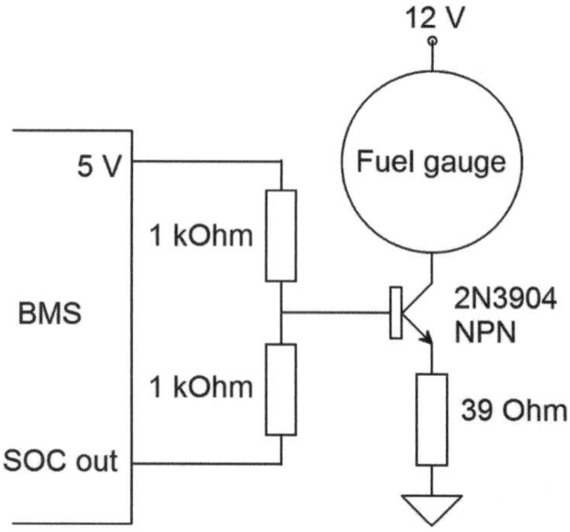

Figure 6.25 Adapter from SOC output to analog fuel gauge.

Many vehicles have a fuel gauge that is connected to 12V at one end, and through a resistive sensor (sender) in the fuel tank, to ground. The sender's resistance is: 4Ω when full, 107Ω when empty. The gauge's resistance is about 127Ω, and will show full with 90 mA through it, and empty with 35 mA.

6.2 Configuring

Many BMSs are preset for a given application (e.g., a protector for eight standard Li-Ion cells may only be used with exactly eight cells, and only with $LiCoO_2$ cells). Others need to be configured for a particular application, such as the number of cells, their chemistry, and how they interface to the external system.

6.2.1 Cell Configuration

Some BMSs may be configured for a particular cell, typically through a computer interface or by adjusting a trimmer. This configuration entails setting voltage thresholds for the particular cell used. At the very least, there are three thresholds (though some BMS can have as many as seven different settings): minimum cell voltage, balance voltage, and maximum cell voltage. A few BMSs also allow you to set the temperature range for charging and discharging. It is up to you to study the cell and its spec sheet to decide the best thresholds for it and for your application. Some suggested values are in Table 6.1.

6.2.2 Pack Configuration

Some BMSs require you to set the number of cells used, and possibly the number of batteries of banks in which the pack is divided. If the pack uses multiple batteries in parallel, the BMS needs to be configured accordingly, so that it may calculate the correct pack voltage when adding the individual cell voltages. If the BMS is able to estimate pack SOC, the nominal capacity of the pack must be specified. You may

Table 6.1 Suggested Settings for Various Cell Chemistries

	Minimum [V]	*Balance [V]*	*Maximum [V]*	*Charge [°C]*	*Discharge [°C]*
Li-Titanate (Altair Nano)	1.5	2.7	2.8	20~55	–40~55
$LiFePO_4$ (A123, K2...)	2.0	3.4	3.6	0~40	–30~60
$LiFeYPO_4$ (Thundersky)	2.8	3.4	4.0	–25~75	–25~75
LiPo (Kokam...)	2.7	4.0	4.2	0~40	–20~60
$LiCoO_2$ (Gaia...)	2.7	4.0	4.2	0~40	–30~60

also be able to configure the maximum charging and discharging current (nominal, peak, or both, depending on the BMS).

6.2.3 System Configuration

The BMS may be configurable to work best with the other devices in the system (e.g., load, charger), to be able to control them. For example, the polarity of the BMS's limit outputs may be selectable (normally open or normally closed). And you may be able to set a reaction delay, to avoid nuisance tripping of the limits. The BMS may allow you to specify the messages present on its data links, and the format of the data on those messages, for compatibility with the other devices in the system. For example, the speed of its CAN bus may be set to 125, 250, or 500 kHz.

A CCCV charger should be set for a voltage that is exactly equal to the cell top cutoff voltage (e.g., 4.2V) multiplied by the number of cells. That way, the BMS and the CCCV charger will work in harmony, allowing the BMS to do cell balancing, and then allow the charger to fully charge all the cells (Figure 6.26).

- *Stage 1:* The charger is fully on, charging all the cells, until one of them reaches the top cutoff.
- *Stage 2:* The BMS cycles the charger on (at full current) and off, as it balances the pack, until all the cells reach the same SOC, though they won't yet be completely full.
- *Stage 3:* The cells are a hair below the top cutoff voltage, so the BMS allows the charger to remain on. As all the cells are at the same voltage, and nearly full, the pack voltage reaches the CCCV charger's constant voltage. The cells draw less and less current, as they become completely charged.

6.3 Testing

Just because a BMS is physically located in a pack does not mean that the pack is protected. You need to be able to prove that the BMS is indeed able to do its job. Before you start using the product, do test that the pack current is shut off if any cell is about to be operated outside its SOA. Go through the following procedure, while watching the conditions of the cells like a hawk. At this point you can't trust the BMS.

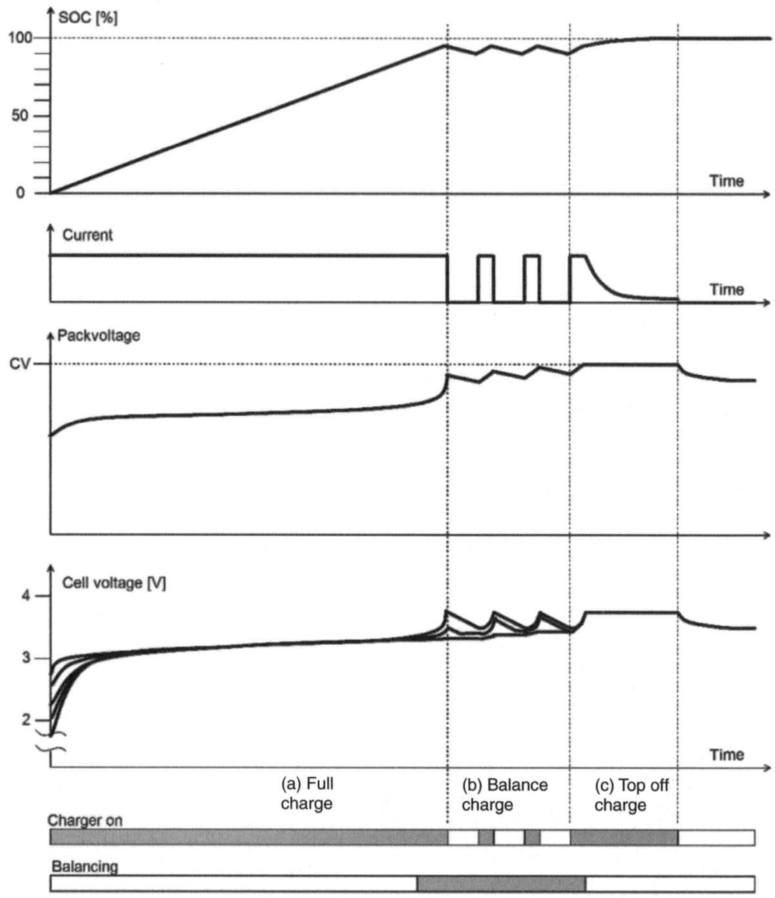

Figure 6.26 Three stages of charging: (a) charging at full current; (b) balancing, with charger cycling on and off; and (c) topping off the charge.

Test the charging:

1. Turn on power to the charger.
2. Note that the BMS appears to be operational. If a digital BMS, note the cell voltages.
3. Note that current is flowing into the pack.
4. Note that the BMS is aware of the current flowing into the pack, and that the SOC value (if available) is increasing.
5. Note the cell temperatures, making sure they are not excessive. If the BMS can read temperatures, note that it is seeing the correct values.
6. Note that balancing starts occurring on the most charged cells when their voltage reaches a certain threshold (if a top balancing BMS).
7. Note that, as soon as any cell's voltage reaches the maximum threshold, charging is interrupted.
8. When the pack is fully charged, turn off power to the charger.

Test the discharging:

1. Turn on the load (e.g., the ignition).
2. Note that the BMS appears to be operational. If a digital BMS, note the cell voltages.
3. Note that current is flowing out of the pack.
4. Note that the BMS is aware of the current flowing out of the pack, and that the SOC value (if available) is decreasing.
5. Note the cell temperatures, making sure they are not excessive.
6. Note that, as soon as any cell's voltage reaches the minimum threshold, discharging is interrupted.
7. Turn off the load.

If any of the above steps fails, identify and correct the problem. Do not trust the BMS until you have proved that it is doing its job.

6.4 Troubleshooting

Troubleshooting depends a lot on the particular BMS. However, here are some general pointers.

You may be able to do some quick tests of the BMS functions by forcing the BMS's limits. For example, you may temporarily set the maximum cell voltage to a low value, just to see that charging does stop.

Communication problems may be due to poor immunity to electrical noise. If the BMS starts reporting bad data or not seeing some cell boards whenever the charger is on, or whenever the load is drawing high currents, poor noise immunity is the likely culprit. Often, minimizing emissions at their source is not an option, so that leaves trying to improve the BMS's noise immunity. Here are some pointers, though not all of them will be applicable to your system.

6.4.1 Grounding

A short, low-inductance conductor between the BMS case (or its ground plane) to the chassis may help. Cut the ground loop between a BMS and a laptop computer or display that is using an RS232 cable, by removing the local ground on that device.

6.4.2 Shielding

Use of shielded cables, and/or twisted pair, for communication cables, may reduce data errors.

6.4.3 Filtering

Try ferrite clamps on data cables, CAN filters on the CAN bus, and a bypass capacitor to ground on DC lines (signal and power). Don't bother with the power wires to the load: you are not likely to achieve anything with those.

6.4.4 Wire Routing

Route high-voltage communication wires between cell boards along the power conductor (do not wrap it around the power conductor) and away from chassis ground. Route low-voltage communication cables to cell boards away from power conductors and close to chassis ground.

6.4.5 Nuisance Cutouts

If the BMS shuts down the pack when a large pulse of current is drawn from it (e.g., hard accelerations, load inrush current) especially after the pack is partially depleted, the battery design may not be appropriate for the application (e.g., cells have too high a resistance, too few cells are in parallel). You can try to add more cells in parallel, use cells with lower resistance, or reduce the maximum current that the load can draw, or you could try to "kill the messenger" by reducing the low-voltage cutout threshold, or by adding a delay to the low cutoff voltage response to make it more forgiving.

6.5 Using

A BMS should not require a section in this book on using it, because its should operate reliably in the background, much like, say, antivirus software. However, not all BMS can operate independently of the user. We saw that meters (Section 2.1.3) require that the user be an integral part of the BMS, by constantly paying attention to a display, or listening for a sound, and taking appropriate action in case of abnormal conditions. Only a trained user should be allowed to operate a product that is equipped with a meter-type BMS.

Similarly, we saw that some BMSs (regulators, CCCV chargers) provide no pack protection whatsoever. Only people of faith should operate a product that is equipped with such an ineffective BMS. Having read this book, you certainly understand that the only BMS that is worth installing on a pack is one that is able to protect that pack without any human or divine intervention.

Reference

[1] Beauregard, Garrett, Report of investigation, *Hybrids Plus Plug In Hybrid Electric Vehicle,* National Rural Electric Cooperative Association, Inc. and U.S. Department of Energy, Idaho National Laboratory, ETEC, June 26, 2008.

List of Acronyms and Abbreviations

μA	microAmp
μF	microFarad
μV	Microvolt
$\mu\Omega$	Micro-ohm
Ω	Ohm
A	Amp
AC	Alternating current
Ah	Amp-hour
ASIC	Application-specific integrated circuit
BJT	Bipolar junction transistor
BMS	Battery management system
C	Coulomb, or charge
CAN	Controller area network
CCCV	Constant current/constant voltage
CCL	Charge current limit
DAC	Digital-to-analog converter
DC	Direct current
DCL	Discharge current limit
DOD	Depth of discharge
DVM	Digital voltmeter
ECU	Electronic control unit
EEM	Electrical equivalent model
EMI	Electromagnetic interference
EV	Electric vehicle
F	Farad
FPGA	Field programmable gate array
HEV	Hybrid electric vehicle
HLIM	High limit
HVC	High voltage cutoff
HVL	High voltage limit
Hz	Hertz
I2C	Inter IC
IC	Integrated circuit

IGBT	Insulated gate bipolar transistor
IR	Current times resistance
J	Joule
kHz	kilohertz
kW	kilowatt
kWh	kilowatt-hour
LA	Lead acid
LC	Inductor-capacitor
Li-Ion	Lithium ion
$Li_4Ti_5O_{12}$	Lithium-titanate
$LiCoO_2$	Standard lithium-cobalt-oxide
$LiFePO_4$	Nano-phosphate/lithium-iron-phosphate/lithium-ferro-phosphate
$LiMn_2O_4$	Lithium-manganese-oxide
LiMnNiCo	Lithium-manganese-nickel-cobalt
$LiMnO_2$	Lithium-manganese-oxide
$LiNiO_2$	Lithium-nickel-oxide
LiPo	Lithium polymer
LLIM	Low limit
LT	Linear technology
LVC	Low voltage cutoff
LVL	Low voltage limit
mA	milliAmp
Min	Minute
MOSFET	Metal oxide semiconductor—field effect transistor
ms	millisecond
mΩ	milli-ohm
NiMH	Nickel metal hydride
OBD	Onboard diagnostics
OCV	Open circuit voltage
OOB	Out of bounds
PCM	Protection circuit module
PCM	Powertrain control module
PHEV	Plug-in hybrid electric vehicle
PID	Parameter identifier
Pot	Potentiometer
PWM	Pulse width modulation
RC	Resistor-capacitor
Regen	Regenerative braking
RF	Radio frequency
RX	Receive, receiver
s	Second

SAE	Society of Automotive Engineers
SMB	System management bus
SOA	Safe operating area
SOC	State of charge
SOH	State of health
Spec	Specification
SPI	Serial peripheral interface
SSR	Solid state relay
TI	Texas Instruments
TVS	Transient voltage suppressor
TX	Transmit, transmitter
UART	Universal asynchronous receiver transmitter
V	Volt
VCU	Vehicle control unit
VIM	Vehicle information management
VMS	Voltage management system
W	Watt
Wh	Watt-hour

Glossary

AC impedance Complex resistance, which includes a real part (the resistance) and an imaginary part (inductance or capacitance).

AC motor inverter Device that converts DC to AC to transfer energy from a battery to an AC induction motor.

Alternating current Electrical current that switches direction rapidly and regularly.

Ammeter Current meter, measuring amps.

Amp Unit of measure for current; 1 amp is the current that flows through a 1-ohm resistor when 1 volt is applied to it.

Amp-hour Unit of measure for charge; 1 Ah is the charge stored in a cell or battery by a 1A current for 1 hour.

Analog switch Electronic component that makes or breaks a connection between two points, based on the state of its digital control input.

Asynchronous Communication link with no separate clock line: the clock is embedded in the data line.

Backup Device that stores energy for use later when the main source of power is not available.

Balanced The state of a battery in which all the cells are at the same state of charge.

Battery A collection of cells (or blocks) wired in series, and constituting a single physical module, providing a higher voltage.

Battery management system A device or system whose purpose is to monitor, control, and/or optimize a battery.

Battery pack A collection of batteries, arranged in any series and/or parallel combination.

Block A collection of cell wired directly in parallel, also providing 3V to 4V in the case of Li-Ion.

Bottom balancing Balancing a battery so that all of its cells are at 0% SOC.

Calendar life A measure of the loss of capacity of a cell as it sits unused, measured in percent loss over years.

CAN bus A multipoint communication bus standard in industrial and vehicular applications.

Capacitance A measure of the ability of an electrical component (specifically a capacitor) to store charge.

Capacitor An electrical component whose capacitance is characterized.

Capacity A measure of the ability of a cell or battery to accept charge.

Cell The most basic element of a battery (providing 3V to 4V in the case of Li-Ion).

Charge A quantity of electrons.

Charge current limit Maximum charging current allowed into the battery according to the BMS.

Charger Device that converts AC to DC (not necessarily filtered) to transfer energy from the grid to charge a battery.

Chassis The metal body of a product, especially a vehicle, usually considered to the be local ground, or common.

Clock Electronic circuit that generates a signal with a stable frequency.

Comparator Electronic component that compares the voltages on its two inputs, and whose output is high if the voltage on the positive input is higher than the voltage on its negative input.

Contactor High-power relay.

Converter, EV A person who converts standard cars to electric vehicles, or a company that does so.

Coulomb Unit of measure for the charge that flows through a conductor in 1 second.

Counter Electronic component that receives and counts clock pulses, generating a set of binary outputs whose value represents the number of pulses received.

Current The flow of electrons through a conductor, measured in Amp.

Current source A theoretical device that sinks or sources a given current regardless of the voltage across it.

Cutoff voltage Voltage below which a cell should not be discharged, or above which a cell should not be charged.

Cycle life A measure of the loss of capacity of a cell as it is fully charged and discharged numerous times, measured in percent loss over number of cycles.

Daisy chain A link that goes from A to B, then from B to C, and so forth.

Darligton Two transistors, the first one driving the second one, to increase the gain.

DC motor controller Device that converts DC to DC at a lower voltage to transfer energy from a battery to a DC motor.

Decoder Electronic component that receives an n-bit binary address and asserts the corresponing one of its outputs.

Depth of discharge A measure of the charge out of a battery, measured in Ah (may be measured in %).

Direct current Electrical current that is always in the same direction, even though it may vary over time.

Discharge current limit Maximum discharging current allowed out of the battery according to the BMS.

Duplex An asynchronous link in which communication occurs in both directions simultaneously, as opposed to simplex.

Efficiency The ratio of two parameters using the same units, such as energy out over energy in.

Electrical equivalent model An electrical circuit that closely approximates the behavior of a cell.

Electrolyte The chemical between the cell's electrodes.

Energy A measure of the work performed by an entity.

Equalizer *See* Regulator, definition (a).

Estimated OCV The theoretical voltage of the voltage source inside the cell, calculated through IR compensation.

EV converter *See* Converter, EV.

Farads Unit of measure for capacity: 1F is the capacity of a capacitor that is charged to 1V when 1A current flows into it for 1 second.

Fuel gauge An alternate name for the SOC evaluation function, especially when done in electric vehicles.

Hertz Unit of measure for frequency.

High cutoff voltage Voltage above which a cell should not be charged.

High limit BMS line that is on or off depending on whether the battery may or may not receive charge.

High voltage limit *See* High limit.

I2C bus InterIntegrated circuit bus, a two-wire bus used between onboard peripherals.

Impedance *See* AC impedance.

Inductance Property of electrical components, especially inductors.

Integrator, vehicle *See* Vehicle integrator.

Inverter Device that converts DC to AC, for example, to transfer energy from a battery to the grid or to an AC induction motor.

IR compensation Calculation that allows to determine the theoretical OCV given the current through the resistance, and the value of that resistance.

Joules Unit of measure for energy: 1J is the energy dissipated in heat by a 1-ohm resistor when 1-A current flows through it for 1 second.

Lead acid Chemistry of standard car batteries.

Leakage Current inside a cell that depletes its charge.

Lithium ion Chemistry of Li-Ion cells.

Loop antenna A radio antenna formed by a loop of wire, which is best to detect the magnetic field component of electromagnetic waves, especially at low frequencies.

Low cutoff voltage Voltage below which a cell should not be discharged.

Low limit BMS line that is on or off depending on whether the battery may or may not be discharged.

Low voltage limit *See* Low limit.

Model *See* Electrical equivalent model.

Motor controller *See* DC motor controller.

Motor inverter *See* AC motor inverter.

Multiplexer Electronic component that selects one of multiple inputs, as specified by its address inputs.

Nickel metal hydride The chemistry of a type of cell commonly used in HEVs.

Off the shelf Commercially available, as opposed to custom.

Ohm A measure of resistance.

Op-Amp Electronic component that amplifies the voltages difference between its two inputs.

Open circuit voltage The voltage of a cell that has been sitting unused for a prolonged time, or the theoretical voltage of the voltage source inside the cell, calculated through IR compensation.

Overdischarge The state of a cell from which more charge has been taken than could be done safely.

Overcharge The state of a cell to which more charge has been added than could be done safely.

Pack Collection of batteries, arranged in any series and/or parallel combination.

Parameter ID Protocol to request the value of a parameter from an ECU, and receiving it from that ECU.

Peak shaving The process of supplementing the power grid with energy from a battery when the demand is particularly high.

Potentiometer Variable resistor.

Power A measure of the energy per second produced or consumed by a device.

Power supply Device that converts AC to filtered DC to transfer energy from the grid to a load.

Powertrain control module *See* Vehicle control unit.

Protection circuit module Protector-type battery management system.

RC circuit An electrical circuit that uses a resistor and a capacitor, whose values determine the circuit's time constant ($T = R * C$).

Real part The resistive component of an AC impedance.

Reference *See* Voltage reference.

Regenerative braking Braking that recovers mechanical energy and converts it to electrical energy, which may be stored into a pack.

Regulator (a) Device placed across a cell to limit its voltage. (b) *See* Voltage regulator.

Resistance A measure of a device's inability of carrying current, determined as the ration of the voltage across it over the current through it ($R = V/I$).

Resistor A device that has a well-defined resistance.

Safe operating area Range of values of particular parameters within which a device can operate safely.

Self-leakage *See* Leakage.

Self-discharge *See* Leakage.

Simplex A asynchronous link in which communication occurs in a single direction at a time, as opposed to duplex.

SMB A superset of the I2C bus, used with smart batteries.

Solid state relay Electronic device that mimics some of the functions of a mechanical relay, but using an opto-isolator and some type of transistor.

Spec sheet Manufacturer's data on a component.

SPI bus Serial peripheral interface bus, a three-wire bus used between onboard peripherals.

State of charge Proportion of the charge in a cell or a battery, compared to its capacity.

State of health An arbitrary measure of a battery's condition with respect to its nominal conditions.

Synchronous Communication link with a clock line separate from the data line.

Thermal runaway A vicious cycle that is the result of high temperature and that produces more heat, in turn resulting in higher temperature, in an accelerating fashion.

Thermistors Temperature sensing resistor.

Time constant The time that a parameter that is varying asymptotically takes to reach $1/e$ of its final value.

Top balancing Balancing a battery so that all of its cells are at 100% SOC.

Traction pack Large battery pack used in vehicles.

Trimmer Small potentiometer for occasional adjustments.

Unbalance State of a battery in which the cells are not all at the same state of charge.

Underdischarge The state of a cell which has stopped giving charge even though it could give more.

Undercharge The state of a cell which has stopped receiving charge even though it could receive more.

Vehicle control unit Central computer that manages the various devices in a vehicle.

Vehicle information management *See* Vehicle control unit.

Vehicle integrator Company that integrates various devices into a new or converted vehicle.

Volt Unit of measure for voltage.

Voltage Electrical potential difference between 2 points, measured in volts.

Voltage drop A change in voltage due to a current flowing through a resistance.

Voltage management system Battery management system.

Voltage reference Electronic component that receives a certain voltage and generates a low-power output with a lower, accurate voltage.

Voltage regulator Electronic component that receives a certain voltage and generates a fixed voltage able to power other circuits.

Voltage source A theoretical device that produces a given voltage regardless of the current it sinks or sources.

Watt Unit of measure for power; 1W is the power dissipated in heat by a 1-ohm resistor when 1-A current flows through it.

Watt-hour Unit of measure for energy: 1 Wh is the energy dissipated in heat by a 1-ohm resistor when 1-A current flows through it for 1 hour.

Zener diode Electronic component that conducts in one direction, while in the other direction it only conducts if the voltage across it reaches its breakdown voltage.

About the Author

Davide Andrea is the owner of Elithion LLC, founded in 2008 to make his lithium-ion battery management system technology available to the industry. Previously, Mr. Andrea cofounded Hybrids Plus, where, using his BMS technology, he developed Prius and Escape PHEV conversions, as well as the Inverger, a bidirectional charger/inverter for V2G applications. Previously, Mr. Andrea offered electronic consulting services through DAVIDE, DBA. He has 6 years of experience in Li-Ion BMS design, 25 years of experience in electronics design, and a B.S. in electrical engineering and computer science from the University of Colorado.

Index

Q

R

T

U